Advances in Experimental Medicine and Biology

Volume 675

For further volumes:
http://www.springer.com/series/5584

Patrick C. Hallenbeck
Editor

Recent Advances in Phototrophic Prokaryotes

Springer

Editor
Patrick C. Hallenbeck
Département de Microbiologie et
Immunologie
Université de Montréal
H3C 3J7 Montréal, Québec
Succursale Centre-ville
Canada
patrick.hallenbeck@umontreal.ca

ISBN 978-1-4419-1527-6 e-ISBN 978-1-4419-1528-3
DOI 10.1007/978-1-4419-1528-3
Springer New York Dordrecht Heidelberg London

Library of Congress Control Number: 2010925847

Printed on acid-free paper

Springer is part of Springer Science+Business Media (www.springer.com)

Preface

ISPP2009, the 13th International Symposium on Phototrophic Prokaryotes, was held in Montreal, Canada, from August 9 to August 14. This was only the second time that the ISPP series was in North America. ISPP2009 was well attended with about 280 registered participants from over 30 countries. A stimulating and informative program showcased the recent developments in this ever-evolving field. This is always one of my favourite conference series to attend because not only does it inform my specific research passions, it broadly educates me in ways that improve my teaching and increase my breadth of understanding in a variety of outside areas. Indeed, the ISPP series brings together a broad spectrum of interests, techniques, and disciplines. Both established researchers and newcomers to this field gave oral presentations in a large number (80) of plenary and parallel symposia sessions which proved to have active audience participation and lively discussions. A large number of excellent poster presentations supplemented the oral program. I think that the high quality of the scientific presentations, as well as the enjoyable social events, was widely appreciated. Things ran very smoothly, from the original registration to the closing ceremony, thanks to Isabel Stengler and her team at IS Event Solutions.

This volume is based on ISPP2009 and presents selected invited works, reviews, and research articles covering the spectrum of the diverse subject matter covered by this symposium series. Here the authors have greatly expanded on what was presented in Montreal to ensure complete coverage of each topic, covering the gamut from ecology, physiology, proteins and genomics, signal transduction to applied aspects. These articles not only are timely, presenting up-to-date published and unpublished results, but, since they are extensively referenced, should serve as valuable source material for some time to come. I hope that you enjoy and find this volume useful, covering the range of topics in which these organisms play an important role, one that is ever relevant to our understanding of the natural world around us, the biogeochemical cycles that shape our environment, and the impact of the microbial world on climate change.

Best wishes, I hope to see you at a future ISPP conference which will without a doubt be as interesting, exciting, and illuminating as ISPP2009.

Montréal, QC, Canada Patrick C. Hallenbeck

Contents

Contributors

Richard M. Alvey Department of Biochemistry and Molecular Biology, The Pennsylvania State University, University Park, PA 16802, USA, rma14@psu.edu

Jens Appel School of Life Science, Arizona State University, P.O. Box 874501, Tempe, AZ 85287, USA, jens.appel@asu.edu

Semra Aygun-Sunar Department of Biology, University of Pennsylvania, Philadelphia, PA 19104, USA, fdaldal@sas.upenn.edu

Carl E. Bauer Biology Department, Indiana University, Bloomington, IN 47405, USA, bauer@indiana.edu

Hermann Bauwe Institute of Biological Sciences and Plant Physiology, University of Rostock, Albert-Einstein-Strasse 3, D-18051 Rostock, Germany, hermann.bauwe@biologie.uni-rostock.de

J. Thomas Beatty Department of Microbiology and Immunology, University of British Columbia, Vancouver, BC, Canada, V6T 1Z3, jbeatty@interchange.ubc.ca

Avijit Biswas Department of Biological Sciences, University of New Orleans, New Orleans, LA 70148, USA, abiswas@uno.edu

H. Bolhuis Department of Marine Microbiology, Netherlands Institute of Ecology NIOO-KNAW, P.O. Box 140, 4400 AC Yerseke, The Netherlands, h.bolhuis@nioo.knaw.nl

Donald A. Bryant Department of Biochemistry and Molecular Biology, The Pennsylvania State University, University Park, PA 16802, USA, dab14@psu.edu

Leong-Keat Chan Department of Marine Sciences, University of Georgia, Athens, GA 30602, USA, lkchan@uga.edu

Julius T. Csotonyi Department of Microbiology, University of Manitoba, Winnipeg, MB R3T 2N2, Canada, umcsoton@cc.umanitoba.ca

Fevzi Daldal Department of Biology, University of Pennsylvania, Philadelphia, PA 19104, USA, fdaldal@sas.upenn.edu

Andrew W. Dangel Department of Microbiology, The Ohio State University, 484 West 12th Avenue, Columbus, OH 43210-1292, USA, dangel.2@osu.edu

Marion Eisenhut Department of Biology, Plant Physiology and Molecular Biology, University of Turku, FI-20014, Turku, Finland, mareis@utu.fi

Enrique Flores Instituto de Bioquímica Vegetal y Fotosíntesis, C.S.I.C. and Universidad de Sevilla, Américo Vespucio 49, E-41092 Seville, Spain, flores@cica.es

Sarah M. Florizone Department of Microbiology & Immunology, University of British Columbia, Vancouver, BC V6T 1Z3, Canada, sflorizone@interchange.ubc.ca

Karl Forchhammer Institut für Mikrobiologie der Eberhard-Karls-Universität Tübingen, Auf der Morgenstelle 28, D-72076 Tübingen, Germany, karl.forchhammer@uni-tuebingen.de

Claudia Hackenberg Institute of Biological Sciences and Plant Physiology, University of Rostock, Albert-Einstein-Strasse 3, D-18051 Rostock, Germany, claudia.hackenberg@uni-rostock.de

Martin Hagemann Institute of Biological Sciences and Plant Physiology, University of Rostock, Albert-Einstein-Strasse 3, D-18051 Rostock, Germany, martin.hagemann@biologie.uni-rostock.de

Patrick C. Hallenbeck Département de Microbiologie et Immunologie, Université de Montréal, CP 6128, succursale Centre-Ville, Montréal, QC H3C 3J7, Canada, patrick.hallenbeck@umontreal.ca

Thomas E. Hanson College of Earth, Ocean and Environment and Delaware Biotechnology Institute and Department of Biological Sciences, University of Delaware, Newark, DE 19711, USA, tehanson@udel.edu

Jennifer Hiras College of Earth, Ocean and Environment and Delaware Biotechnology Institute, University of Delaware, Newark, DE 19711, USA, jhiras@udel.edu

Dörte Hoffmann School of Life Science, Arizona State University, P.O. Box 874501, Tempe, AZ 85287, USA, dorte.hoffmannl@asu.edu

Kazuhito Inoue Research Institute for Photobiological Hydrogen Production and Department of Biological Sciences, Kanagawa University, 2946 Tsuchiya, Hiratsuka Kanagawa 259-1293, Japan, inoue@waseda.jp

Gauri Joshi Department of Microbiology The Ohio State University 484 West 12th Avenue, Columbus, OH 43210-1292, USA, joshi.67@osu.edu

Diana Kirilovsky Commissariat à l'Energie Atomique (CEA), Institut de Biologie et Technologies de Saclay (iBiTecS), and Centre National de la Recherche Scientifique (CNRS), 91191 Gif sur Yvette, France, diana.kirilovsky@cea.fr

Masaharu Kitashima Department of Biological Sciences, Kanagawa University, 2946 Tsuchiya, Hiratsuka, Kanagawa 259-1293, Japan, sakurai@waseda.jp

Rick Laguna Department of Microbiology, The Ohio State University, 484 West 12th Avenue, Columbus, OH 43210-1292, USA, laguna.2@osu.edu

Andrew S. Lang Department of Biology, Memorial University of Newfoundland, St. John's, NL A1B 3X9, Canada, aslang@mun.ca

Molly M. Leung Department of Microbiology & Immunology, University of British Columbia, Vancouver, BC V6T 1Z3, Canada, mleung@interchange.ubc.ca

Keran Li Biology Department, Indiana University, Bloomington, IN 47405, USA, kerli@indiana.edu

Amanda Luther Department of Microbiology, The Ohio State University, 484 West 12th Avenue, Columbus, OH 43210-1292, USA, luther.29@osu.edu

Vicente Mariscal Instituto de Bioquímica Vegetal y Fotosíntesis, C.S.I.C. and Universidad de Sevilla, Américo Vespucio 49, E-41092 Seville, Spain, vicente.mariscal@ibvf.csic.es

Bernd Masepohl Lehrstuhl für Biologie der Mikroorganismen Fakultät für Biologie und Biotechnologie Ruhr-Universität Bochum, 44780 Bochum, Germany, bernd.masepohl@rub.de

Hajime Masukawa Research Institute for Photobiological Hydrogen Production Kanagawa University, 2946 Tsuchiya, Hiratsuka, Kanagawa, 259-1293, Japan, masukawa@waseda.jp

Crystal A. Miller Department of Biological Sciences, University of New Orleans, New Orleans, LA 70148, USA, camille2@uno.edu

Hongtao Min Department of Biological Sciences, Purdue University, West Lafayette, IN 47907, USA, hmin@purdue.edu

Rachael M. Morgan-Kiss Department of Microbiology, Miami University, Oxford, OH 45056, USA, morganr2@muohio.edu

Robert A. Niederman Department of Molecular Biology and Biochemistry, Rutgers University, Piscataway, NJ 08854, USA, rniederm@rci.rutgers.edu

Ozlem Onder Department of Biology, University of Pennsylvania, Philadelphia, PA 19104, USA, ozlem@sas.upenn.edu

Jörg Overmann Bereich Mikrobiologie, Department Biologie I, Ludwig-Maximilians-Universität München, Großhadernerstrasse 2-4, D-82152 Planegg-Martinsried, Germany, joerg.overmann@dsmz.de

Himadri B. Pakrasi Department of Biology, Washington University, St. Louis, MO 63130, USA, pakrasi@wustl.edu

Zach Poss School of Life Science, Arizona State University, P.O. Box 874501, Tempe, AZ 85287, USA, zach.poss@asu.edu

Hidehiro Sakurai Research Institute for Photobiological Hydrogen Production Kanagawa University, 2946 Tsuchiya, Hiratsuka, Kanagawa, 259-1293, Japan, sakurai@waseda.jp

Nicolle A. Saunée Department of Biological Sciences, University of New Orleans, New Orleans, LA 70148, USA, nsaunee@chnola-research.org

Wendy M. Schluchter Department of Biological Sciences, University of New Orleans, New Orleans, LA 70148, USA, wschluch@uno.edu

Christoph Schwarz School of Life Science, Arizona State University, P.O. Box 874501, Tempe, AZ 85287, USA, christoph.schwarz@asu.edu

Nur Selamoglu Department of Biology, University of Pennsylvania, Philadelphia, PA 19104, USA, nursela@sas.upenn.edu

Ina Severin Department of Marine Microbiology, Netherlands Institute of Ecology, NIOO-KNAW, P.O. Box 140, 4400 AC Yerseke, The Netherlands, i.severin@nioo.knaw.nl

Gaozhong Shen Department of Biochemistry and Molecular Biology, The Pennsylvania State University, University Park, PA 16802, USA, GXS22@psu.edu

Louis A. Sherman Department of Biological Sciences, Purdue University, West Lafayette, IN 47907, USA, lsherman@purdue.edu

Erko Stackebrandt DSMZ-Deutsche Sammlung von Mikroorganismen und Zellkulturen GmbH, Mascheroder Weg 1b, 38124 Braunschweig, Germany, erko@dsmz.de

Lucas J. Stal Department of Marine Microbiology, Netherlands Institute of Ecology, NIOO-KNAW, P.O. Box 140, 4400 AC Yerseke, The Netherlands, l.stal@nioo.knaw.nl

Jolantha Swiderski DSMZ-Deutsche Sammlung von Mikroorganismen und Zellkulturen GmbH, Mascheroder Weg 1b, 38124 Braunschweig, Germany, jswi@dsmz.de

F. Robert Tabita, Department of Microbiology, The Ohio State University, 484 West 12th Avenue, Columbus, OH 43210-1292, USA, tabita.1@osu.edu

Terumi A. Taylor Department of Microbiology & Immunology, University of British Columbia, Vancouver, BC, V6T 1Z3, Canada, ttaylor@interchange.ubc.ca

Jörg Toepel Department of Biological Sciences, Purdue University, West Lafayette, IN 47907, USA, lsherman@purdue.edu

Shervonda R. Williams Department of Biological Sciences, University of New Orleans, New Orleans, LA 70148, USA, shervondawilliams@ups.com

Kamil Woronowicz Department of Molecular Biology and Biochemistry, Rutgers University, Piscataway, NJ 08854, USA, kworonow@rci.rutgers.edu

Vladimir Yurkov Department of Microbiology, University of Manitoba, Winnipeg, MB R3T 2N2, Canada, vyurkov@cc.umanitoba.ca

Sébastien Zappa Biology Department, Indiana University, Bloomington, IN 47405, USA, szappa@indiana.edu

Part I
Diversity and Ecology

Chapter 1
A New Extreme Environment for Aerobic Anoxygenic Phototrophs: Biological Soil Crusts

Julius T. Csotonyi, Jolantha Swiderski, Erko Stackebrandt, and Vladimir Yurkov

Abstract Biological soil crusts improve the health of arid or semiarid soils by enhancing water content, nutrient relations and mechanical stability, facilitated largely by phototrophic microorganisms. Until recently, only oxygenic phototrophs were known from soil crusts. A recent study has demonstrated the presence of aerobic representatives of Earth's second major photosynthetic clade, the evolutionarily basal anoxygenic phototrophs. Three Canadian soil crust communities yielded pink and orange aerobic anoxygenic phototrophic strains possessing the light-harvesting pigment bacteriochlorophyll *a*. At relative abundances of 0.1–5.9% of the cultivable bacterial community, they were comparable in density to aerobic phototrophs in other documented habitats. 16S rDNA sequence analysis revealed the isolates to be related to *Methylobacterium*, *Belnapia*, *Muricoccus* and *Sphingomonas*. This result adds a new type of harsh habitat, dry soil environments, to the environments known to support aerobic anoxygenic phototrophs.

1.1 Introduction

Arid soils worldwide are threatened by erosion because their hydrologically stressed environment makes soil-binding plant roots relatively scarcer than in mesic habitats, a problem compounded by mechanical disturbances such as trampling by livestock. Global climate change may expand such fragile habitats in North America (Schlesinger et al. 1990; Belnap et al. 2004), triggering ancillary problems such as contamination of groundwater by erosion-mobilized nitrate from subsurface reservoirs (Walvoord et al. 2003). Biological soil crusts (BSC) are drought-tolerant communities composed primarily of microorganisms that form a cohesive, erosion-resistant organic veneer on the surface of arid soils (Belnap et al. 2001). They colonize large tracts of otherwise bare ground between vascular plants (Belnap et al. 2001; Reddy and Garcia-Pichel 2007) (Fig. 1.1), making the importance of their protective role obvious. Microorganisms, especially phototrophs, are the primary BSC

V. Yurkov (✉)
Department of Microbiology, University of Manitoba, Winnipeg, MB R3T 2N2, Canada
e-mail: vyurkov@cc.umanitoba.ca

P.C. Hallenbeck (ed.), *Recent Advances in Phototrophic Prokaryotes*, Advances in Experimental Medicine and Biology 675, DOI 10.1007/978-1-4419-1528-3_1,

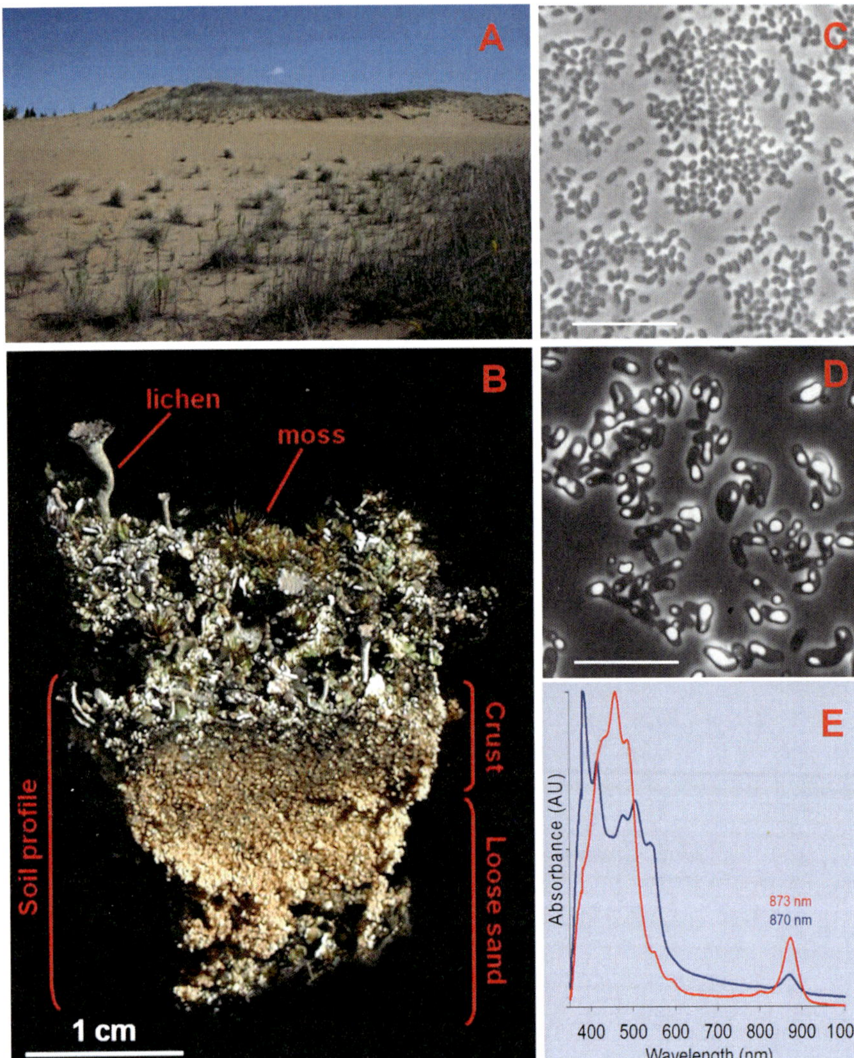

Fig. 1.1 Biological soil crust environment and characteristics of isolated phototrophs. (**a**) Source of samples S1, S2, S3: sand dune near Brandon, Manitoba. (**b**) Profile of BSC (sample R4). (**c**) and (**d**) Phase-contrast micrographs of ovoid to short rod-shaped strain JO5 and pleomorphic strain P194, respectively. Scale bars: 10 μm. (**e**) In vivo absorbance spectra of strains JO5 (*red trace*) and P194 (*blue trace*), highlighting BChl *a* incorporated into LH complexes I absorbing at 873 and 870 nm, respectively. Peaks at 400–550 nm and the small 800 nm peak are due to carotenoids and to BChl *a* integrated into the reaction centre, respectively

components that enhance soil moisture, nutrient content, structural development (via, e.g., adhesive polysaccharide production) and biogeochemical cycling of elements (Belnap et al. 2001). Consequently, understanding BSC microbial distribution and function will aid in the development of strategies to preserve soils that are crucial to agriculture and ranching. Several elegant and groundbreaking reports on BSC microbiological communities have been published recently (Nagy et al. 2005; Gundlapally and Garcia-Pichel 2006). Their application of genotypic analysis indicates that BSC contain a large reservoir of unfamiliar microbial taxa distantly related to known species (Nagy et al. 2005; Gundlapally and Garcia-Pichel 2006). However, these studies have only scratched the surface of a vast field of knowledge, and the subject remains poorly understood.

Phototrophic microorganisms are crucial to carbon cycling in BSC, because some transduce light to chemical energy and fix inorganic carbon, whereas others can accelerate its conversion to different forms. Oxygen-producing cyanobacteria are the most visible and hence best studied phototrophs in many crusts (Belnap et al. 2001; Nagy et al. 2005). By contrast, Earth's second major branch of photosynthetic organisms, the evolutionarily basal anoxygenic phototrophs, which generate no oxygen as a byproduct of photosynthesis, were unreported from BSC prior to the work of Csotonyi et al. (2010). Usually requiring anoxic conditions to harvest light, these "living fossils" are relicts of a 2 billion-year-old chemically reducing Earth. Therefore, most species are not expected to thrive in the aerobic BSC habitat. The obligately aerobic anoxygenic phototrophs (AAP), a recently discovered subgroup (Yurkov and Csotonyi 2003; Yurkov 2006; Yurkov and Csotonyi 2009), are exceptional in this regard, for they are well suited to the oxygen- and light-rich BSC. Although the AAP are primarily chemoheterotrophic, they nevertheless produce a functional photosynthetic reaction centre (RC) and one or more peripheral light-harvesting (LH) complexes with which they can amend heterotrophic energy generation by up to 20% (Yurkov and Van Gemerden 1993; Kolber et al. 2001). Five determinative traits distinguish AAP from classical anoxygenic phototrophs: (1) a requirement of O_2 for photosynthesis, freeing them from the need for restrictive anaerobic illuminated habitats; (2) the inhibition of bacteriochlorophyll (BChl) synthesis by light, which minimizes the formation of toxic triplet BChl in illuminated aerobic environments; (3) a low number of photosynthetic units (PSU) per cell; but (4) a great abundance of carotenoids; and (5) the absence of the Calvin cycle and inability to subsist on inorganic carbon (Yurkov and Csotonyi 2003; Rathgeber et al. 2004; Yurkov and Csotonyi 2009).

Unknown prior to 30 years ago (Shiba et al. 1979), research increasingly demonstrates the global importance of AAP (Yurkov and Csotonyi 2009). They are abundant, comprising up to about 20% of the photic marine microbial community (Yurkov and Csotonyi 2009), and the distribution of AAP even reaches into the deep ocean (Rathgeber et al. 2008). Their biogeochemical potential is also high: because AAP can rely on sunlight for 15–20% of their cellular energy, they accelerate rates of nutrient cycling over that of strict heterotrophs (Yurkov and Van Gemerden 1993; Kolber et al. 2001). Marine members of the abundant *Roseobacter* clade may even play a part in global climate change: by participating in the regulation of volatile cloud-seeding sulphur compounds (Wagner-Döbler and Biebl 2006), they may contribute to changes in Earth's albedo (Yurkov and Csotonyi 2009).

Biological soil crusts may be considered as a type of extreme environment, especially for microorganisms, which cannot easily escape physically from extreme soil microenvironments. The unsheltered exposure, and low heat and water capacity, of sandy BSC microhabitats results in a soil–atmosphere interface that intermittently reaches high temperature (alternating with extremely low winter temperature at high latitude), high ultraviolet radiation exposure, considerable desiccation by both sun and wind and oxidizing O_2-rich conditions. However, AAP are expected to survive such conditions because most species are exceptionally extremotolerant (Yurkov and Csotonyi 2003, 2009), constituting 30–36% of cultivable bacteria in some extreme environments (Yurkov and Beatty 1998b; Csotonyi et al. 2008). The first non-marine thermotolerant species were isolated from Russian thermal springs in 1990 (Yurkov and Beatty 1998a). Hypersaline environments and acidic coal mine tailings have also yielded numerous halotolerant and acidotolerant AAP (Yurkov and Csotonyi 2003). Perhaps the presence of an alternative mode of energy generation, namely phototrophy, has driven many AAP to evolve extremotolerance or extremophily as a selective advantage in competition with other species, for it would enhance their fitness when extreme conditions most severely compromise the vigour of cohabitants. Halotolerance is especially frequently observed in AAP, and a host of species inhabit meromictic lakes (Yurkova et al. 2002; Karr et al. 2003), salt flats (Yurkov and Csotonyi 2003) and hypersaline springs (Csotonyi et al. 2008). Yurkov and Csotonyi (2003) comprehensively reviewed the ubiquity of AAP in extreme environments. However, aside from enumerating and characterizing novel AAP, microbial ecologists have barely scratched the surface of the biology of AAP in extreme environments, and new types of environments are constantly found to yield novel AAP. For example, cultivation of anoxygenic phototrophs from hydrologically stressful environments implies possession of survival strategies against drought, an important trait not demonstrated for AAP prior to their isolation from BSC (Csotonyi and Yurkov unpublished).

As microbiologists converge on a better understanding of marine AAP ecology, attention is turning to environments in which AAP are more likely to interact with humans: soil, freshwater lakes and rivers. Interest is both academic and conservationist. First, the only known betaproteobacterial AAP representative (*Roseateles*) is riverine, and two AAP genera (*Craurococcus* and *Paracraurococcus*) were isolated from mesic urban soils, implying a large store of diversity (Yurkov and Csotonyi 2003). Second, enumeration of AAP in ecologically sensitive habitats would be a valuable contribution to catalogues of biodiversity in these regions of increasing human-induced environmental damage. An unsuccessful attempt to detect their diagnostic BChl pigment via HPLC suggested the absence of anoxygenic phototrophs from Sonoran BSC samples (Nagy et al. 2005), but this result may be attributed to characteristically low aerobic pigment production of AAP (Yurkov and Csotonyi 2009). Until recently, no other reports on assays of BSC habitats for anoxygenic phototrophs have been published.

These facts spurred a cultivation-based spectrophotometrically intensive search for anoxygenic phototrophs in three Canadian sites harbouring BSC communities. The basis of the research was the detection of absorbance of light in the

800–1,000 nm region in vivo and 768 nm region in acetone/methanol [7:2] extract, followed by 16S rRNA gene sequencing of BChl *a*-positive isolates. A sand dune (Fig. 1.1a) covered intermittently by drought-tolerant taxa of cyanobacteria (*Nostoc*), mosses (*Tortula*) and lichens was sampled near Brandon, Manitoba. Sandy soil near Marchand, Manitoba, capped by either a thin veneer of microorganisms at an early successional stage following logging or a well-developed BSC community about 0.5 cm thick, dominated by lichens (mainly *Cladonia*) and mosses (Fig. 1.1b), formed a second set of samples. A third set came from an established sand dune inhabited by cyanobacteria (*Nostoc*), mosses (*Ceratodon*) and lichens near the shores of Jasper Lake, Alberta.

1.2 Enumeration of AAP in Biological Soil Crusts

AAP were initially enumerated by culturing on rich organic media (Yurkov and Beatty 1998a). This technique effectively assesses the abundance of a physiological subset of the community and offers the benefit that all enumerated organisms can subsequently be examined in detail phenotypically and phylogenetically. However, the miniscule proportion of microorganisms that is amenable to cultivation spurs a growing interest in census techniques that circumvent the need to culture organisms. Regrettably, most culture-independent methods possess significant shortfalls, and measuring the abundance of AAP in nature is no trivial task. Based on the strategy employed, AAP candidates may be confused with either purple nonsulphur bacteria or closely related non-phototrophs (Yurkov and Csotonyi 2009). Thus far, three preferable techniques exist: (1) infrared fast repetition rate fluorometry, which unfortunately requires state-of-the-art instruments, but can measure and distinguish between aerobic and anaerobic photosynthetic electron transport; (2) quantitative polymerase chain reaction of *pufLM* (and of genes that can be used to discriminate between aerobic and anaerobic photosynthesis); and (3) cultivation, followed by 16S rDNA genetic analysis of cultured organisms. The recent Canadian BSC study successfully utilized the third strategy (Csotonyi et al. 2010).

Nine out of 12 samples (Table 1.1) yielded many carotenoid-rich bright pink and orange strains that possessed BChl *a* absorbing light at 870 or 873 nm in vivo (Fig. 1.1e) and at 770 nm in an acetone/methanol (7:2 v/v) extract. BChl *a* was found in strains from all three locations sampled, at rates of 13.9% (Marchand locality), 13.6% (Brandon sites) and 4.3% (Alberta location) of pigmented strains. Anoxygenic phototrophs constituted 0.1–5.9% of the total cultivable bacterial community (Table 1.1), comparable to the range of relative AAP abundance determined by several methods for other systems (Yurkov and Csotonyi 2009). Even so, the densities of AAP measured in the Marchand and Brandon sites ($\sim 10^5$–10^7 CFU/cm^3; Table 1.1) are likely only lower limits of annual population sizes, since sampling was performed during a dry period. Indeed, isolation of sand dune strains O47, P85 and P110 from sample dilutions of 10^8–10^{10} implies heterogeneous distribution and possibly extraordinarily high densities.

Table 1.1 Abundance of anoxygenic phototrophs in BSC

		Anoxygenic phototrophs		
Sample[a]	Medium of isolation	CFU/cm^3 of sample	Percent of all colonies	Strains[b]
C1	B	2.00×10^{6}	0.47	P4, P8
C2	B	3.57×10^{6}	2.28	**P12**, P13, P16, P18, P194
R1	B	1.00×10^{7}	1.03	P198
R3	A	8.33×10^{4}	0.16	*O11*
R4	B	1.11×10^{6}	5.88	P233
S1	B	4.76×10^{5}	0.0034	**P40**
	B	4.76×10^{7}	0.34	**P73**
	C	4.76×10^{7}	4	P67
S2	B	3.47×10^{9}	0.74	**O47**
	B	6.94×10^{11}	1.65	P110, P112
	B	3.47×10^{6}	0.0056	P44
	A	3.47×10^{10}	1.69	P85
S3	B	6.25×10^{5}	0.66	P132
N	A	2.50×10^{6}	0.63	**JO1**
	A	2.50×10^{5}	0.12	*JO5*

[a]C1–C2 and R1–R4, Marchand, Manitoba; S1–S3, Brandon, Manitoba; N, Jasper Lake, Alberta
[b]Strain name font face denotes proteobacterial subclass of phylogenetic affiliation: **bold**, alpha-1; underscored, alpha-2; *italic*, alpha-4; plain, not sequenced

1.3 Phylogenetic Analysis of Biological Soil Crust AAP

16S rDNA is currently actively used for identification of purified species and natural samples. This genetic standard is therefore a worthwhile basis on which to determine the relatedness of newly found organisms to characterized species. For natural environments, phylogenetic analysis provides us with an appreciation of the diversity and approximate number of novel species expected to be described.

Therefore, 16S rDNA analysis was chosen to analyse the collection of novel phototrophic BSC strains from Alberta and Manitoba (Csotonyi et al. 2010). Phylogenetic 16S rDNA sequencing of 16 representative strains (Table 1.1, Fig. 1.2) provided the first insight into the genotypic composition of anoxygenic phototrophic BSC communities. Interestingly, the nearest relative of each strain was non-phototrophic. This reflects the frequently encountered wide phylogenetic distribution of AAP (Yurkov and Csotonyi 2009). The aerobic phototrophic bacteria are closely allied with the purple nonsulphur bacteria, according to 5S and 16S rDNA analyses, DNA–DNA hybridization and 16S–23S internal-transcribed spacer sequence analysis (Yurkov and Csotonyi 2003). Also similar to purple nonsulphur bacteria, the AAP are not a homogeneous group, but instead are phylogenetically interspersed with phototrophic and heterotrophic proteobacteria (Yurkov and Beatty 1998a; Yurkov and Csotonyi 2009).

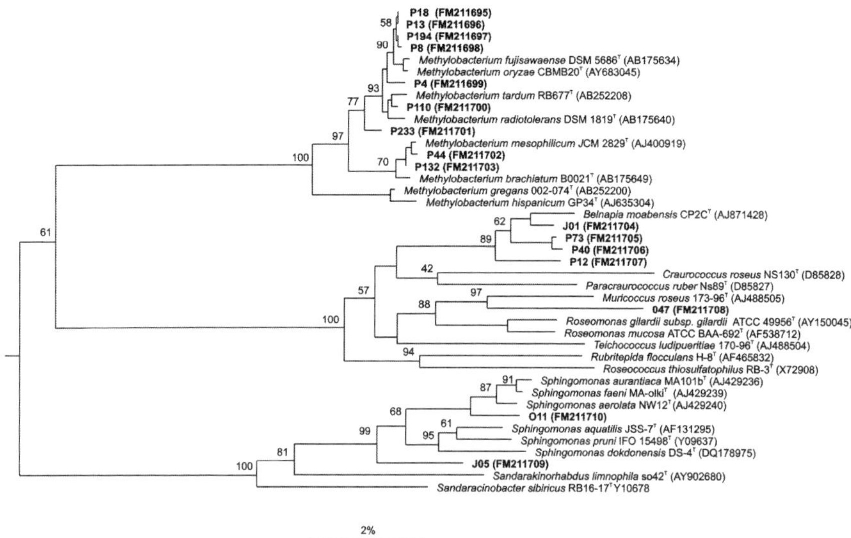

Fig. 1.2 Neighbour-joining phylogenetic tree showing relatedness of 16 BSC isolates based on 16S rDNA sequences more than 1,400 nucleotides long. The tree is rooted in *Agrobacterium tumefasciens* (AM181758). Accession numbers follow strain names. *Bootstrap values* are indicated at branch points

About half of the strains (P4, P8, P13, P18, P44, P110, P132, P194, P233) were pleomorphic anoxygenic phototrophs related to species of the order Rhizobiales, specifically the genus *Methylobacterium* (98.8–99.5% sequence similarity) (Fig. 1.2). The isolation of strains P44 and P110 from moss-rich BSC is consistent with species of the *Rhizobiales* often associating closely with plants (Fleischman and Kramer 1998). Five coccoidal alpha-1-proteobacterial strains (JO1, O47, P12, P40, P73) were recovered from all three locations, suggesting a widespread group of BSC anoxygenic phototrophs. Three of these (JO1, P40, P73) originated from *Nostoc*-dominated crusts, implying a cyanobacterial association (Table 1.1). Strain O47 was marginally related (94.4% similarity; Fig. 1.2) to the non-phototrophic *Muricoccus roseus* (Kämpfer et al. 2003), which suggests a new genus affiliation. The remaining strains were most closely affiliated (97.4–98.8% sequence similarity; Fig. 1.2) with *Belnapia moabensis*, the first heterotrophic bacterium described from BSC. This non-phototroph is related to the AAP *Craurococcus* and *Paracraurococcus*, described from mesic soils (Reddy et al. 2006). The Marchand and Jasper Lake sites also yielded rod-shaped alpha-4-proteobacteria (JO5, O11) related to non-phototrophic species of *Sphingomonas* (96.6–98.5% similarity) (Fig. 1.2).

Although no phototrophs were recovered from Colorado Plateau BSC, 4.8% of all PCR-amplified sequences from DGGE fingerprints of those samples were alphaproteobacterial (Gundlapally and Garcia-Pichel 2006). The Canadian phototrophic isolates were all alphaproteobacterial. Of course, further investigation and more extensive sampling will be required to determine whether BSC phototrophs are restricted to the Alphaproteobacteria or if they are members of other classes as well.

1.4 Photosynthetic Pigments of Newly Isolated AAP

Thus far, all AAP are known to produce BChl *a*, but not BChl *b*, found in some purple nonsulphur bacteria. All known AAP incorporate BChl into a reaction centre (RC) (absorbing light primarily of 800 nm) and at least one light-harvesting (LH) complex, typically absorbing at about 870 nm, although some interesting exceptions exist, as reviewed by Yurkov and Csotonyi (2009). Some AAP also produce a second LH complex, such as the B832-type complex of *Erythromicrobium* and *Porphyrobacter* species (named for its peak absorbance at 832 nm; Yurkov and Csotonyi 2009) or the B806 found in *Roseobacter* (Shimada et al. 1985) and *Roseicyclus* (Rathgeber et al. 2005).

The photosynthetic pigment analysis of BSC AAP did not produce any surprises or novelties and confirmed that they all produce BChl *a*. Absorbance of light by cells at 870 and 873 nm implied incorporation of BChl *a* into a single LH complex I, whereas a peak at 800 nm indicated the presence of a functional photosynthetic RC (Fig. 1.1e). Typically, AAP synthesize about 0.05–2 nmol of BChl per mg dry weight (Yurkov and Csotonyi 2003). All isolated BSC strains synthesized BChl *a* in quantities at the low end of this range. Strains JO1, JO5, O47, P40, P110 and P194 produced only 0.07, 0.19, 0.07, 0.06, 0.16 and 0.19 nmol/mg dry weight, respectively.

Low BChl *a* levels in pure cultures might underestimate the pigment's relative importance in nature. Biosynthesis of BChl in AAP is sometimes finely tuned to physico-chemical environmental cues that may be lacking in the laboratory (Biebl and Wagner-Döbler 2006; Yurkov and Csotonyi 2009). For example, whereas marine *Erythrobacter* relative NAP1 grows optimally at 32.5°C, BChl production is maximal at 22.5°C and halted at temperatures above 30°C (Koblížek et al. 2003). Pigment expression in *Thalassobacter stenotrophicus*, *Hoeflea phototrophica* and *Citromicrobium bathyomarinum* is salt sensitive, despite the marine habitat of each species; *T. stenotrophicus* is unpigmented at 7% salinity (Macián et al. 2005), *H. phototrophicum* produces most BChl at 0.6% salinity but no BChl at 3.5% salinity (Biebl et al. 2006) and *C. bathyomarinum* generates the most BChl in the absence of NaCl (Rathgeber et al. 2004). Studies have demonstrated that AAP tend to upregulate pigment synthesis under suboptimal nutrient conditions, implying that photosynthesis proves especially advantageous under oligotrophic conditions (Yurkov and Csotonyi 2009). Sometimes, other stressful conditions can enhance BChl production as well. For instance, thermophilic *R. flocculens* synthesizes BChl at 30°C, but not at its preferred growth temperature of 50°C (Alarico et al. 2002). Indeed, AAP may produce much more BChl under very specific conditions of native habitats, and competitive superiority may depend on a very small energetic advantage.

Low BChl content in actively growing cultures may also be explained by alternative roles of BChl in AAP, in addition to classic photosynthetic light-assimilating function. For instance, light-induced cyclic electron flow in *Roseateles depolymerans* was hypothesized to help maintain the electrochemical proton gradient of the plasma membrane during starvation conditions rather than contribute to biomass production (Suyama et al. 2002). BChl may thus contribute to survival under stressful conditions. Similar strategies could help explain the peculiar abundance of AAP in extreme or harsh environments (Yurkov and Csotonyi 2009), including BSC.

In contrast to their low cellular levels of BChl, BSC anoxygenic phototrophs produced copious pink and orange carotenoids, absorbing in the 400–550 nm range. Profuse synthesis of carotenoids is a determinative trait of AAP, conferring on them a diverse range of intense colours, including red, brown, orange, yellow, pink, purple and intermediates (Yurkov 2006; Yurkov and Csotonyi 2009). Interestingly, contrary to the light-harvesting function of carotenoids in anaerobic purple phototrophs, the majority of AAP carotenoids are not incorporated into an energy transduction pathway, but are distributed evenly throughout the cytoplasmic membrane, cytoplasm and cell wall (Yurkov and Beatty 1998a; Koblížek et al. 2003). Although the primary purpose of this photosynthetically disengaged pool of carotenoids is uncertain, a leading suggestion is preventative or remediative amelioration of phototoxicity (Fraser et al. 2001; Beatty 2002), which is an especially great risk in the ultraviolet-rich aerobic environment of BSC.

1.5 Conclusions and Prospective Research

Oxygenic phototrophs such as cyanobacteria, algae and mosses have been well known from BSC since this type of biological community was studied from a soil management perspective nearly a century ago. Although the second of the two branches of phototrophs on Earth, anoxygenic phototrophs, has been known to science for a relatively long time, their occurrence in BSC has only now been demonstrated (Csotonyi et al. 2010). The presence of APP implies a higher efficiency of light harvesting by soil crust organisms than previously realized. Their utilization of the near-IR radiation not used by oxygenic phototrophs facilitates light-accelerated turnover of soil organic carbon content by these unusual heterotrophs.

An area requiring further study is the enhancement of soil physical structure by anoxygenic phototrophs, which may be important because of the aggregative growth habit of some species. Several strains of AAP from meromictic lakes and salt springs produce holdfast structures or copious extracellular adhesive materials that facilitate intense flocculation (Yurkov and Csotonyi 2009). Production of adhesive substances for anchorage would also be advantageous for survival on loose windy sand dunes as it would increase soil tensile strength and facilitate establishment of BSC, promoting soil structural development following mechanical disturbance. Erosion-reducing capacity is especially important because it would circumvent the loss of a recently discovered large subsurface nitrate pool that exists beneath arid soils, and its consequent contamination of groundwater (Walvoord et al. 2003).

AAP species are also particularly resistant to numerous toxic metal oxides and are therefore potentially key players in the biogeochemical cycling of these elements (Yurkov and Csotonyi 2003, 2009). AAP can resist and chemically reduce very high concentrations of metal or metalloid oxyanions such as tellurite, selenite, vanadate and arsenate (Yurkov and Csotonyi 2003), even when their immediate habitat is known to lack significant concentrations of these elements (Csotonyi et al. 2008). For example, they can reduce tellurite to elemental Te at concentrations up to 2,700 μg/ml (Yurkov et al. 1996). On soils contaminated by tailings from metal (e.g. gold) mines, soil crusts may be critical to the structural integrity of soils that are relatively depleted in vascular plants. The AAP component in BSC may therefore fill an important role in detoxifying hazardous pollutant elements mobilized through soils by erosion or evaporation.

Yet another burgeoning industrial application of AAP from BSC is the environmentally sustainable manufacture of biodegradable plastics, the subject of much ongoing research (Nath et al. 2008). The pleomorphic cells of pink *Methylobacterium* relatives from Canadian BSC possessed very large quantities of intracellular polyhydroxyalkanoate inclusions (Fig. 1.1d), the raw materials for bioplastic production (Nath et al. 2008).

Biogeographically, AAP are very widely distributed and have been found in many types of environments, from ocean depths to high-altitude lakes, saline springs and acidic mine drainage (Csotonyi et al. 2008; Jiang et al. 2009; Yurkov and Csotonyi 2009). Therefore, it will also be interesting to determine by cultivation and genotypic studies whether the presence of anoxygenic phototrophs is a cosmopolitan feature of BSC or whether they are characteristic only of high-latitude locations such as the Canadian sites from 50 to 53°N. The frequently demonstrated extremotolerance of AAP suggests that at least this particular component of the anoxygenic BSC community may increase in proportion with decreasing latitude and escalating physical stress. The role and distribution of anoxygenic phototrophs such as AAP in the pivotally important BSC biological communities is certain to be the centre of extensive research in the near future.

Acknowledgements This research was supported by a Natural Science and Engineering Research Council (NSERC) operating grant to V.Yurkov.

References

Alarico S, Rainey FA, Empadinhas N, Schumann P, Nobre MF, Da Costa MS (2002) *Rubritepida flocculans* gen. nov., sp. nov., a new slightly thermophilic member of the α-1 subclass of the *Proteobacteria*. Syst Appl Microbiol 25:198–206

Beatty JT (2002) On the natural selection and evolution of the aerobic phototrophic bacteria. Photosynth Res 73:109–114

Belnap J, Kaltenecker JH, Rosentreter R, Williams J, Leonard S, Eldridge D (2001) Biological soil crusts: ecology and management (BLM Technical Reference 1730–1732). United States Bureau of Land Management, Washington, DC

Belnap J, Phillips SL, Miller, ME (2004) Response of desert biological soil crusts to alterations in precipitation frequency. Oecologia 141:306–316

Biebl H, Wagner-Döbler I (2006) Growth and bacteriochlorophyll *a* formation in taxonomically diverse aerobic anoxygenic phototrophic bacteria in chemostat culture: influence of light regimen and starvation. Proc Biochem 41:2153–2159

Biebl H, Tindall BJ, Pukall R, Lünsdorf H, Allgaier M, Wagner-Döbler I (2006) *Hoeflea phototrophica* sp. nov., a novel marine aerobic alphaproteobacterium that forms bacteriochlorophyll *a*. Int J Syst Evol Microbiol 56:821–826

Csotonyi JT, Swiderski J, Stackebrandt E, Yurkov V (2008) Novel halophilic aerobic anoxygenic phototrophs from a Canadian hypersaline spring system. Extremophiles 12:529–539

Csotonyi JT, Swiderski J, Stackebrandt E, Yurkov V (2010) A new environment for anoxygenic phototrophic bacteria: biological soil crusts. Environ Microbiol Rep. doi: 10.1111/j.1758–2229.2010.00151.x

Fleischman D, Kramer D (1998) Photosynthetic rhizobia. Biochim Biophys Acta 1364:17–36

Fraser NJ, Hashimoto H, Cogdell RJ (2001) Carotenoids and bacterial photosynthesis: the story so far.... Photosynth Res 70:249–256

Gundlapally SR, Garcia-Pichel F (2006) The community and phylogenetic diversity of biological soil crusts in the Colorado plateau studied by molecular fingerprinting and intensive cultivation. Microb Ecol 52:345–357

Jiang H, Dong H, Yu1 B, Lv G, Deng S, Wu Y, Dai M, Jiao N (2009) Abundance and diversity of aerobic anoxygenic phototrophic bacteria in saline lakes on the Tibetan plateau. FEMS Microbiol Ecol 67:268–278

Kämpfer P, Andersson MA, Jäckel U, Salkinoja-Salonen M (2003) *Teichococcus ludipueritiae* gen. nov. sp. nov., and *Muricoccus roseus* gen. nov. sp. nov. representing two new genera of the α-1 subclass of the *Proteobacteria*. Syst Appl Microbiol 26: 23–29

Karr EA, Sattley WM, Jung DO, Madigan MT, Achenbach LA (2003) Remarkable diversity of phototrophic purple bacteria in a permanently frozen Antarctic lake. Appl Environ Microbiol 69:4910–4914

Koblížek M, Béjà O, Bidigare RR, Christensen S, Benetiz-Nelson B, Vetriani C, Kolber MK, Falkowski PG, Kolber ZS (2003) Isolation and characterization of *Erythrobacter* sp. strains from the upper ocean. Arch Microbiol 180:327–338

Kolber ZS, Plumley FG, Lang AS, Beatty JT, Blankenship RE, Van Dover CL, Vetriani C, Koblížek M, Rathgeber C, Falkowski PG (2001) Contribution of aerobic photoheterotrophic bacteria to the carbon cycle in the ocean. Science 292:2492–2495

Macián MC, Arahal DR, Garay E, Ludwig W, Schleifer KH, Pujalte MJ (2005) *Thalassobacter stenotrophicus* gen. nov., sp. nov., a novel marine α-proteobacterium isolated from Mediterranean sea water. Int J Syst Evol Microbiol 55:105–110

Nagy ML, Pérez A, Garcia-Pichel F (2005) The prokaryotic diversity of biological soil crusts in the Sonoran Desert (Organ Pipe Cactus National Monument, AZ). FEMS Microbiol Ecol 54:233–245

Nath A, Dixit M, Bandiya A, Chavda S, Desai AJ (2008) Enhanced PHB production and scale up studies using cheese whey in fed batch culture of *Methylobacterium* sp. ZP24. Bioresour Technol 99:5749–5755

Rathgeber C, Beatty JT, Yurkov V (2004) Aerobic phototrophic bacteria: new evidence for the diversity, ecological importance and applied potential of this previously overlooked group. Photosynth Res 81:113–128

Rathgeber C, Yurkova N, Stackebrandt E, Schumann P, Beatty JT, Yurkov V (2005) *Roseicyclus mahoneyensis* gen nov., sp. nov., an aerobic phototrophic bacterium isolated from a meromictic lake. Int J Syst Evol Microbiol 55:1597–1603

Rathgeber C, Lince MT, Alric J, Lang AS, Humphrey E, Blankenship RE, Verméglio A, Plumley FG, Van Dover CL, Beatty JT, Yurkov V (2008) Vertical distribution and characterization of aerobic phototrophic bacteria at the Juan de Fuca Ridge in the Pacific Ocean. Photosynth Res 97:235–244

Reddy GS, Garcia-Pichel F (2007) *Sphingomonas mucosissima* sp. nov. and *Sphingomonas desiccabilis* sp. nov., from biological soil crusts in the Colorado plateau, USA. Int J Syst Evol Microbiol 57:1028–1034

Reddy GSN, Nagy M, Garcia-Pichel F (2006) *Belnapia moabensis* gen. nov., sp. nov., an alphaproteobacterium from biological soil crusts in the Colorado plateau, USA. Int J Syst Evol Microbiol 56:51–58

Schlesinger WH, Reynolds JF, Cunningham GL, Huenneke LF, Jarrell WM, Virginia RA, Whitford WH (1990) Biological feedbacks in global desertification. Science 247:1043–1048

Shiba T, Simidu U, Taga N (1979) Distribution of aerobic bacteria which contain bacteriochlorophyll *a*. Appl Environ Microbiol 38:43–45

Shimada K, Hayashi H, Tasumi M (1985) Bacteriochlorophyll-protein complexes of aerobic bacteria, *Erythrobacter longus* and *Erythrobacter* species OCh 114. Arch Microbiol 143:244–247

Suyama T, Shigematsu T, Suzuki T, Tokiwa Y, Kanagawa T, Nagashima KVP, Hanada S (2002) Photosynthetic apparatus in *Roseateles depolymerans* 61A is transcriptionally induced by carbon limitation. Appl Environ Microbiol 68:1665–1673

Wagner-Döbler I, Biebl H (2006) Environmental biology of the marine *Roseobacter* lineage. Annu Rev Microbiol 60:255–280

Walvoord MA, Phillips FM, Stonestrom DA, Evans RD, Hartsough PC, Newman BD, Striegl RG (2003) A reservoir of nitrate beneath desert soils. Science 302:1021–1024

Yurkov V (2006) Aerobic phototrophic bacteria. In: Dworkin M, Falkow S, Rosenberg E, Schleifer K-H, Stackebrandt E (eds) Prokaryotes, 3rd edn. Springer, Berlin

Yurkov VV, Beatty JT (1998a) Aerobic anoxygenic phototrophic bacteria. Microbiol Mol Biol Rev 62:695–724

Yurkov V, Beatty JT (1998b) Isolation of aerobic anoxygenic photosynthetic bacteria from black smoker plume waters of the Juan de Fuca Ridge in the Pacific Ocean. Appl Environ Microbiol 64:337–341

Yurkov VV, Csotonyi JT (2003) Aerobic anoxygenic phototrophs and heavy metalloid reducers from extreme environments. In: Pandalai SG (ed) Recent research developments in bacteriology, vol 1. Transworld Research Network, Trivandrum

Yurkov V, Csotonyi JT (2009) New light on aerobic anoxygenic phototrophs. In: Hunter N, Daldal F, Thurnauer MC, Beatty JT (eds) The purple phototrophic bacteria. Springer, New York

Yurkov V, Van Gemerden H (1993) Impact of light/dark regimen on growth rate, biomass formation and bacteriochlorophyll synthesis in *Erythromicrobium hydrolyticum*. Arch Microbiol 159:84–89

Yurkov V, Jappe J, Vermeglio A (1996) Tellurite resistance and reduction by obligately aerobic photosynthetic bacteria. Appl Environ Microbiol 62:4195–4198

Yurkova N, Rathgeber C, Swiderski J, Stackebrandt E, Beatty JT, Hall KJ, Yurkov V (2002) Diversity, distribution and physiology of the aerobic phototrophic bacteria in the mixolimnion of a meromictic lake. FEMS Microbiol Ecol 40:191–204

Chapter 2
The Phototrophic Consortium "*Chlorochromatium aggregatum*" – A Model for Bacterial Heterologous Multicellularity

Jörg Overmann

Abstract Phototrophic consortia currently represent the most highly developed interspecific association between prokaryotes and consist of green sulfur bacterial epibionts which surround a central, motile, chemotrophic bacterium. Several independent experimental findings indicate that a rapid signal transfer occurs between the epibionts and the central bacterium. First, the cell division of the partner bacteria occurs in a highly coordinated fashion. Second, consortia accumulate scotophobotactically in the light, whereby the central bacterium confers motility to the consortium and the epibionts act as light sensors. Third, the organic carbon uptake of the central bacterium seems to be controlled by the epibiont. A decade ago, a laboratory culture of the phototrophic consortium "*Chlorochromatium aggregatum*" could be established and maintained. Using "*C. aggregatum*," recent genomic, transcriptomic, and proteomic studies have started to unravel the molecular basis of prokaryotic heterologous multicellularity in this model system.

2.1 Introduction

Consortia are close associations of bacteria maintaining a permanent cell-to-cell contact and an organized arrangement of cells (Schink 1991, 2002). To date, 19 different morphological types of bacterial consortia have been recognized, including, among others, the "corn-cob" bacterial formations in the human oral cavity, the anaerobic methane-oxidizing consortia in deep sea sediments, and the consortia of methanogenic archaea surrounding endospore-forming bacteria in the termite hindgut (Overmann 2001). Monospecific cell–cell interactions have also been documented between cells of a single *Deltaproteobacterium* species which constitute highly structured magnetotactic multicellular prokaryotes (Wenter et al. 2009). Microbial consortia are not only relevant for maintaining biogeochemical cycles in different environments but also of medical (e.g., in dental plaque;

J. Overmann (✉)
Bereich Mikrobiologie, Department Biologie I, Ludwig-Maximilians-Universität München, Großhadernerstrasse 2-4 D-82152, Planegg-Martinsried, Germany
e-mail: joerg.overmann@dsmz.de

P.C. Hallenbeck (ed.), *Recent Advances in Phototrophic Prokaryotes*, Advances in Experimental Medicine and Biology 675, DOI 10.1007/978-1-4419-1528-3_2,

Whittaker et al. 1996) and technological significance (e.g., in wastewater treatment; De Bok et al. 2004). Still, previous research on symbiotic interactions involving prokaryotes has largely focused on their associations with higher eukaryotes. It is likely that purely bacterial interactions like those in consortia preceded pathogenic or symbiotic relationships with eukaryotes. Therefore, a better understanding of the molecular mechanisms of bacterial heterologous multicellularity will have implications for the elucidation of the evolution and mechanistic basis of human or plant pathogens.

Phototrophic consortia are tightly packed cell aggregates formed by colorless bacterial cells and green sulfur bacteria (Chlorobiaceae). The spatial arrangement of the cells is not at random but rather always occurs in a highly ordered fashion. In the natural environment, the most frequently observed morphological type of phototrophic consortia is barrel shaped and motile (Overmann 2006). This type was already discovered more than a century ago (Lauterborn 1906) and consists of up to 69 green sulfur bacterial epibionts which surround a spindle-shaped, motile, and colorless bacterium in the center (Overmann 2001, 2006). In "*Chlorochromatium aggregatum*," the colorless bacterium is surrounded by green-colored, rod-shaped bacteria, while brown epibionts are found in "*Pelochromatium roseum.*" These smaller consortia are spindle shaped and typically harbor up to 20 epibiont cells. Significantly larger are "*Chlorochromatium magnum*," "*Pelochromatium latum*," and "*Pelochromatium roseo-viride*," which display a globular shape and contain $\geq$40 epibionts. The epibionts of "*C. magnum*" are green colored, whereas those in "*P. latum*" are brown. Interestingly, "*P. roseo-viride*" contains an inner layer of brown-colored and an outer layer of green-colored epibionts. The green epibionts in "*Chlorochromatium glebulum*" contain gas vesicles. Crescent-shaped green or brown epibionts are found in "*Chlorochromatium lunatum*" and "*Pelochromatium selenoides*," respectively. Phototrophic consortia of the type "*C. glebulum*" are bent and contain green epibionts. Finally, two non-motile morphotypes of phototrophic consortia are known which exhibit a different arrangement of the two associated bacteria. In "*Chloroplana vacuolata*," long slender colorless and gas-vacuolated rods alternate with chains of rod-shaped and gas-vacuolated green cells, thereby forming a sheath-like structure with up to 400 cells of green bacteria. "*Cylindrogloea bacterifera*" consists of green-colored bacteria surrounding a central chain of colorless bacteria with thick capsules. Among the known consortia, these phototrophic consortia probably represent the most highly specialized type of association and have reached the highest degree of mutual interdependence between non-related bacteria.

Phototrophic consortia have been found in the chemocline of many freshwater lakes worldwide (Caldwell and Tiedje 1975; Croome and Tyler 1984; Gorlenko 1988; Overmann and Tilzer 1989; Eichler and Pfennig 1990; Gasol et al. 1995; Glaeser and Overmann 2004). Consortia with brown-colored epibionts are usually found at greater water depths than their green-colored counterparts. Similar to free-living green sulfur bacteria, the phototrophic consortia carrying brown-colored epibionts thus seem to have a selective advantage at greater depths due to their increased absorption in the blue-green or green portion of the wavelength spectrum which typically reaches these deeper layers (Overmann et al. 1998). In certain lakes, the biomass of phototrophic consortia amounts to two-thirds of the total bacterial biomass which is present in the chemocline (Gasol et al. 1995), suggesting a significance of these associations for the biogeochemical cycles in these ecosystems. Under certain conditions, up to 88% of all green sulfur bacteria occur as epibionts in phototrophic consortia rather than as free-living cells (Glaeser and Overmann 2003a). This, together with the large diversity of phototrophic consortia maintained in nature, indicates that an association with the colorless bacterium provides the green sulfur bacterial epibiont with a selective advantage over an independent lifestyle.

Based on the stable and highly ordered structure, the two non-related bacterial partners in phototrophic consortia must have developed specific recognition mechanism, specific means of intercellular communication, and a high degree of mutual physiological interdependence. Elucidating these processes in phototrophic consortia thus would contribute toward a general understanding of interorganismic interactions among prokaryotes.

Over 90 years after the discovery of the phototrophic consortia, the first enrichment culture of "*C. aggregatum*" could be established employing an anoxic mineral medium supplemented with 2-oxoglutarate (Fröstl and Overmann 1998). This culture represents the first laboratory model system allowing detailed studies of the physiological interactions and the molecular basis of bacterial heterologous multicellularity.

2.2 Identification of the Partner Bacteria in Phototrophic Consortia

2.2.1 The Epibiont

Based on their color and the presence of chlorosomes as documented by electron microscopy, it had been concluded that the epibionts of phototrophic consortia belong to the green sulfur bacteria (Caldwell and Tiedje 1975). In fact, fluorescent in situ hybridization with an oligonucleotide probe specific for green sulfur bacteria (Chlorobiaceae) confirmed the phylogenetic affiliation of the epibiont cells (Tuschak et al. 1999). However, epibionts of different phototrophic consortia represent novel and unique 16S rRNA gene sequence types within the radiation of green sulfur bacteria (Fröstl and Overmann 2000; Glaeser and Overmann 2004).

A subsequent worldwide investigation of the 16S rRNA gene sequences of epibionts from natural populations of phototrophic consortia in 14 different lakes revealed that all epibionts in a particular type of phototrophic consortium taken from the same lake invariably belonged to one single sequence type. Each sequence type represented a distinct and novel branch within the radiation of green sulfur bacteria (Glaeser and Overmann 2004). Interestingly, morphologically indistinguishable phototrophic consortia, when collected from different lakes, were found to harbor genetically different epibionts. As an example, "*C. aggregatum*" consortia sampled from European and North American lakes contained seven different epibiont phylotypes. It is to be concluded that the epibionts of phototrophic consortia are significantly more diverse than judged just from the consortia morphotypes as recognized by light microscopy (see Section 2.1). Based on these 16S rRNA gene sequence analyses, the current estimate amounts to 19 different sequence types of epibionts.

Phylogenetic analyses also provide some insights into the specificity and evolution of the bacterial interaction in phototrophic consortia. If phototrophic consortia would form randomly from bacterial cells which encounter each other just by chance, morphologically identical consortia from the same lake would be expected to harbor different sequence types of green sulfur bacteria, and the same sequence types should also be detectable as free-living cells in the environment. Moreover, even one single phototrophic consortium could contain different sequence types of green sulfur bacteria. However, none of the above three predictions could be verified. Rather, all phototrophic consortia with the same morphology that share the same habitat contain only a single type of epibiont and none of the 16S rRNA gene sequences of epibionts could be detected in free-living green sulfur bacteria. In addition, epibiont sequences do not form a monophyletic group within the Chlorobiaceae (Glaeser and Overmann 2004). These phylogenetic analyses suggest (1) that the epibionts always occur associated with the central bacterium under natural conditions and (2) that the ability to form symbiotic associations either arose independently from different ancestors or was present in a common ancestor prior to the radiation of green sulfur bacteria and the transition to the free-living state in independent lineages. The possibility of multiple origins of phototrophic consortia then raises a question as to the mechanism by which different lineages of green sulfur bacteria could independently acquire the genetic determinants for heterologous multicellularity (compare Section 2.5).

Recently, one type of epibiont from the phototrophic consortium "*C. aggregatum*" could be isolated in pure culture using anoxic media supplemented with dithionite and fermented rumen extract (Vogl et al. 2006). Similar to other green sulfur bacteria, the isolate *Chlorobium chlorochromatii* CaD is non-motile, obligately anaerobic, and photolithoautotrophic with sulfide as electron donor in the free-living state. Exhaustive physiological testing of this strain *Chl. chlorochromatii* CaD did not reveal any unusual capabilities as compared to known Chlorobiaceae. Acetate and peptone are photoassimilated in the presence of sulfide and bicarbonate. Furthermore, in situ measurements of light-dependent $H^{14}CO_3^-$ fixation in a natural population of phototrophic consortia and determination of the stable carbon isotope ratios ($\delta^{13}C$) of their bacteriochlorophyll molecules indicated that photoautotrophic growth of the epibionts also occurs in the associated state, under natural conditions, and employing the reverse tricarboxylic acid cycle (Glaeser and Overmann 2003a). The capability of the isolate *Chl. chlorochromatii* strain CaD to grow in pure culture indicates that it is not an obligately symbiotic bacterium.

2.2.2 The Central Bacterium

The central bacterium of the barrel-shaped phototrophic consortia typically is a rod-shaped bacterium with tapered ends that typically exhibits only low phase contrast

(Overmann et al. 1998). Electron microscopy revealed that the central bacterium is monopolarly monotrichously flagellated (Glaeser and Overmann 2003b). While the central bacterium is assumed to grow chemoheterotrophically, no firm conclusions can be drawn with respect to its physiology since this bacterium cannot so far be cultured separately from its epibionts.

It had been suggested earlier that the central bacterium might be a sulfur- or sulfate-reducing bacterium (see Section 2.4). Since sulfur- or sulfate-reducing bacteria typically belong to either the Deltaproteobacteria or the Firmicutes, it came as a surprise when the colorless central bacteria in "*C. aggregatum*" as well as "*C. magnum*" obtained from two lakes in Eastern Germany and the United States, respectively, was identified as a member of the Betaproteobacteria based on fluorescent in situ hybridization (Fröstl and Overmann 2000).

Chemotactic enrichment of "*C. magnum*" and sequencing of the 16S rRNA gene revealed that the central bacterium of this consortium represents a so far isolated phylogenetic lineage distantly related to *Rhodoferax* spp., *Polaromonas vacuolata*, and *Variovorax paradoxus* within the family Comamonadaceae (Kanzler et al. 2005). The majority of the relatives of this lineage are not yet cultured and are found in low-temperature aquatic environments and/or aquatic environments containing xenobiotics or hydrocarbons.

Subsequent molecular analyses of purified genomic DNA from the central bacterium of "*C. aggregatum*" demonstrated the presence of two *rrn* operons which are arranged in a tandem fashion and separated by an unusually short intergenic region of only 195 base pairs. When this gene order was exploited to screen natural communities by PCR, a previously unknown and diverse subgroup of Betaproteobacteria was detected in the chemocline of stratified freshwater lakes. This group was absent in soil microbial communities or other aquatic ecosystems. Using fluorescent in situ hybridization with oligonucleotide probes specific for this subgroup, some of the sequences obtained can be attributed to the central bacteria of different consortia. Most notably, phylogenetic analyses reveal that the central bacteria of other populations of phototrophic consortia are only distantly related to the colorless bacterium in "*C. magnum*" (Pfannes et al. 2007). This indicates that the chemotrophic symbionts in phototrophic consortia have a polyphyletic origin similar to the epibionts.

2.3 Evidence for a Direct Interaction Between the Two Partners in Phototrophic Consortia

2.3.1 Aggregate Structure and Cell Division

In natural populations as well as laboratory cultures of the motile, barrel-shaped phototrophic consortia, the number of epibiont cells per central bacterium shows a nonrandom frequency distribution. Depending on the type and origin of the phototrophic consortium, the most frequent number of epibionts determined per consortium was 11 ("*C. aggregatum*" from Lake Cisó; Gasol et al. 1995), 13 ("*C. aggregatum*" from Lake Grünenplan; Overmann et al. 1998), 20 ("*P. roseum*" from Lake

Dagow; Overmann et al. 1998), or 36 ("*C. glebulum*" from Lake Dagow; Overmann et al. 1998). The nonrandom number of epibiont cells which is consistently observed in each population of phototrophic consortia strongly suggests that the number of epibiont cells is tightly controlled during the growth and division of individual consortia. The more numerous epibionts in the consortia "*C. glebulum*," "*C. magnum*," and "*P. roseo-viride*" form two consecutive layers, with cells of the outer layer being not directly attached to their heterotrophic partner bacterium. This fact clearly shows that the number of epibiont cells is not simply controlled by the area provided by the cell surface of the central bacterium for attachment of epibionts.

When consortia were partially disaggregated, either during the preparation of samples for electron microscopy or in squash preparations of wet mounts on agar-coated slides for light microscopy, the central bacterium was often found to exhibit a pronounced invagination and hence to be in a stage of cell division (Overmann et al. 1998; Fig. 2.1). This observation, together with the nonrandom frequency distribution of epibionts in phototrophic consortia, strongly suggests that the cell division of all epibionts of the same consortium proceeds in a highly coordinated fashion and parallel to that of the central bacterium.

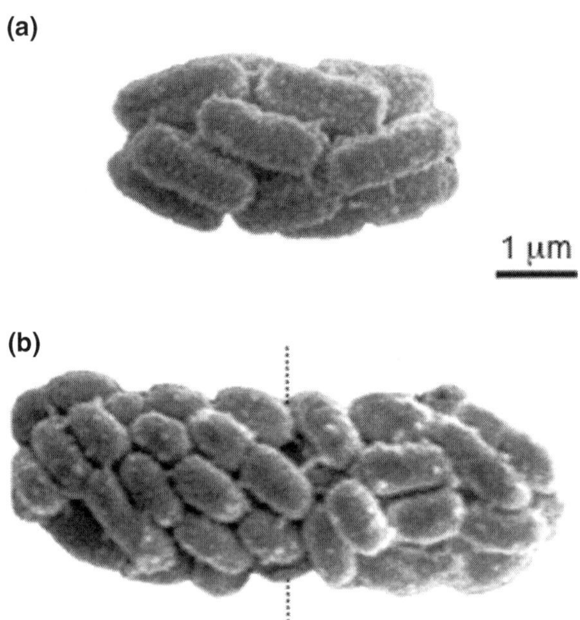

Fig. 2.1 Two different stages in the life cycle of the phototrophic consortium "*P. roseum*" from Lake Dagow. (**a**) Typical morphology of a consortium with *rod-shaped* epibiont cells. (**b**) Consortium shortly after synchronous cell division of all epibionts, which display an almost *coccoid shape*. *Dashed line* indicates the plane of division during the subsequent separation of the two daughter consortia

Intact "*C. aggregatum*" consortia can be disaggregated by the addition of EGTA or pyrophosphate but stay intact in the presence of various reducing agents, ionic or non-ionic detergents, denaturing agents, lectin-binding sugars, or proteolytic enzymes (Vogl et al. 2008). Apparently, Ca^{2+} ions are involved in the specific aggregate formation of phototrophic consortia.

2.3.2 Ultrastructure

Comparative electron microscopy studies of epibiont cells in pure culture and in "*C. aggregatum*" yielded further insights into the subcellular processes which are involved in maintaining the heterologous multicellularity. Epibionts are interconnected and also (to a lesser extent) connected to the central bacterium by means of electron-dense, hair-like filaments. Furthermore, a connection between the partners is established by numerous periplasmic tubules which extend from the outer membrane of the central bacterium and form direct contact with the outer membrane of the epibiont cells. Based on detailed ultrastructural studies combining high-resolution analytical scanning electron microscopy and transmission electron microscopy with three-dimensional reconstruction and image analysis of the topology of the periplasmic tubules, it has been suggested that the central bacterium and the epibionts share a common periplasmic space (Wanner et al. 2008).

In cells from pure cultures, chlorosomes are equally distributed along the inner face of the cytoplasmic membrane. In contrast, the distribution of the chlorosomes in symbiotic epibiont cells is uneven, with chlorosomes being entirely absent at the site of attachment to the central bacterium. Instead, a conspicuous additional 17-nm-thick layered structure was found at the attachment site of the symbiotic epibiont cells (Vogl et al. 2006; Wanner et al. 2008). The asymmetric distribution of chlorosomes and the altered architecture of the adhesion site are a specific feature of the symbiotic state and were previously unknown for any of the other, free-living, green sulfur bacteria. This intracellular sorting and differentiation processes of the epibionts thus provide interesting targets for future studies of reciprocal signal exchange in phototrophic consortia.

2.3.3 Scotophobic Response

Another line of evidence for a close signal exchange between the partners in phototrophic consortia stems from the behavior of intact consortia toward light. Motile "*C. aggregatum*" in enrichment cultures exhibit a scotophobic accumulation in dim ($\leq$ 1.5 μmol quanta$\cdot m^{-2} \cdot s^{-1}$) light (Buder 1914; Fröstl and Overmann 1998). Unlike the phototaxis of, e.g., unicellular algae, the scotophobic reaction of bacteria is solely controlled by changes in light intensity but not by the direction of light. Consortia incubated in microcuvettes on a microscope stage and exposed to a microspectrum accumulated at a wavelength of 740 nm (Fröstl and Overmann 1998). Since this wavelength corresponds to the position of the absorption maximum of bacteriochlorophyll *c*, the photosynthetic pigments of the epibiont most likely act as the photoreceptor of the scotophobic response. Yet, only the central bacterium carries a flagellum. Therefore, it has been concluded that the partner bacteria maintain an efficient interspecific communication within the consortia.

2.4 Physiology and Possible Selective Advantage of the Interaction

2.4.1 Motility Acquired by the Epibiont

Typically, phototrophic consortia are found in freshwater lakes at depths which receive low light intensities (<5 μmol quanta·m^{-2}·s^{-1}) and that contain low concentrations of sulfide. In addition, low concentrations of dissolved oxygen have often been detected in these layers (Overmann et al. 1998). The buoyant density of consortia (1,046.8 kg·m^{-3} determined for "*P. roseum*" in Lake Dagow) can significantly surpass the density of the ambient chemocline water (995.8 kg·m^{-3}) (Overmann et al. 1998). If they were non-motile, the resulting sedimentation rate (up to 3 cm·s^{-1}) thus would lead to a rapid sedimentation of phototrophic consortia into the dark deeper water layers where the obligately (see Section 2.2.1) phototrophic epibionts would ultimately die. Clearly, the association of the immotile epibionts with a motile central bacterium could counteract sedimentation, provided phototrophic consortia as a whole could orient themselves properly in light and sulfide gradients. It is likely that the scotophobic response toward dim light described in the previous section could serve this purpose under natural conditions.

In addition to the scotophobic response, phototrophic consortia have been shown to exhibit a pronounced chemotactic response, both in laboratory enrichments (Fröstl and Overmann 1998) and in situ (Glaeser and Overmann 2003b). Sulfide proved to consistently act as a strong attractant and would lead to a rapid accumulation of intact consortia at hot spots of sulfate reduction (e.g., sinking organic particles) in the chemocline. Actually, enrichments of phototrophic consortia were only successful at low sulfide concentrations of ≤300 μM, whereas other, free-living green sulfur bacteria outcompeted the phototrophic consortia at higher sulfide concentrations. Obviously, a strong limitation of external sulfide is of competitive advantage for intact consortia. In pure culture, the epibiont *Chl. chlorochromatii* CaD isolated from "*C. aggregatum*" is strictly dependent on sulfide as electron donor and sulfur source. Future research will show whether sensing of sulfide occurs in the green sulfur bacterial epibiont and the signal is then communicated to the central bacterium or whether the central bacterium is capable of sensing sulfide itself.

Due to the rapid orientation toward light and electron-donating substrates, the motility acquired in association provides the epibionts with a selective advantage over its non-motile green sulfur bacterial relatives competing for sulfide as electron donor. Some green sulfur bacteria form gas vesicles, which decrease the buoyant density of the cells. Typically, such changes in buoyant density occur over a time period of several days (Overmann et al. 1991) and thus are much slower than the diurnal fluctuations in light intensities and sulfide concentrations. By comparison, the vertical migration of phototrophic consortia, mediated by flagellar motility, is orders of magnitude faster and hence is expected to lead to increased sulfide utilization and light utilization efficiency by the epibionts in phototrophic consortia.

2.4.2 Reciprocal Control of the Metabolism of the Two Partner Bacteria

Interestingly, 2-oxoglutarate also serves as a chemoattractant in situ as well as in laboratory cultures. Like all free-living green sulfur bacteria, the epibiont *Chl. chlorochromatii* CaD does not utilize 2-oxoglutarate for growth, suggesting that 2-oxoglutarate is taken up and utilized by the central bacterium. Microautoradiography was used to investigate whether intact consortia could assimilate 2-[^{14}C(U)]-oxoglutarate. The major fraction of a natural population of consortia incorporated this carbon compound, but exclusively in the presence of light and sulfide (Glaeser and Overmann 2003b). These results indicate that the incorporation of 2-oxoglutarate by the central bacterium is regulated by the physiological state of the green sulfur bacterial epibiont.

2.4.3 Physiological Interactions

Little is known so far of the putative physiological coupling between the two partners in phototrophic consortia. In analogy to the syntrophic growth of green sulfur bacteria in coculture with sulfur- and sulfate-reducing bacteria, it had initially been speculated that a sulfur cycle also occurs in phototrophic consortia, in which the colorless central bacterium rereduced the elemental sulfur or sulfate formed during anoxygenic photosynthesis of the green sulfur bacterial partner (Pfennig 1980. Based on its phylogenetic affiliation, the central bacterium does not represent a typical sulfur- or sulfate-reducing bacterium, however.

In green sulfur bacteria, the cell yield increases significantly in the presence of assimilable organic carbon substrates (Overmann and Pfennig 1989). Therefore it would appear possible that epibionts grow mixotrophically using organic carbon compounds excreted by the central chemotrophic bacterium. However, the stable carbon isotope discrimination values determined for epibionts in natural populations of phototrophic consortia clearly indicate that the epibionts grow photoautotrophically in situ rather than incorporating organic substrates (Glaeser and Overmann 2003a).

While the association with a motile partner represents a clear advantage for the immotile epibiont (see Section 2.4.1), the selective advantage of the partnership for the central bacterium has remained rather unclear. Green sulfur bacteria are known to excrete considerable amounts of photosynthetically fixed carbon (Czeczuga and Gradski 1972), which has also been confirmed for the epibiont (Pfannes 2007). It had been suggested that the epibiont supplies the central bacterium with such organic carbon excretion products (Glaeser and Overmann 2003b). A recent series of experiments in fact provide evidence for a rapid transfer of photosynthetically fixed carbon from the epibiont to the central bacterium (Johannes Müller and Jörg Overmann, unpublished observations).

Theoretically, the tight arrangement of the cells in phototrophic consortia (Fig. 2.1) could result in a diffusion limitation and hence a physiological isolation of the central bacterium. However, transmission electron studies of cryofixed and cryosubstituted preparations show that considerable intercellular space exists between the cells and that the central bacterium only occupies 25% of the volume available (Wanner et al. 2008). Since gaps are also left by the epibiont cells forming the cortex of phototrophic consortia, this extracellular space probably prevents diffusion limitation of the central bacterium in phototrophic consortia. This conclusion is supported by experiments with fluorogenic substrate analogs of intracellular esterases which lead to a rapid accumulation of fluorescent signal in the central bacterium within less than 2 min and before the surrounding epibionts (Bayer 2007).

2.5 The Molecular Basis for Microbial Symbiosis in Phototrophic Consortia

Possibly the most fascinating – and so far unique – aspect of the biology of phototrophic consortia is the specificity of the interorganismic interaction which manifests itself in a coadaptation as well as in particular mechanisms of mutual recognition and signal exchange of the partner organisms. Based on the unique lifestyle and considerable phylogenetic distance of *Chl. chlorochromatii* CaD to other green sulfur bacteria, niche-specific genes were expected to occur in the genome of the epibiont. Recent efforts have yielded first insights into the molecular mechanisms underlying the interaction between the epibionts and the central bacterium in phototrophic consortia by unraveling a series of unique genes and an unusual regulation of central metabolic functions in the epibiont.

2.5.1 Several Candidate Symbiosis Genes in the Epibiont of **"C. aggregatum"** *Resemble Bacterial Virulence Factors*

The isolation of *Chl. chlorochromatii* CaD in pure culture enabled the sequencing of its genome (http://genome.jgi-psf.org/finished_microbes/chlag/chlag.home.html). In parallel, the genome sequence of 11 strains of free-living green sulfur bacteria became publicly available (http://img.jgi.doe.gov/cgi-bin/geba/main.cgi?section=TaxonList&page=taxonListPhylo&pidt=14955.1250667420). This information provided the opportunity to identify niche-specific genes in the epibiont genome. In two recent studies, suppression subtractive hybridization and dot blot hybridization of genomic DNA were combined with comparative bioinformatic approaches to identify candidate symbiosis genes which are unique to the epibiont genome (Vogl et al. 2008; Wenter et al. 2010).

Initially, four candidate symbiosis genes were recovered (Vogl et al. 2008). The two hemagglutinin-like open reading frames (ORFs) Cag0614 and Cag0616 are exceptionally large and code for 36,805 and 20,646 amino acid-long gene products, respectively. Therewith they represent the largest ORFs known to date and are only surpassed by the human titin gene. Three sequence regions were almost identical between both genes and both contain repetitive regions of high-sequence similarity, suggesting that the two ORFs arose through a gene duplication event. Furthermore, Cag0614 and Cag0616 encode RGD tripeptides known from filamentous hemagglutinin which in bacterial pathogens participate in binding to the mammalian cells. In addition, Cag0616 harbors two $\beta\gamma$-crystalline Greek key motifs containing all established elements of this motif, like the conserved sequence $(Y/F/W)X_6GX_{28\text{–}34}S$, two Ca^{2+}-binding sites, and three to four beta strands. Bacterial proteins containing this motif typically participate in the response to stress conditions. The third gene Cag1920 encodes a 3,834 amino acid-long putative hemolysin, whereas the gene product of Cag1919 is a 1,526 amino acid-long putative RTX-like protein. All four genes are transcribed constitutively and do not occur in any of the 16 free-living green sulfur bacterial relatives of the epibiont tested so far. In light of the fact that epibionts seem to be specifically adapted to life in association with the central bacterium and so far have not been detected in a free-living state in nature, it has been hypothesized that a regulatory mechanism for the expression of the potential symbiosis genes may be dispensable.

The RTX-toxin-like gene product of Cag1919 was studied in more detail. A 100 amino acid-long region of its C-terminus comprises six repetitions of the consensus motif GGXGXD and, based on its high similarity with the corresponding region in the alkaline protease of *Pseudomonas aeruginosa*, is predicted to form a Ca^{2+}-binding beta roll. In this beta roll structure, Ca^{2+} ions are coordinated at the turns between the two strands by the aspartic acid residues of the repeat sequence. In contrast to the RTX toxins known from pathogenic Proteobacteria, however, the gene product of Cag1919 could not be detected by $^{45}Ca^{2+}$ autoradiography and therefore occurs only at a low abundance in epibiont cells. Furthermore, no secretion signal typical for RTX toxins could be identified. The RTX-type terminus coded by Cag1919 exhibited a significant sequence similarity to the RTX modules found in virulence factors of pathogenic Proteobacteria like *Erwinia carotovora*, *Pseudomonas fluorescens*, *Ralstonia solanacearum*, or *Vibrio vulnificus*. This putative symbiosis gene Cag1919 thus may have been acquired via horizontal gene transfer from a proteobacterium. In the Gram-negative bacterial pathogens, the RTX module is necessary for binding to the target cell. In a similar fashion, the Gag1919 gene product may be involved in the cell–cell binding within the phototrophic consortium "*C. aggregatum*."

Subsequent bioinformatic analyses yielded 186 ORFs that were unique to the epibiont genome as compared to the 11 genomes of free-living relatives (Wenter et al. 2010). The fact that the main fraction (99) of these ORFs code for hypothetical proteins with unknown function indicates that the cell–cell interactions in phototrophic consortia involve numerous novel modules. In contrast to this limited number of unique ORFs, genome comparisons among closely related bacteria have demonstrated that a much higher fraction, up to 1,387 genes, of the bacterial genome may encode niche-specific functions. Low numbers of niche-specific genes comparable to those in the epibiont have only been reported for pathogenic bacteria which evolved from ancestors already adapted to interactions with animal cells, e.g., *Salmonella enterica* and *Bacillus anthracis*. As a consequence, additional genes which are not unique to the epibiont genome may still be involved in the cell–cell interaction of phototrophic consortia.

In order to search for the presence of such non-unique ORFs with potential relevance to the symbiosis, recent bioinformatic and transcriptomic analyses targeted ORFs with similarity to known bacterial virulence factors (Wenter et al. 2010). Several additional genes were detected, among them an outer membrane efflux protein that contains a conserved TolC-like domain and typically is part of bacterial type I secretion systems (Cag0615); Cag1408 which is related to the *Escherichia coli* membrane fusion protein HlyD, mediating the transport of hemolysin across the periplasm as part of the type I secretion system; and Cag1570 that is related to VapD, a putative toxin of a toxin–antitoxin pair found in many pathogenic bacteria. Furthermore, four different VCBS (*Vibrio*, *Colwellia*, *Bradyrhizobium*, *Shewanella*) domain proteins (Cag0738, 1239, 1242, 1560), three additional hemagglutinin-like proteins (Cag1053, 1055, 1512), and two hemolysin activation/secretion proteins (Cag1054, 1056) were documented.

Thus, a total of eight unique epibiont genes and eight non-unique genes matching known bacterial virulence factors have been detected. Hence, the low overall number of epibiont-specific ORFs and the significant fraction of non-unique genes with homologues in free-living green sulfur bacteria support the above hypothesis that the ancestor of the epibiont was preadapted to a symbiotic lifestyle in a manner comparable to that seen in the evolution of human pathogens.

2.5.2 Regulation Involves Central Metabolic Pathways Rather than Symbiosis Genes

As a second approach toward the elucidation of the genetic basis of the symbiotic interaction in phototrophic consortia, a recent study employed proteomics, cDNA suppression subtractive hybridization, and transcriptomics to compare the expression in free-living (pure culture of *Chl. chlorochromatii*CaD) and symbiotic ("*C. aggregatum*") epibiont cells (Wenter et al. 2010).

Two-dimensional differential gel electrophoresis of the cytoplasmic proteome covered 78% of all expected gene proteins of the epibiont. Fifty-four of the proteins were detected exclusively in consortia but not in pure epibiont cultures. Among them, the proteins identified as a nitrogen regulatory protein P-II, a 2-isopropylmalate synthase, and a glutamate synthase are likely to be involved in central amino acid metabolism. Two proteins involved in sugar metabolism (glyceraldehyde-3-phosphate dehydrogenase, phosphotransferase protein IIA), as well as a porphobilinogen deaminase and an UDP-3-*O*-[3-hydroxymyristoyl] glucosamine *N*-acyltransferase were identified. Subsequent RT-qPCR analyses showed that the greatest change occurred for the gene encoding nitrogen regulatory protein P-II, for which a pronounced 189-fold increase in transcript abundance was found.

Analyses of the membrane proteins indicated that a branched chain amino acid ABC transporter binding protein is only expressed in the symbiotic state of the epibiont, and cross-linking experiments indicated that this membrane protein is possibly located at the cell contact site to the central bacterium. Whereas supplementation of pure epibiont cultures with peptone or branched chain amino acids (leucine, valine, or isoleucine in non-inhibitory concentrations of 0.1 or 1 mM) did not affect the pattern of membrane proteins, cultivation of the epibiont with consortia culture supernatant stimulated the expression of the ABC transporter binding protein. Furthermore, analyses of chlorosomal proteins revealed that an uncharacterized 97 amino acid-long, 11-kDa epibiont protein (encoded by Cag1285) is only expressed during symbiosis. This protein was distantly related to putative uncharacterized proteins of four green sulfur bacteria *Chlorobium phaeovibrioides* DSM 265,. *Chlorobium luteolum*DSM273^T, *Chlorobium phaeobacteroides*BS1, and *Prosthecochloris aestuarii* DSM271^T as well as to proteins of human and phytopathogenic Betaproteobacteria belonging to the genera *Ralstonia*, *Burkholderia*, and *Neisseria*. Since its differential expression correlates with the distinct changes in the intracellular distribution of chlorosomes in the epibiont, Cag1285 may be involved in the intracellular sorting of chlorosomes (Wenter et al. 2010).

Finally, a comparison of the transcriptomes of symbiotic and free-living epibionts indicated that 328 genes were differentially transcribed. In this context, and in agreement with earlier results, mRNAs of the four putative symbiosis genes Cag0614, 0616, 1919, and 1920 (see Section 2.5.1) were found to be present both in the "*C. aggregatum*" consortium and in the free-living *Chl. chlorochromatii*.

Similar to other green sulfur bacteria (Eisen et al. 2002), the genome of the epibiont encodes only a few (56) proteins for environmental sensing and regulatory responses. Taken together, the results of the above three molecular approaches were largely complementary and yielded a first inventory of 352 genes which were differentially expressed in the symbiotic and free-living state and hence likely to be involved in the bacterial interactions in "*C. aggregatum.*" Extrapolating the results of the transcriptome analysis toward the entire genome, on the order of 700 (35%) of the protein coding genes would be expected to be differentially regulated. This high fraction of regulated genes is comparable to that observed during a heat-shock response of the sulfate-reducing *Desulfovibrio vulgaris*, but higher than the fraction found in many other bacteria, like *S. enterica*, during a switch between the extracellular and the intravacuolar environment (20.6%), *Listeria monocytogenes* during temperature shock (maximum 25%), or *Thermotoga maritima* switching between biofilm and planktonic populations (6%).

Most of the regulated genes encoded components of central metabolic pathways whereas only very few (7.5%) of the unique "symbiosis genes" turned out to be regulated under the experimental conditions tested. While the comparison between a symbiotic and a free-living state was chosen in order to study the most pronounced response, such a switch under natural conditions is unlikely. Yet, even this most pronounced change in environmental conditions did not elicit a regulation of most of the unique symbiosis genes. Hence, such a regulation seems to be dispensable for symbiotic interaction, or the symbiotic association has been stable in nature for a sufficient time that regulation of the potential symbiosis genes no longer confers any selective advantage. Rather, the regulatory response pertains to the central metabolic pathways. Since a switch between non-symbiotic and symbiotic lifestyle is unlikely, the observed regulation may serve the purpose of fine-tuning the physiological coupling between the epibiont and the central bacterium if the phototrophic consortium as a whole experiences changes in environmental conditions.

Acknowledgments Several motivated students have participated in our research on phototrophic consortia: Jürgen Fröstl, Jens Glaeser, Kajetan Vogl, Martina Schlickenrieder, Martina Müller, Birgit Kanzler, Kristina Pfannes, Katharina Hütz, Anne Bayer, Dörte Dibbern, and Johannes Müller. Their contributions were decisive for the success of the project. Support by the Deutsche Forschungsgemeinschaft (grants Ov20/3-1, Ov20/3-2, Ov20/3-3, Ov20/10-1, Ov20/10-2) is gratefully acknowledged.

References

Bayer A (2007) Neue Ansätze zur Analyse der bakteriellen Interaktionen in phototrophen Konsortien. Diplomarbeit University of Munich, Munich, 104 pp.

Buder J (1914) Chloronium mirabile. Berichte deutsche botanische Gesellschaft 31:80–97

Caldwell DE, Tiedje JM (1975) A morphological study of anaerobic bacteria from the hypolimnia of two Michigan lakes. Can J Microbiol 21:362–376

Croome RL, Tyler PA (1984) Microbial microstratification and crepuscular photosynthesis in meromictic Tasmanian lakes. Verh Int Verein Limnol 22:1216–1223

Czeczuga B, Gradski F (1972) Relationship between extracellular and cellular production in the sulphuric green bacterium *Chlorobium limicola* Nads. as compared to primary production of phytoplankton. Hydrobiologia 42:85–95

De Bok FA, Plugge CM, Stams AJ (2004) Interspecies electron transfer in methanogenic propionate degrading consortia. Water Res 38:1368–1375

Eichler B, Pfennig N (1990) Seasonal development of anoxygenic phototrophic bacteria in a holomictic drumlin lake (Schleinsee, F.R.G.). Arch Hydrobiol 119:369–392

Eisen JA, Nelson KE, Paulsen IT et al (2002) The complete genome sequence of *Chlorobium tepidum* TLS, a photosynthetic, anaerobic, green-sulfur bacterium. Proc Natl Acad Sci USA 99:9509–9514

Fröstl J, Overmann J (1998) Physiology and tactic response of "*Chlorochromatium aggregatum*". Arch Microbiol 169:129–135

Fröstl JM, Overmann J (2000) Phylogenetic affiliation of the bacteria that constitute phototrophic consortia. Arch Microbiol 174:50–58

Gasol JM, Jürgens K, Massana R, Calderón-Paz JI, Pedrós-Alió C (1995) Mass development of Daphnia pulex in a sulfide-rich pond (Lake Cisó). Arch Hydrobiol 132:279–296

Glaeser J, Overmann J (2003a) Characterization and in situ carbon metabolism of phototrophic consortia. Appl Environ Microbiol 69:3739–3750

Glaeser J, Overmann J (2003b) The significance of organic carbon compounds for in situ metabolism and chemotaxis of phototrophic consortia. Environ Microbiol 5:1053–1063

Glaeser J, Overmann J (2004) Biogeography, evolution and diversity of the epibionts in phototrophic consortia. Appl Environ Microbiol 70:4821–4830
Gorlenko VM (1988) Ecological niches of green sulfur bacteria. In: Olson JM, Ormerod JG, Amesz J, Stackebrandt E, Trüper HG (eds) Green photosynthetic bacteria. Plenum, New York, pp 257–267
Kanzler B, Pfannes KR, Vogl K, Overmann J (2005) Molecular characterization of the non-photosynthetic partner bacterium in the consortium "*Chlorochromatium aggregatum*". Appl Environ Microbiol 71:7434–7441
Lauterborn R (1906) Zur Kenntnis der sapropelischen Flora. Allg Bot 2:196–197
Overmann J (2001) Phototrophic consortia: a tight cooperation between non-related eubacteria. In: Seckbach J (ed) Symbiosis: mechanisms and model systems. Kluwer, Dordrecht, pp 239–255
Overmann J (2006) The symbiosis between nonrelated bacteria in phototrophic consortia. In: Overmann J (ed) Molecular basis of symbiosis. Progress in molecular subcellular biology, Chapter II. Springer, Berlin, pp 21–37
Overmann J, Lehmann S, Pfennig N (1991) Gas vesicle formation and buoyancy regulation in *Pelodictyon phaeoclathratiforme* (Green sulfur bacteria). Arch Microbiol 157:29–37
Overmann J, Pfennig N (1989) *Pelodictyon phaeoclathratiforme*sp. nov., a new brown-colored member of the Chlorobiaceae forming net-like colonies. Arch Microbiol 152:401–406
Overmann J, Tilzer MM (1989) Control of primary productivity and the significance of photosynthetic bacteria in a meromictic kettle lake (Mittlerer Buchensee, West Germany). Aquat Sci 51:261–278
Overmann J, Tuschak C, Fröstl J, Sass H (1998) The ecological niche of the consortium "*Pelochromatium roseum*". Arch Microbiol 169:120–128
Pfennig N (1980) Syntrophic mixed cultures and symbiotic consortia with phototrophic bacteria: a review. In: Gottschalk G, Pfennig N, Werner H (eds) Anaerobes and anaerobic infections. Fischer, Stuttgart, New York, pp 127–131
Pfannes K (2007) Characterization of the symbiotic bacterial partners in phototrophic consortia. Dissertation, University of Munich, 180p.
Pfannes KR, Vogl K, Overmann J (2007) Heterotrophic symbionts of phototrophic consortia: members of a novel diverse cluster of *Betaproteobacteria* characterised by a tandem *rrn* operon structure. Environ Microbiol 9:2782–2794
Schink B (1991) Syntrophism among prokaryotes. In: Balows A, Trüper HG, Dworkin M, Harder W, Schleifer K-H (eds) The prokaryotes, 2nd edn. Springer, Berlin, pp 276–299
Schink B (2002) Synergistic interactions in the microbial world. Ant van Leeuwenhoek 81:257–261
Tuschak C, Glaeser J, Overmann J (1999) Specific detection of green sulfur bacteria by in situ hybridization with a fluorescently labeled oligonucleotide probe. Arch Microbiol 171:265–272
Vogl K, Glaeser J, Pfannes KR et al (2006) *Chlorobium chlorochromatii* sp. nov., a symbiotic green sulfur bacterium isolated from the phototrophic consortium "*Chlorochromatium aggregatum*". Arch Microbiol 185:363–372
Vogl K, Wenter R, Dressen M, et al (2008) Identification and analysis of four candidate symbiosis genes from '*Chlorochromatium aggregatum*', a highly developed bacterial symbiosis. Environ Microbiol 10:2842–2856
Wanner G, Vogl K, Overmann J (2008) Ultrastructural characterization of the prokaryotic symbiosis in '*Chlorochromatium aggregatum*'. J Bacteriol 190:3721–3730
Wenter R, Hütz K, Dibbern D, Reisinger V, Li T, Plöscher M, Eichacker L, Eddie B, Hanson T, Bryant D, Overmann J (2010) Expression-based identification of genetic determinants of the bacterial symbiosis in '*Chlorochromatium aggregatum*'. Environ Microbiol (in press; Ms. No. EMI-2009-0895)
Wenter R, Wanner G, Schüler D, Overmann J (2009) Ultrastructure, phylogeny and tactic behaviour of a novel multicellular magnetotactic prokaryote from North Sea sediments. Environ Microbiol 11:1493–1505
Whittaker CJ, Klier CM, Kolenbrander PE (1996) Mechanisms of adhesion by oral bacteria. Annu Rev Microbiol 50:513–552

Chapter 3
The Ecology of Nitrogen Fixation in Cyanobacterial Mats

Lucas J. Stal, Ina Severin, and H. Bolhuis

Abstract All cyanobacterial mats that have been investigated have been proven to be diazotrophic, i.e., use atmospheric dinitrogen (N_2) as the source of nitrogen. Many cyanobacteria possess the capacity to fix N_2 and different species have evolved various ways to cope with the sensitivity of nitrogenase toward oxygen which is produced by these oxygenic phototrophs. These different strategies give rise to complex patterns of nitrogenase activity in microbial mats. Nitrogenase activity may exhibit complex variations over a day–night cycle but different types of microbial mats may also have their own characteristic patterns. Besides the cyanobacteria, numerous other members of the Bacteria as well as some Archaea are known to be diazotrophic. The complexity of the microbial community and of the observed patterns of nitrogenase activity makes it difficult to understand how the different groups of organisms contribute to N_2 fixation in microbial mats. Cyanobacteria have ample access to energy (sunlight) and reducing equivalents (water) and therefore easily satisfy the demands of nitrogenase. As well, since they also fix CO_2, they are able to synthesize the acceptor molecules for the fixed nitrogen. However, it is also feasible that other diazotrophs in a joint venture with cyanobacteria are responsible for the bulk of the fixed nitrogen. In this review we discuss the importance of cyanobacteria as diazotrophs in microbial mats, their interactions with other potential N_2-fixing microorganisms, and the factors that control their activities.

3.1 Introduction

Microbial mats are vertically stratified benthic communities of microorganisms, frequently with cyanobacteria as the main component and as the pioneers (Stal 1995). These small-scale microbial ecosystems often form a tough and coherent

L.J. Stal (✉)
Department of Marine Microbiology, Netherlands Institute of Ecology, NIOO-KNAW, 4400 AC, Yerseke, The Netherlands
e-mail: l.stal@nioo.knaw.nl

P.C. Hallenbeck (ed.), *Recent Advances in Phototrophic Prokaryotes*, Advances in Experimental Medicine and Biology 675, DOI 10.1007/978-1-4419-1528-3_3,

structure as the result of entangled filamentous organisms and of the extracellular polymeric substances (EPS) that the microbial community exudes. The EPS forms the matrix in which the microorganisms are embedded. The typical coherent structure of this microbial ecosystem, which sometimes can be lifted off from the sediment substrate on which it thrives, is the reason for typifying it as a 'mat' (like a 'doormat').

Microbial mats have been considered as analogues to the most ancient ecosystems on Earth known to us. Lithified microbial mats, or stromatolites, are known from the fossil record dating back almost 3.5 billion years (Schopf 2000). However, non-lithifying microbial mats have also been detected in the fossil record, known as microbially induced sediment structures (MISS), as early as the Precambrian (Noffke et al. 2006).

Modern microbial mats develop in environments characterized by extreme conditions that largely exclude grazing organisms (Fenchel 1998). Only in the absence of grazing pressure are the mat-forming microorganisms capable of accumulating and forming the macroscopic structure of a microbial mat. In the vast majority of cases the pioneering organisms are cyanobacteria. Cyanobacteria are oxygenic phototrophic Bacteria that use light, water, and inorganic carbon (CO_2) as ubiquitous sources of energy, electrons, and carbon, respectively. Moreover, many species are capable of fixing atmospheric dinitrogen (N_2), making them the organisms with minimum growth requirements (Stewart 1980). For that reason cyanobacteria are capable of colonizing low-nutrient environments. Cyanobacteria are further characterized by a wide range of tolerances toward salt, temperature (low and high), and drought (Seckbach 2007). Cyanobacteria are metabolically versatile and react quickly to changing conditions. As far as is known, all cyanobacterial mats are characterized by N_2 fixation (Severin and Stal 2010). However, it is unclear whether cyanobacteria are the only organisms contributing to N_2 fixation in microbial mats.

3.2 The Physiological Ecology of Microbial Mats

Cyanobacteria enrich the sediment with organic carbon and nutrients (particularly nitrogen, but also with phosphate and other micronutrients) through their photosynthetic fixation of carbon and nitrogen. This organic matter is released by a variety of processes including exudation of EPS, low-molecular fermentation products, glycolate produced during photorespiration, and death and cell lysis (Stal 1995). All these components can serve as substrates for a range of other microorganisms, which mostly degrade these under anaerobic conditions. While during the day the top layer of the mat is enriched with oxygen by cyanobacterial photosynthesis, this is respired within minutes once it is dark (Jørgensen et al. 1983). Below the zone of oxygenic photosynthesis the mat is permanently anoxic. Hence, anaerobic processes form an important component of the physiology of microbial mats.

Sulfate-reducing bacteria (SRB) are an important functional group in marine microbial mats, mainly because of the high concentration of sulfate in seawater (28 mM) (Baumgartner et al. 2006). SRB are Bacteria that anaerobically mineralize organic matter by using sulfate (SO_4^{2-}) as the terminal electron donor, reducing it to sulfide (S^{2-}). SRB use predominantly low-molecular organic matter (e.g., acetate, ethanol, lactate) but some species have been shown to also degrade more complex organic matter including EPS. When iron is available, the sulfide produced by SRB reacts with it to form the black precipitate FeS. This FeS can react further with elemental sulfur to form the more stable mineral pyrite (FeS_2). Sulfide also reacts with elemental sulfur to form polysulfides (Visscher et al. 1990). Sulfide is only stable under anaerobic conditions since it reacts chemically with oxygen. However, colorless sulfur bacteria (CSB) will use the sulfide as electron donor and respire it with (low concentrations of) oxygen (and, if available, with nitrate). These autotrophic (CO_2-fixing) Bacteria are very efficient and overrule chemical oxidation (Visscher et al. 1992). The second functional group of microorganisms that are involved in the oxidation of sulfide are the anaerobic anoxygenic phototrophic bacteria, notably the purple sulfur bacteria and to a lesser extent the green sulfur bacteria (Visscher et al. 1992). These Bacteria use sulfide as electron donor in photosynthesis. The colorless and photosynthetic sulfur bacteria eventually oxidize the sulfide back to sulfate, whether or not with elemental sulfur as an intermediate.

The daily cycle results in steep and contrasting millimeter scale gradients of oxygen and sulfide in the microbial mat (Revsbech et al. 1983). These physicochemical gradients cause the formation of vertically stratified communities of functional groups of microorganisms (Stal 2000). In the green top layer cyanobacteria are present. Sometimes this layer is mixed or overlaid by epipelic diatoms or by a thin layer of sand. Diatoms belong to the domain Eukarya and are oxygenic phototrophic microalgae with a siliceous cell wall. Also, the top layer may consist of dead cyanobacteria of which the remaining pigmented polysaccharide sheaths serve as protection of the underlying viable cyanobacteria from excessive radiation ('sunglasses'). Underneath the green layer of cyanobacteria, a purple layer of purple sulfur bacteria may be present. Below the layer of purple sulfur bacteria, usually a deep horizon of black FeS is found, which is permanently anoxic. This is often confused as the layer of the SRB. However, SRB occur throughout the mat and take advantage of the organic matter produced directly or indirectly by photosynthetic activity (Canfield and DesMarais 1991). Some SRB are remarkably tolerant toward oxygen and in exceptional cases they can even use it to some extent as electron acceptor.

Sometimes the layers of the cyanobacteria and the purple sulfur bacteria are separated by a layer of oxidized iron (Stal 2001). It is possible that this layer indicates the presence of anoxygenic phototrophic Bacteria that utilize ferrous iron as electron donor (Widdel et al. 1993). This layer may also protect the anaerobic Bacteria from oxygen diffusion from the top layer by reacting with ferrous iron or preventing the toxic sulfide from diffusing upward into the cyanobacterial layer by reacting with ferric iron. However, these hypotheses await scientific investigation and proof.

3.3 N_2 Fixation and N_2-Fixing Organisms

All of the above-mentioned functional groups of organisms have representatives that are capable of fixing N_2. Biological N_2 fixation is confined to Bacteria and to some Archaea. No member of the Eukarya possesses the capacity to fix atmospheric N_2 except in symbiotic associations with Bacteria. Diazotrophic (growth at the expense of dinitrogen as sole source of nitrogen) organisms have the genetic potential to synthesize nitrogenase, a complex consisting of two enzymes: dinitrogenase reductase and dinitrogenase (Zehr et al. 2003). The former reduces the latter using low-potential electrons (reduced ferredoxin) while hydrolyzing 4 ATP per 2 electrons transported. The fixation of N_2 to $2NH_3$ is accompanied with the evolution of one molecule of H_2. Hence, N_2 fixation costs 16 ATP and 8 low-potential electrons, which makes it energetically expensive. Another characteristic of nitrogenase is its extreme sensitivity toward O_2 (Gallon 1992). Hence, diazotrophic organisms need to provide nitrogenase with an anaerobic environment. Many diazotrophic microorganisms are in fact anaerobes. It is not always clear whether anaerobic diazotrophs actually utilize N_2 because it would draw considerably on their energy supplies and in many cases such organisms may live under energy-limiting conditions, while anaerobic environments are often not depleted in nitrogen.

3.4 N_2-Fixing Cyanobacteria

Many cyanobacteria possess the capacity to synthesize nitrogenase and fix N_2 (Bergman et al. 1997). Using sunlight, these organisms may meet the energy demand of nitrogenase. However, this is paradoxical since cyanobacteria not only usually thrive in oxygenated environments, they also evolve O_2 through their photosynthetic metabolism. Cyanobacteria have evolved various means to accommodate N_2 fixation in an oxygen-rich environment, although except for the heterocystous cyanobacteria, this has not been extensively investigated in all cyanobacteria that possess the genetic capacity for N_2 fixation (Stal and Zehr 2008). Many species of cyanobacteria accomplish the strict separation of nitrogenase from oxygen by avoiding O_2 through temporally living anaerobically, either in the light by anoxygenic photosynthesis or by fermentative metabolism in the dark.

Cyanobacteria of sections IV and V comprise filamentous species that differentiate specialized cells, heterocysts, which are the site of N_2 fixation (Adams 2000). Heterocysts have lost the oxygenic photosystem II but retain the anoxygenic photosystem I and are therefore capable of harvesting light energy but do not evolve O_2 and cannot fix CO_2. Therefore, heterocysts depend upon the neighboring vegetative cells for reducing equivalents that are imported as carbohydrate, probably as sucrose (Curatti et al. 2002). Heterocysts also possess a thick glycolipid cell wall that serves as a gas diffusion barrier. This cell wall limits the flux of O_2 into the heterocyst and, as well, of N_2 (Walsby 2007). Hence, there must be a trade-off between minimizing the inward diffusion of O_2 while permitting the entry of sufficient N_2 into the heterocyst. The O_2 that enters the heterocyst is consumed by a high-affinity respiratory system that maintains the interior of the heterocyst virtually anoxic, a prerequisite for nitrogenase activity. Thus, heterocystous cyanobacteria are capable of simultaneous oxygenic photosynthesis and N_2 fixation, which are spatially separated into two different cell types.

Another group of diazotrophic cyanobacteria is composed of unicellular and filamentous non-heterocystous types. Their strategy is to separate N_2 fixation and oxygenic photosynthesis temporally (Bergman et al. 1997). Basically, cyanobacteria belonging to this group fix N_2 during the night and obviously confine photosynthesis to the daytime. Dark nitrogenase activity in these cyanobacteria is supported by aerobic respiration which should also be sufficiently fast in order to maintain anaerobicity in the N_2-fixing cells. If these cyanobacteria are confronted with anoxic conditions at night, nitrogenase activity is supported by ATP generated by substrate-level phosphorylation in fermentative pathways (Stal and Moezelaar 1997). However, the fact that many potentially diazotrophic cyanobacteria are unable to express nitrogenase aerobically when exposed to alternating light–dark cycles means that species that fix N_2 through temporal separation of N_2 fixation and photosynthesis must have additional mechanisms to render the diazotrophic cells anoxic, mechanisms that are not available to the 'anaerobic N_2-fixing' cyanobacteria. This may be associated with the characteristics of their respiratory systems and/or with the properties of gas diffusion into the diazotrophic cell, but thus far this aspect has not been studied.

It should be noted that several (and perhaps all) of the aerobic N_2-fixing cyanobacteria with the 'temporal separation' strategy seem, in fact, to be not really 'aerobic N_2 fixing.' Using an online, real-time setup for acetylene reduction assay (ARA) to monitor nitrogenase activity, with precisely controlled O_2 concentrations, it was discovered that several strains, previously reported to be 'aerobic' N_2-fixing cyanobacteria, did in fact not show any nitrogenase activity at atmospheric levels of O_2. For instance, the filamentous non-heterocystous mat-forming cyanobacterium *Lyngbya aestuarii* did not show nitrogenase activity above 5% O_2, even though the organism could be grown in aerobic batch cultures (Ferreira et al. 2009). The unicellular *Gloeothece* sp. and *Crocosphaera watsonii* did not fix N_2 above 10% O_2, although both strains grow well in aerobic cultures and it is known that *C. watsonii* thrives as plankton in the ocean (Compaoré and Stal 2009). We suspect that the same applies to the unicellular *Cyanothece*. Cyanobacteria such as *L. aestuarii* and *Gloeothece* also grow in continuous light and they still seem to maintain the separation of oxygenic photosynthesis and N_2 fixation. This may be different in *C. watsonii*. We were unable to grow this strain diazotrophically under conditions of continuous light. When grown in continuous culture *Gloeothece* was even shown to confine nitrogenase activity largely to the light period (Ortega-Calvo and Stal 1991). It was argued that this was necessary in order to fix sufficient N_2 to maintain a growth rate equal to the dilution rate of the culture. However, the mechanism was not elucidated.

The type of N_2 fixation observed with continuous cultures of *Gloeothece* resembles to some extent that of the oceanic, tropical filamentous diazotroph *Trichodesmium*. This cyanobacterium fixes N_2 exclusively during the day (Capone et al. 1990), thus resembling the pattern by heterocystous cyanobacteria. However, while the latter also fix N_2 during the dark, albeit usually at a lower rate, *Trichodesmium* is unable to fix N_2 in the dark. Using the online ARA setup it was shown that *Trichodesmium* is a real aerobic diazotroph (Staal et al. 2006) and was capable of maintaining nitrogenase activity at up to 25% O_2. The strategy by which *Trichodesmium* fixes N_2 in the light is most likely a combination of spatial and temporal separation from oxygenic photosynthesis (Berman-Frank et al. 2001). In some strains nitrogenase seems to be confined to special cells, in the literature sometimes termed 'diazocytes' (Fredriksson and Bergman 1997). In other cases, nitrogenase was present in all cells (Ohki 2008), but activity may have been limited to selected cells, which presumably have then switched off oxygenic photosynthesis. There is even evidence that cells may switch quickly between oxygenic photosynthesis and N_2 fixation (Küpper et al. 2004). There is little doubt that both processes are incompatible and can therefore not take place simultaneously. Since *Trichodesmium* thrives in the tropical and subtropical oceans at temperatures well above 25°C it has been proposed that a gas diffusion barrier such as glycolipid cell wall of the heterocyst is not required because O_2 solubility at full salinity seawater is 30% less than in freshwater and metabolic processes such as respiration are fast at elevated temperature (Staal et al. 2003; Stal 2009). A cell wall acting as a gas diffusion barrier would in fact represent a disadvantage because it would also limit the flux of N_2. *Trichodesmium* is unable to maintain active nitrogenase during the dark. Obviously, light is required to provide sufficient reducing equivalents to scavenge O_2, and light-dependent processes such as the Mehler reaction may be needed to render the diazotrophic cell anoxic (Kana 1993).

Other filamentous, non-heterocystous diazotrophic cyanobacteria that may resemble *Trichodesmium* with respect to the daily pattern of N_2 fixation are *Symploca* and *Lyngbya*. *Symploca* PCC8002 was originally isolated from tidal sediment and assigned as *Microcoleus chthonoplastes*. Fredriksson et al. (1998) reported that nitrogenase activity in this organism was confined to the light and showed that nitrogenase was present in a subset of the cells similar to the supposed diazocytes of *Trichodesmium*. However, this pattern may not be a general characteristic of this genus (Kumazawa et al. 2001). Also a tropical benthic marine *Lyngbya* sp. was reported to fix N_2 during the light (Jones 1992). *Trichodesmium*, *Lyngbya,* and *Symploca* are closely related organisms and it is possible they have adopted a similar strategy for aerobic N_2 fixation.

Another group of putative unicellular diazotrophic cyanobacteria seems to be important in the plankton of the oceans. These organisms, of which no cultivated representatives exist, are known as 'Group A' unicellular diazotrophic cyanobacteria. This group is remarkable because it expresses *nifH* during the day and there is also evidence that it actually fixes N_2 during the day. Flow cytometric sorting and subsequent metagenomic sequencing of this organism has revealed that it apparently lacks the genes for the oxygenic photosystem II and it therefore resembles a heterocyst (Zehr et al. 2008). Such an organism would be dependent on externally supplied organic matter and might actually live symbiotically. This would also explain N_2 fixation in this picoplanktonic cell that, due to its small size, would be impossible to render anoxic when free-living as single cells suspended in oxygenated seawater.

3.5 Heterocystous Cyanobacteria in Microbial Mats

Although heterocystous cyanobacteria seem to be best adapted for diazotrophic oxygenic photosynthetic growth, they are absent from full salinity seawater, at least as free-living planktonic organisms. For the tropical waters this can be explained because the glycolipid cell wall of the heterocyst is superfluous and even disadvantageous (see previous paragraph). However, it is unknown why such organisms are absent from the temperate and cold seas and oceans, while they are abundant in freshwater lakes and brackish water basins such as the Baltic Sea. Heterocystous cyanobacteria are also rare in marine microbial mats.

Comparing mats composed of heterocystous cyanobacteria with those dominated by non-heterocystous forms in the same area shed some light on this question. The heterocystous cyanobacterium *Calothrix* grew in Laguna Ojo de Liebre, Baja California, Mexico, on the higher parts of the intertidal flat and formed porous mats that were rarely inundated (Stal 1995). Nitrogenase activity was confined to the daytime, typical for heterocystous cyanobacteria. Depth profiles of oxygen in these mats revealed little dynamics, i.e., the mats were more or less air saturated with O_2 during the day and night, except when the mat was inundated. The mat was also oxygen saturated below the cyanobacterial mat and consequently no sulfide was present. In contrast, the non-heterocystous *Lyngbya* developed mats in the lower reaches of the intertidal flat that were inundated at every high tide. These mats were dense and showed dynamic oxygen depth profiles with oxygen supersaturation during the day (up to pure O_2) and anoxic conditions at night. Below the cyanobacterial layer the mat was permanently anoxic and consequently sulfide was present. N_2 fixation was confined to the night as might be expected, although it must have been supported by fermentative energy generation since aerobic respiration was impossible due to the absence of O_2. It was also concluded that the conditions in this mat type were apparently inadequate for heterocystous cyanobacteria. This might be attributed to the inability of heterocystous cyanobacteria (or of the heterocysts) of anaerobic dark metabolism (fermentation) or their sensitivity to sulfide, or both. O_2 supersaturation during the light might also discriminate against heterocystous cyanobacteria, although this seems less likely because some strains may still fix some N_2 in 100% O_2. *Lyngbya* may have been incapable of coping with air saturation levels of O_2 when fixing N_2 at night. However, other species might have tolerated atmospheric levels of O_2 or their respiratory activity might have sufficiently decreased the O_2 concentration. Non-heterocystous diazotrophic cyanobacteria therefore might not be able to compete with heterocystous types.

Sulfide may select against heterocystous cyanobacteria as has been shown in culture and also in microbial mats in the Bay of Arcachon in France (Villbrandt and Stal 1996). The sediments and waters in this system contained high levels of iron which scavenged any sulfide by precipitating it as FeS and pyrite. These mats contained heterocystous cyanobacteria, while mats in another lagoon where the sediments contained high levels of free sulfide exclusively non-heterocystous diazotrophic cyanobacteria were observed.

Heterocystous cyanobacteria are also encountered in microbial mats that are exposed to lower salinity, e.g., when they receive upwelling freshwater or rain. It is unlikely that it is the sea salt concentration that impairs growth of heterocystous cyanobacteria. It has been suggested that the high sulfate concentration in seawater (28 mM) could interfere with N_2 fixation in heterocysts (Stal et al. 1999). Sulfate is a structural analogue of molybdate, a co-factor for the classical Fe–Mo nitrogenase. The uptake of molybdate might be restricted in the presence of high concentrations of sulfate, although there is no conclusive evidence for this mechanism. In microbial mats the sulfate concentration in the pore water may be lowered through the activity of sulfate-reducing bacteria.

The heterocystous *Cyanobacterium Rivularia* forms stromatolitic microbial mats in freshwater streams (Sabater et al. 2000). Its occurrence has been associated with nitrogen-depleted environments and high rates of N_2 fixation, although not much research has focused on this aspect. Non-calcifying *Rivularia* colonies have also been found in the rocky shores in the supralittoral of intertidal marine bays (Khoja et al. 1984). *Rivularia* is related to *Calothrix* which is known from marine and brackish environments. N_2 fixation in these cyanobacteria is impaired at full seawater salinity (Reed and Stewart 1983).

Hot springs are another type of environment in which heterocystous cyanobacteria thrive and form microbial mats (Finsinger et al. 2008). Hot spring heterocystous cyanobacteria belong to section V comprising species that exhibit true branching and have been assigned to in the literature as *Mastigocladus*, *Chlorogloeopsis*, and *Fischerella*. They grow and fix N_2 up to temperatures of $\sim$60°C. However, growth and N_2 fixation appeared to be highly sensitive to sulfide and where sulfide occurred in higher concentrations *Oscillatoria* became dominant and presumably thrived by anoxygenic photosynthesis (Castenholz 1976). It is not known whether this cyanobacterium fixes N_2.

Mats that contain heterocystous cyanobacteria may exhibit more complex daily patterns of N_2 fixation. These organisms show highest nitrogenase activities during the day, although they may have varying levels of dark activity. The latter depends, however, on aerobic respiration and when oxygen is unavailable, no N_2 is fixed by the heterocystous cyanobacteria. The daily pattern of nitrogenase activity may be further complicated by the presence of not only other diazotrophs, including non-heterocystous filamentous and unicellular cyanobacteria, but also other Bacteria and Archaea (Severin and Stal 2008).

3.6 Other Diazotrophs in Microbial Mats

Sulfate-reducing bacteria and anoxygenic phototrophic bacteria are two functional groups of Bacteria that play important roles in microbial mats and that have representatives capable of diazotrophic growth. Their *nif* genes have been found in microbial mats as well as their transcripts (Olson et al. 1999; Steppe and Paerl 2002; Severin et al. 2010). This has led to a different view on the microbial ecology of N_2 fixation. Considering the supposed difficulties of N_2 fixation in non-heterocystous cyanobacteria, a model was proposed that featured chemotrophic Bacteria in the mat as the actual N_2-fixing organisms (Steppe et al. 1996). They would live in close proximity to the cyanobacteria which would supply the chemotrophic Bacteria with organic substrate, while the latter provide fixed nitrogen, CO_2, and other growth factors. This model was supported by the detection of nitrogenase gene transcripts putatively assigned to chemotrophic Bacteria and the inability of detecting any of such genes attributable to cyanobacteria. However, the latter finding may be the result of a bias introduced by the nucleic acid extraction methods that were used as many benthic cyanobacteria are not easily extracted due to their rigid extracellular polysaccharide sheaths.

3.7 *Microcoleus chthonoplastes*: A Diazotroph?

Many cyanobacterial mats are formed by the cosmopolitan filamentous non-heterocystous cyanobacterium *M. chthonoplastes*. This species form bundles of trichomes enclosed by a common sheath, although this property is often lost upon cultivation in the laboratory. Originally, it was thought that *M. chthonoplastes* was the N_2-fixing organisms in the mats in which it was the dominant cyanobacterium. It was assumed that the bundle formation of *M. chthonoplastes* facilitated N_2 fixation by providing micro-zones with low oxygen. However, N_2-fixing strains that were subsequently isolated from microbial mats and assigned to *M. chthonoplastes* appear to have been misidentified. For instance, the aerobically N_2-fixing *M. chthonoplastes* isolated by Pearson (Pearson et al. 1979; Malin and Pearson 1988) was later identified as *Symploca* sp. (Janson et al. 1998) and the anaerobic N_2-fixing *M. chthonoplastes* 'strain 11' is related to *Geitlerinema* (Siegesmund et al. 2008), a genus which also includes the famous anoxygenic phototrophic and anaerobic N_2-fixing Solar Lake strain '*Oscillatoria limnetica*.' Dubinin et al. (1992) and Sroga (1997) also reported on N_2-fixing *Microcoleus* but their correct assignments have not been confirmed.

Using strictly anaerobic conditions, Rippka et al. (1979) were unable to detect nitrogenase activity in the type strain of *M. chthonoplastes* PCC7420 and Villbrandt and Stal (unpublished results) were unable to induce nitrogenase activity in the collection of the 'true' *M. chthonoplastes* of Garcia-Pichel et al. (1996). Moreover, Steppe et al. (1996) were unable to detect cyanobacterial *nifH* in four strains of *M. chthonoplastes*, including the type strain PCC7420. However, they did amplify from these cultures *nifH* belonging to γ-proteobacteria and these sequences were also found in microbial mats dominated by *M. chthonoplastes*. These authors concluded subsequently that the cultures were contaminated and that diazotrophic heterotrophic Bacteria live in a consortium with *M. chthonoplastes* in the microbial mat. However, if this model is correct they should have been able to grow this consortium with N_2, but that was not the case.

Also Zehr et al. (1995) did not find cyanobacterial *nifH* genes in mats dominated by *M. chthonoplastes*. The *nifH* sequences retrieved in the *M. chthonoplastes* mats belonged to anaerobic heterotrophic Bacteria most closely related to γ-proteobacteria. Olson et al. (1999) studied three distant mats of *M. chthonoplastes* and found only heterotrophic *nifH* sequences which were all closely related notwithstanding their large geographic distances. They concluded that cyanobacteria were not important for N_2 fixation and that the same type of γ-proteobacteria was the diazotroph in these mats. Similarly, Steppe and Paerl (2002) found that sulfate-reducing bacteria were potentially important dinitrogen fixers and that most of the *nifH* genes expressed clustered with the δ-proteobacteria and other anaerobic Bacteria. Omoregie et al. (2004a) investigated a mat showing high rates of nitrogenase activity during the night but they did not find any *nifH* sequences related to filamentous cyanobacteria. The *nifH* sequences they found belonged to the γ-proteobacteria and other novel sequences, and therefore the authors concluded that cyanobacteria were not involved in N_2 fixation in this mat. Nevertheless, in a follow-up study Omoregie et al. (2004b) found the expression of *nifH* belonging to unicellular cyanobacteria that were not found in the clone libraries of this gene. But still many expressed sequences belonged to anaerobic Bacteria.

Within the sheath of *M. chthonoplastes*, a new gliding filamentous purple sulfur phototrophic bacterium has been found (D'Amelio et al. 1987). It is not known whether this organism could be responsible for N_2 fixation. D'Amelio et al. (1987) proposed that in *M. chthonoplastes*, when thriving in the zone below maximum oxygenic photosynthesis, the associated bacterium could carry out sulfide-dependent anoxygenic photosynthesis or grow photoheterotrophically using organic matter excreted by the cyanobacterium. Moreover, *M. chthonoplastes* may be capable of anoxygenic photosynthesis oxidizing sulfide to elemental sulfur (Garlick et al. 1977). The purple sulfur bacterium could then oxidize the elemental sulfur further to sulfate.

Recently, the genome of *M. chthonoplastes* PCC7420 was sequenced. Remarkably, a complete nitrogenase operon was found to be present in this genome and the same *nif*genes were detected in a collection of other axenic *M. chthonoplastes* strains (Bolhuis et al. 2009). Phylogenetic analysis of *nifHDK* revealed that they cluster with the δ-proteobacteria and that they are closely related to sulfate-reducing bacteria *Desulfovibrio* and it was proposed that this cluster was obtained by the cyanobacterium through lateral gene transfer. This of course sheds a totally new light on the previous reports of the inability of *M. chthonoplastes* to fix N_2 and the fact that the *nif* genes were incorrectly not attributed to *M. chthonoplastes*. It has hitherto proven impossible to induce nitrogenase activity in *M. chthonoplastes* or even express the *nif* genes. It is not clear whether this is due to inadequate conditions or whether the gene cluster is in fact inactive although this is unlikely as it is otherwise difficult to explain why this cluster is maintained in *M. chthonoplastes* strains found all over the world. Moreover, the expression of this typical *nifHDK* cluster was seen in samples of *M. chthonoplastes* mats. This suggests that the expression of the *nif* genes and the induction of an active nitrogenase in *M. chthonoplastes* depend on the appropriate conditions. After all, the hypothesis of a consortium of *M. chthonoplastes* with an N_2-fixing chemotroph may have been proven wrong now, since we know that this cyanobacterium is most likely capable of fixing its own N_2.

3.8 Day–Night Variations of Nitrogenase Activity

The day–night variations in nitrogenase activity in microbial mats are the result of the combined action of the microorganisms that contribute to N_2 fixation and the conditions that control N_2 fixation (Severin and Stal 2008). Among the latter the actual environmental concentration of O_2 is important. However, it should be noted that O_2 is not only a negative factor. Admittedly, it will prevent nitrogenase activity in anaerobic organisms or in facultative anaerobes that lack the possibility to furnish nitrogenase with anaerobic conditions in an otherwise aerobic environment. However, O_2 might be essential for nitrogenase activity in the dark since aerobic respiration may be the only process that supplies nitrogenase with energy. For instance, this is the case in heterocystous cyanobacteria. This group of cyanobacteria is, however, not homogenous with respect to the light–dark variations of nitrogenase activity they exhibit. Some species confine virtually all nitrogenase activity to the light period, while others have varying levels of nitrogenase activity in the dark, including some where this activity is basically not different from the levels of activity in the light (Evans et al. 2000). These differences may be species specific or dependent on the light history of the organism. Dark nitrogenase activity in heterocystous cyanobacteria not only depends on respiratory activity, and therefore on diffusion of O_2 into the heterocyst, but also depends on the import of reducing equivalents from the neighboring vegetative cells which are influenced by the light history of the organism. When the O_2 diffusion exceeds the import of reducing equivalents it will eventually lead to the inactivation of nitrogenase. Also, in the light the diffusion of O_2 will draw upon the availability of reducing equivalents, even though heterocystous cyanobacteria may cope with 100% O_2, the fixation of N_2 will be lower due to the fact that a substantial amount of the electrons are used for respiration. All this will lead to complex daily variations of nitrogenase activity that are difficult to interpret.

Non-heterocystous diazotrophic cyanobacteria in microbial mats are mostly of the type that separates nitrogenase activity temporally from photosynthesis and thus confine it to the dark period. Hence, the cessation of oxygenic photosynthesis may be the trigger for the induction of nitrogenase. Dark nitrogenase activity depends on aerobic respiration as a source of energy, but because many microbial mats turn anoxic in the dark, fermentation and substrate-level phosphorylation may be the only way for these cyanobacteria to generate energy for N_2 fixation. This presumes that these cyanobacteria are capable of fermentation, but, in fact, this is likely since they are living in such environments. However, the low ATP yield of fermentative energy generation will support only a limited amount of N_2 fixation. In some microbial mats this leads to a peculiar daily pattern of nitrogenase activity (Villbrandt et al. 1990). At sunset, when the light intensity is low and photosynthesis ceases, nitrogenase is induced, taking advantage of the low rate of photosynthesis and the low levels of O_2, while some light is available. This activity decreases to very low levels after the sun has set and the mat has turned anoxic. Nitrogenase then becomes fully expressed, unhindered by the presence of O_2, and at sunrise a huge peak of nitrogenase activity is supported by the first light, while the mat is still anoxic. This activity subsequently rapidly disappears when photosynthesis starts and oxygenates the mat.

Much less is known about the contribution of diazotrophic Bacteria other than cyanobacteria. It is anticipated that such organisms should also exhibit daily variations of nitrogenase activity. For instance, sulfate-reducing bacteria are dependent for their substrate mainly on the cyanobacteria. However, their stimulation may be both during the daytime, when cyanobacteria may provide photosynthate (glycolate, EPS), or during the night, when they release low-molecular fermentation products. It is also feasible that different species are involved in the decomposition of different substrates and these different species may or may not be diazotrophs. Not much is known about their contribution. The sulfide produced by the sulfate-reducing bacteria is used by anoxygenic phototrophic bacteria. It is probable that diazotrophic anoxygenic phototrophic bacteria depend on light for fixing N_2. However, also in this case, no information is available on the contribution of this group of Bacteria to N_2 fixation in microbial mats.

The daily patterns of nitrogenase activity in microbial mats are complex and poorly understood. The analyses of specific nitrogenase gene expression patterns are poorly correlated to measured nitrogenase activity. More detailed knowledge is required on the different functional groups of diazotrophs in microbial mats and the factors that control their (diazotrophic) activity.

References

Adams DG (2000) Heterocyst formation in cyanobacteria. Curr Opin Microbiol 3:618–624

Baumgartner LK, Reid RP, Dupraz C, Decho AW, Buckley DH, Spear JR, Przekop KM, Visscher PT (2006) Sulfate reducing bacteria in microbial mats: changing paradigms, new discoveries. Sediment Geol 185:131–145

Bergman B, Gallon JR, Rai AN, Stal LJ (1997) N_2 fixation by non-heterocystous cyanobacteria. FEMS Microbiol Rev 19:139–185

Berman-Frank I, Lundgren P, Chen Y-B, Küpper H, Kolber Z, Bergman B, Falkowski P (2001) Segregation of nitrogen fixation and oxygenic photosynthesis in the marine cyanobacterium *Trichodesmium*. Science 294:1534–1537

Bolhuis H, Severin I, Confurius-Guns V, Wollenzien UIA, Stal LJ (2009) Horizontal transfer of the nitrogen fixation gene cluster in the cyanobacterium *Microcoleus chthonoplastes*. ISME J. doi:10.1038/ismej.2009.99

Canfield DE, Des Marais DJ (1991) Aerobic sulfate reduction in microbial mats. Science 251:1471–1473

Capone DG, O'Neil JM, Zehr J, Carpenter EJ (1990) Basis for diel variation in nitrogenase activity in the marine planktonic cyanobacterium *Trichodesmium thiebautii*. Appl Environ Microbiol 56:3532–3536

Castenholz RW (1976) The effect of sulfide on the bluegreen algae of hot springs. I. New Zealand and Iceland. J Phycol 12:54–68

Compaoré J, Stal LJ (2009) Oxygen and the light-dark cycle of nitrogenase activity in two unicellular cyanobacteria. Environ Microbiol. doi:10.1111/j.1462-2920.2009.02034.x

Curatti L, Flores E, Salerno G (2002) Sucrose is involved in the diazotrophic metabolism of the heterocyst-forming cyanobacterium *Anabaena* sp. FEBS Lett 513:175–178

D'Amelio ED, Cohen Y, Des Marais DJ (1987) Association of a new type of gliding, filamentous, purple phototrophic bacterium inside bundles of *Microcoleus chthonoplastes* in hypersaline cyanobacterial mats. Arch Microbiol 147:213–220

Dubinin AV, Gerasimenko LM, Zavarzin GA (1992) Nitrogen fixation by cyanobacterium *Microcoleus chthonoplastes* from hypersaline lagoons of lake Sivash. Microbiology 61: 593–597

Evans AM, Gallon JR, Jones A, Staal M, Stal LJ, Villbrandt M, Walton TJ (2000) Nitrogen fixation by Baltic cyanobacteria is adapted to the prevailing photon flux density. New Phytol 147: 285–297

Fenchel T (1998) Formation of laminated cyanobacterial mats in the absence of benthic fauna. Aquat Microb Ecol 14:235–240

Ferreira D, Stal LJ, Moradas-Ferreira P, Mendes MV, Tamagnini P (2009) The relation between N_2 fixation and H_2 metabolism in the marine filamentous nonheterocystous cyanobacterium *Lyngbya aestuarii* CCY 9616. J Phycol 45:898–905

Finsinger K, Scholz I, Serrano A, Morales S, Uribe-Lorio L, Mora M, Sittenfeld A, Weckesser J, Hess WR (2008) Characterization of true-branching cyanobacteria from geothermal sites and hot springs of Costa Rica. Environ Microbiol 10:460–473

Fredriksson C, Bergman B (1997) Ultrastructural characterisation of cells specialised for nitrogen fixation in a non-heterocystous cyanobacterium, *Trichodesmium* spp. Protoplasma 197:76–85

Fredriksson C, Malin G, Siddiqui PJA, Bergman B (1998) Aerobic nitrogen fixation is confined to a subset of cells in the non-heterocystous cyanobacterium *Symploca* PCC 8002. New Phytol 140:531–538

Gallon JR (1992) Reconciling the incompatible: N_2 fixation and O_2. New Phytol 122:571–609

Garcia-Pichel F, Prufert-Bebout L, Muyzer G (1996) Phenotypic and phylogenetic analyses show *Microcoleus chthonoplastes*to be a cosmopolitan cyanobacterium. Appl Environ Microbiol 62:3284–3291

Garlick S, Oren A, Padan E (1977) Occurrence of facultative anoxygenic photosynthesis among filamentous and unicellular cyanobacteria. J Bacteriol 129:623–629

Janson S, Matveyev A, Bergman B (1998) The presence and expression of *hetR* in the non-heterocystous cyanobacterium *Symploca* PCC 8002. FEMS Microbiol Lett 168:173–179

Jones K (1992) Diurnal nitrogen fixation in tropical marine cyanobacteria: a comparison between adjacent communities of non-heterocystous *Lyngbya* sp. and heterocystous *Calothrix* sp. Br Phycol J 27:107–118

Jørgensen BB, Revsbech NP, Cohen Y (1983) Photosynthesis and structure of benthic microbial mats: micro-electrode and SEM studies of four cyanobacterial communities. Limnol Oceanogr 28:1075–1093

Kana TM (1993) Rapid oxygen cycling in *Trichodesmium thiebautii*. Limnol Oceanogr 38:18–24

Khoja TM, Livingstone D, Whitton BA (1984) Ecology of a marine *Rivularia*population. Hydrobiologia 108:65–73

Kumazawa S, Yumura S, Yoshisuji H (2001) Photoautotrophic growth of a recently isolated N_2-fixing marine non-heterocystous filamentous cyanobacterium *Symploca* sp. (Cyanobacteria). J Phycol 37:482–487

Küpper H, Ferimazova N, Šetlík I, Berman-Frank I (2004) Traffic lights in *Trichodesmium*. Regulation of photosynthesis for nitrogen fixation studied by chlorophyll fluorescence kinetic microscopy. Plant Physiol 135:2120–2133

Malin G, Pearson HW (1988) Aerobic nitrogen fixation in aggregate-forming cultures of the nonheterocystous cyanobacterium *Microcoleus chthonoplastes*. J Gen Microbiol 134: 1755–1763

Noffke N, Beukes N, Gutzmer J, Hazen R (2006) Spatial and temporal distribution of microbially induced sedimentary structures: a case study from siliciclastic storm deposits of the 2.9 Ga Witwatersrand Supergroup, South Africa. Precambrian Res 146:35–44

Ohki K (2008) Intercellular localization of nitrogenase in a non-heterocystous cyanobacterium (Cyanophyte) *Trichodesmium* sp. NIBB1067. J Oceanogr 64:211–216

Olson JB, Litaker RW, Paerl HW (1999) Ubiquity of heterotrophic diazotrophs in marine microbial mats. Aquat Microbiol Ecol 19:29–36
Omoregie EO, Crumbliss LL, Bebout BM, Zehr JP (2004a) Comparison of diazotroph community structure in *Lyngbya* sp. and *Microcoleus chthonoplastes* dominated microbial mats from Guerrero Negro, Baja, Mexico. FEMS Microbiol Ecol 47:305–318
Omoregie EO, Crumbliss LL, Bebout BM, Zehr JP (2004b) Determination of nitrogen-fixing phylotypes in *Lyngbya*sp. and *Microcoleus chthonoplastes* cyanobacterial mats from Guerrero Negro, Baja California, Mexico. Appl Environ Microbiol 70:2119–2128
Ortega-Calvo JJ, Stal LJ (1991) Diazotrophic growth of the unicellular cyanobacterium *Gloeothece*sp. PCC 6909 in continuous culture. J Gen Microbiol 137:1789–1797
Pearson HW, Howsley R, Kjeldsen CK, Walsby AE (1979) Aerobic nitrogenase activity associated with a non-heterocystous filamentous cyanobacterium. FEMS Microbiol Lett 5:163–169
Reed RH, Stewart WDP (1983) Physiological responses of *Rivularia atra* to salinity: osmotic adjustment in hyposaline media. New Phytol 95:595–603
Revsbech NP, Jørgensen BB, Blackburn TH, Cohen Y (1983) Microelectrode studies of the photosynthesis and O_2, H_2S and pH profiles of a microbial mat. Limnol Oceanogr 28: 1062–1074
Rippka R, Deruelles J, Waterbury JB, Herdman M, Stanier RY (1979) Generic assignments strain histories and properties of pure cultures of cyanobacteria. J Gen Microbiol 111:1–61
Sabater S, Guasch H, Romaní A, Muñoz I (2000) Stromatolitic communities in Mediterranean streams: adaptations to a changing environment. Biodivers Conserv 9:379–392
Schopf JW (2000) Solution to Darwin's dilemma: discovery of the missing Precambrian record of life. Proc Natl Acad Sci USA 97:6947–6953
Seckbach J (2007) Algae and cyanobacteria in extreme environments. Springer, Dordrecht
Severin I, Acinas SG, Stal LJ (2010) Present *vs*. active nitrogen-fixing microorganisms in a diverse and dynamic cyanobacterial mat community. Environ Microbiol (in press)
Severin I, Stal LJ (2008) Light dependency of nitrogen fixation in a coastal cyanobacterial mat. ISME J 2:1077–1088
Severin I, Stal LJ (2010) Diazotrophic microbial mats. In: Seckbach J (ed) Microbial mats, Springer, New York (in press)
Siegesmund M, Johansen JR, Karsten U, Friedl T (2008) *Coleofasciculus*gen. nov. (Cyanobacteria): morphological and molecular criteria for revision of the genus *Microcoleus* Gomont. J Phycol 44:1572–1585
Sroga GE (1997) Regulation of nitrogen fixation by different nitrogen sources in the filamentous non-heterocystous cyanobacterium *Microcoleus* sp. FEMS Microbiol Lett 153:11–15
Staal M, Meysman FJR, Stal LJ (2003) Temperature excludes N_2-fixing heterocystous cyanobacteria in the tropical oceans. Nature 425:504–507
Staal M, Rabouille S, Stal LJ (2006) On the role of oxygen for nitrogen fixation in the marine cyanobacterium *Trichodesmium* sp. Environ Microbiol 9:727–736
Stal LJ (1995) Physiological ecology of cyanobacteria in microbial mats and other communities. New Phytol 131:1–32
Stal LJ (2000) Cyanobacterial mats and stromatolites. In: Whitton BA, Potts M (eds) The ecology of cyanobacteria. Kluwer, Dordrecht, pp 61–120
Stal LJ (2001) Coastal microbial mats: the physiology of a small-scale ecosystem. S Afr J Bot 67:399–410
Stal LJ (2009) Is the distribution of nitrogen-fixing cyanobacteria in the oceans related to temperature? Environ Microbiol 11:1632–1645
Stal LJ, Moezelaar R (1997) Fermentation in cyanobacteria. FEMS Microbiol Rev 21:179–211
Stal LJ, Staal M, Villbrandt M (1999) Nutrient control of cyanobacterial blooms in the Baltic Sea. Aquat Microbiol Ecol 18:165–173
Stal LJ, Zehr JP (2008) Cyanobacterial nitrogen fixation in the ocean: diversity, regulation, and ecology. In: Flores E, Herrero A (eds) The cyanobacteria: molecular biology, genomics and evolution. Caister Academic Press, Norfolk, UK, pp 423–446

Steppe TF, Olson JB, Paerl HW, Litaker RW, Belnap J (1996) Consortial N_2 fixation: a strategy for meting nitrogen requirements of marine and terrestrial cyanobacterial mats. FEMS Microbiol Ecol 21:149–156

Steppe TF, Paerl HW (2002) Potential N_2 fixation by sulfate-reducing bacteria in a marine intertidal microbial mat. Aquat Microbiol Ecol 28:1–12

Stewart WDP (1980) Some aspects of structure and function in N_2-fixing cyanobacteria. Annu Rev Microbiol 34:497–536

Villbrandt M, Stal LJ, Krumbein WE (1990) Interactions between nitrogen fixation and oxygenic photosynthesis in a marine cyanobacterial mat. FEMS Microbiol Ecol 74:59–72

Villbrandt M, Stal LJ (1996) The effect of sulfide on nitrogen fixation in heterocystous and non-heterocystous cyanobacterial mat communities. Algol Stud 83:549–563

Visscher PT, Nijburg JW, van Gemerden H (1990) Polysulfide utilization by *Thiocapsa roseopersicina*. Arch Microbiol 155:75–81

Visscher PT, van den Ende FP, Schaub BEM, van Gemerden H (1992) Competition between anoxygenic phototrophic bacteria and colorless sulfur bacteria in a microbial mat. FEMS Microbiol Ecol 101:51–58

Walsby AE (2007) Cyanobacterial heterocysts: terminal pores proposed as sites of gas exchange. Trends Microbiol 15:340–349

Widdel F, Schnell S, Heising S, Ehrenreich A, Assmus B, Schink B (1993) Ferrous iron oxidation by anoxygenic phototrophic bacteria. Nature 362:834–836

Zehr JP, Bench SR, Carter BJ, Hewson I, Niazi F, Shi T, Tripp HJ, Affourtit JP (2008) Globally distributed uncultivated oceanic N_2-fixing cyanobacteria lack oxygenic photosystem II. Science 322:1110–1112

Zehr JP, Jenkins BD, Short SM, Steward GF (2003) Nitrogenase gene diversity and microbial community structure: a cross-system comparison. Environ Microbiol 5:539–554

Zehr JP, Mellon M, Braun S, Litaker W, Steppe T, Paerl HW (1995) Diversity of heterotrophic nitrogen fixation genes in a marine cyanobacterial mat. Appl Environ Microbiol 61:2527–2532

Part II
Physiology, Metabolism, and Global Responses

Chapter 4
Nitrogen and Molybdenum Control of Nitrogen Fixation in the Phototrophic Bacterium *Rhodobacter capsulatus*

Bernd Masepohl and Patrick C. Hallenbeck

Abstract The vast majority of the purple nonsulfur photosynthetic bacteria are diazotrophs, but the details of the complex regulation of the nitrogen fixation process are well understood only for a few species. Here we review what is known of the well-studied *Rhodobacter capsulatus*, which contains two different nitrogenases, a standard Mo-nitrogenase and an alternative Fe-nitrogenase, and which has overlapping transcriptional control mechanisms with regard to the presence of fixed nitrogen, oxygen, and molybdenum as well as the capability for the post-translational control of both nitrogenases in response to ammonium. *R. capsulatus* has two PII proteins, GlnB and GlnK, which play key roles in nitrogenase regulation at each of three different levels: activation of transcription of the *nif*-specific activator NifA, the post-translational control of NifA activity, and the regulation of nitrogenase activity through either ADP-ribosylation of NifH or an ADP-ribosylation-independent pathway. We also review recent work that has led to a detailed characterization of the molybdenum transport and regulatory system in *R. capsulatus* that ensures activity of the Mo-nitrogenase and repression of the Fe-nitrogenase, down to extremely low levels of molybdenum.

4.1 Nitrogen Fixation in Purple Nonsulfur Bacteria

4.1.1 Biological Nitrogen Fixation

Ammonium is the preferred inorganic nitrogen source in bacteria, as it can directly be incorporated into organic compounds. The deprotonated form, ammonia (NH_3), diffuses through biological membranes, while the protonated form, ammonium (NH_4^+), does not (Andrade and Einsle 2007). Thus, at least at neutral pHs and

B. Masepohl (✉)
Lehrstuhl für Biologie der Mikroorganismen, Fakultät für Biologie und Biotechnologie, Ruhr-Universität Bochum, 44780, Bochum, Germany
e-mail: bernd.masepohl@rub.de

P.C. Hallenbeck (ed.), *Recent Advances in Phototrophic Prokaryotes*, Advances in Experimental Medicine and Biology 675, DOI 10.1007/978-1-4419-1528-3_4,

moderate external ammonium concentrations, the concentration of ammonia may be sufficiently high to supply cells with nitrogen. At low external concentrations, ammonium is actively taken up by high-affinity transporters like AmtB (see below).

Most bacteria grow with ammonium as nitrogen source, while exclusively diazotrophic species are able to use atmospheric dinitrogen (N_2). All diazotrophs synthesize nitrogenase, a highly conserved enzyme which catalyzes the reduction of N_2 to NH_3 (4.1). Since nitrogen fixation is a highly energy-demanding process, synthesis of nitrogenase is inhibited as long as ammonium is available (see below):

$$N_2 + \geq 16\,ATP + 8\,e^- + 8\,H^+ \rightarrow 2\,NH_3 + H_2 + \geq 16ADP + \geq 16\,P_i \quad (4.1)$$

Nitrogen fixation (*nif*) genes were originally defined for the diazotrophic model bacterium *Klebsiella pneumoniae* (Dixon and Kahn 2004). The structural genes of nitrogenase (*nifHDK*) encode dinitrogenase reductase (Fe protein, $NifH_2$) and dinitrogenase (MoFe protein, $NifD_2$–$NifK_2$). Dinitrogenase contains a unique iron–molybdenum cofactor FeMoco, which is the site of N_2 reduction (Seefeldt et al. 2009). FeMoco biosynthesis requires the products of several genes including *nifU*, *nifS*, *nifB*, *nifQ*, *nifV*, *nifE*, and *nifN* (Hu et al. 2008). While *nifHDK* and most of the above-mentioned FeMoco biosynthesis genes are present in all diazotrophs, the central regulatory gene *nifA* is restricted to proteobacteria.

In addition to molybdenum-dependent nitrogenase (Mo-nitrogenase) containing FeMoco, some diazotrophs synthesize one or two alternative (in the sense of Mo-independent) nitrogenases (Glazer and Kechris 2009). Vanadium-dependent nitrogenase (V-nitrogenase) contains an iron–vanadium cofactor (FeVco), while iron-only nitrogenase contains a heterometal-free cofactor (FeFeco). V-nitrogenase and Fe-nitrogenase are encoded by distinct sets of genes, namely the *vnf* (vanadium-dependent nitrogen fixation) and the *anf* (alternative nitrogen fixation) genes. Alternative nitrogenases are advantageous when molybdenum becomes limiting (Maynard et al. 1994). On the other hand, V- and Fe-nitrogenases exhibit lower specific activities than Mo-nitrogenase, and thus, Mo-nitrogenase is the preferred enzyme as long as molybdenum is available (Schneider et al. 1994). To provide Mo-nitrogenase with molybdenum at low external concentrations, many diazotrophs synthesize ABC-type, high-affinity molybdate uptake systems (ModABC) (Self et al. 2001).

4.1.2 Diazotrophic Purple Nonsulfur Bacteria

N_2-fixing species are found in most bacterial families and in methanogenic archaea, but not in eukaryotes. Besides diazotrophic species, these families include non-diazotrophic species, the latter being generally in the majority. Remarkably, the ability to fix dinitrogen is almost universal among members of the purple nonsulfur bacterial family (Madigan et al. 1984). Diazotrophic growth has been reported for numerous strains belonging to the genera *Rhodobacter*, *Rhodocyclus*,

Rhodomicrobium, *Rhodopseudomonas*, and *Rhodospirillum*. However, to date about 95% of all publications concerning nitrogen fixation in purple nonsulfur bacteria deal with only four model organisms, namely *Rhodobacter capsulatus*, *Rhodobacter sphaeroides*, *Rhodopseudomonas palustris*, and *Rhodospirillum rubrum*. These strains differ considerably regarding generation times under N_2-fixing conditions with the shortest doubling times (about 3 h) being observed for *R. capsulatus* (Madigan et al. 1984).

This review will focus on the regulation of nitrogen fixation in *R. capsulatus*. First, ammonium control of synthesis and activity of Mo- and Fe-nitrogenase will be briefly depicted (Masepohl and Forchhammer 2007; Masepohl and Kranz 2009). Here, emphasis will be on the roles of PII-type signal transduction proteins, GlnB and GlnK, and the ammonium transporter, AmtB (Drepper et al. 2003; Pawlowski et al. 2003; Tremblay and Hallenbeck 2008; Tremblay and Hallenbeck 2009; Tremblay et al. 2007). Second, molybdenum control of *anfA* (encoding the activator of Fe-nitrogenase genes), *modABC* (encoding the high-affinity Mo transporter), and *mop* (encoding a molbindin-like protein) will be described in more detail (Kutsche et al. 1996; Wiethaus et al. 2006; Wiethaus et al. 2009). Finally, selected aspects of nitrogen fixation in *R. sphaeroides*, *R. palustris*, and *R. rubrum* will be addressed.

4.2 Nitrogen Fixation in *R. capsulatus*

4.2.1 Organization of Nitrogen Fixation Genes

Rhodobacter capsulatus harbors more than 50 nitrogen fixation-related genes, most of which are clustered in four unlinked regions of the chromosome designated *nif* region A, B, C, and D (Masepohl and Klipp 1996, www.ergo-light.com). Among the nitrogen fixation genes are the homologues of the "classical" *K. pneumoniae nif* genes mentioned above. Region A contains genes involved in electron supply to nitrogenase (*rnfABCDGEH*, *rnfF*, and *fdxN*), FeMoco biosynthesis (*nifEN*, *nifQ*, *nifSV*, and *nifB1*), and nitrogen regulation (*nifA1*). The *rnf* genes originally defined as *Rhodobacter*-specific nitrogen fixation genes (Schmehl et al. 1993) have meanwhile been detected in several other diazotrophs as well as a variety of other bacteria where they perform metabolic functions unrelated to nitrogen fixation. Region B contains the structural genes of Mo-nitrogenase, *nifHDK*, the nitrogen regulatory genes *rpoN* and *nifA2*, the FeMoco biosynthesis gene *nifB2*, genes involved in molybdenum uptake (*modABC*), and two molybdenum regulatory genes (*mopA*, *mopB*). Region C contains the *ntrBC* genes coding for a two-component regulatory system, which acts on top of the nitrogen regulatory cascade (see below). Region D contains the nitrogen regulatory *anfA* gene and the structural genes of Fe-nitrogenase, *anfHDGK*. Genes coding for PII signal transduction proteins (*glnB*, *glnK*), a high-affinity ammonium transporter (*amtB*), and proteins involved in post-translational regulation of nitrogenase (*draTG*) are located elsewhere in the chromosome (Drepper et al. 2003; Masepohl et al. 1993; Yakunin and Hallenbeck 2002).

Several *R. capsulatus* nitrogen fixation genes exist in duplicate. Among these are the FeMoco biosynthesis genes *nifB1*/*nifB2*, the regulatory genes *nifA1*/*nifA2*, and *mopA*/*mopB* (Masepohl et al. 1988; Wang et al. 1993). Like most bacteria, *R. capsulatus* harbors two PII genes, *glnB* and *glnK* (Drepper et al. 2003). The products of the respective gene pairs can (at least partially) substitute for each other, and thus, clear phenotypes require analysis of double mutants.

Several *nif* genes are not only essential for activity of Mo-nitrogenase but are, in addition, required for activity of Fe-nitrogenase (Schüddekopf et al. 1993). These are the *rnf* genes involved in electron supply, as well as the FeMoco biosynthesis genes *nifV* and *nifB*. The requirement of NifV (homocitrate synthase) and NifB (catalyzing synthesis of NifBco, a Mo-free precursor of FeMoco) together with other findings suggests that FeFeco is structurally similar to FeMoco except for the "substitution" of molybdenum by iron.

4.2.2 Activation of Nitrogen Fixation Genes Under Nitrogen Limitation

Under nitrogen-limiting conditions, the sensor kinase NtrB autophosphorylates and, in turn, serves as phosphodonor for its cognate response regulator, NtrC (Cullen et al. 1996). In its phosphorylated form, NtrC binds to its target promoters and activates transcription of its target genes including *nifA1* and *nifA2* (Fig. 4.1; Cullen et al. 1996; Masepohl et al. 2001). In contrast to NtrC proteins from other diazotrophs, which activate transcription in concert with the nitrogen-specific sigma factor RpoN, *R. capsulatus* NtrC acts together with the housekeeping sigma factor RpoD (Bowman and Kranz 1998; Foster-Hartnett and Kranz 1992; Foster-Hartnett et al. 1994).

NifA1 and NifA2 exhibit 97% identity to each other and differ only in their extreme N-terminal amino acid sequences (Masepohl et al. 1988). The two NifA proteins functionally substitute for each other, and either of them is sufficient to activate transcription of the structural genes of Mo-nitrogenase, *nifHDK*, and all the other *nif* genes (Fig. 4.1; Masepohl et al. 1988). Like NifA proteins from other bacteria, NifA1 and NifA2 activate transcription of their target genes in concert with RpoN. Expression of the *R. capsulatus rpoN* gene itself depends on NifA and RpoN (Cullen et al. 1994; Hübner et al. 1993; Preker et al. 1992). Finally, apo-nitrogenase proteins are synthesized and assembled, prior to insertion of FeMoco.

In addition to *nifA1* and *nifA2*, NtrC activates transcription of *anfA* (Fig. 4.1; Cullen et al. 1998; Kutsche et al. 1996). As outlined below in more detail, *anfA* transcription is repressed by molybdenum (Wang et al. 1993; Wiethaus et al. 2006). Thus, *anfA* expression occurs exclusively in the absence of both ammonium and molybdenum. In turn, AnfA activates transcription of the structural genes of Fe-nitrogenase, *anfHDGK*, a prerequisite for synthesis of active Fe-nitrogenase.

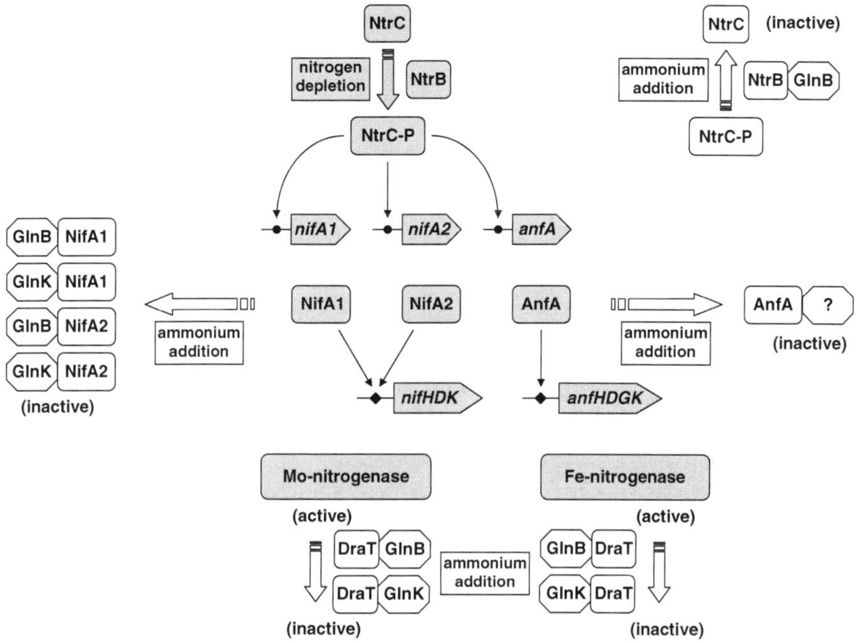

Fig. 4.1 Nitrogen control of nitrogen fixation by GlnB and GlnK in *R. capsulatus*. The regulatory cascade leading to synthesis of Mo and Fe nitrogenase is highlighted in *grey*. Upon nitrogen depletion, NtrC is phosphorylated, and thus activated. NtrC-P activates transcription starting at the RpoD-dependent *nifA1*, *nifA2*, and *anfA* promoters (*filled circles*). For molybdenum repression of *anfA*, see Fig. 4.3. NifA1, NifA2, and AnfA activate transcription starting at RpoN-dependent *nif* and *anf* promoters (*filled diamonds*). Ammonium addition to a nitrogen-fixing culture affects three regulatory levels. First, GlnB interacts with NtrB, and NtrB–GlnB promotes dephosphorylation of NtrC. Second, GlnB and GlnK interact with NifA1 and NifA2, which in turn are no longer able to activate *nif* gene expression. Ammonium addition also inactivates AnfA, but the putative interaction partner of AnfA has not been identified to date. Third, DraT is transiently activated by interaction with GlnB or GlnK, and in turn, DraT modifies nitrogenase reductases by ADP-ribosylation

4.2.3 Nitrogen Control of nifA *Expression by the Ntr System*

As in other free-living diazotrophs, ammonium prevents synthesis of nitrogenase in *R. capsulatus*. In contrast to ammonium, other sources of fixed nitrogen like amino acids and urea do not inhibit nitrogen fixation (Masepohl et al. 2001) suggesting that it is exclusively ammonium availability in the medium that determines whether growth conditions are recognized as nitrogen replete or nitrogen limited. There is growing evidence that the high-affinity transporter AmtB acts as a sensor of extracellular ammonium and, in response, controls nitrogen fixation by reversible membrane sequestration of PII proteins (Tremblay and Hallenbeck 2008; see below).

Sensing of the intracellular nitrogen status involves the Ntr system, which is well-characterized for *Escherichia coli* (Jiang et al. 1998a, b, c). As in *E.*

coli, the *R. capsulatus* Ntr system consists of the bifunctional enzyme GlnD (uridylyltransferase/UMP-removing enzyme), two PII-like signal transduction proteins, GlnB and GlnK, and the two-component regulatory system NtrB–NtrC. In contrast to membrane-bound AmtB, the cytoplasmic Ntr system does not sense ammonium but responds to the ammonium assimilation product glutamine (Gln), the carbon skeleton 2-oxoglutarate, and the energy source ATP. These metabolites bind synergistically to GlnD, GlnB, and GlnK, thus modifying the ability of these proteins to interact with and control the activity of different target proteins. Among these target proteins are NtrB, NifA1, NifA2, and DraT (Fig. 4.1; see below; Pawlowski et al. 2003).

At low intracellular Gln concentrations (corresponding to nitrogen-limiting conditions), GlnD uridylylates GlnB at a conserved tyrosine residue within the T-loop (Forchhammer 2008). Upon modification, GlnB no longer interacts with the sensor kinase NtrB, which can now autophosphorylate. As described above, transfer of the phosphoryl group from NtrB to NtrC activates the response regulator, and thus starts the regulatory cascade leading to *nifA* activation, expression of all the other *nif* genes, and finally, synthesis of nitrogenase.

At high intracellular Gln concentrations (representing nitrogen-replete conditions), GlnD removes the UMP moiety from GlnB, which in its unmodified form interacts with NtrB, preventing autophosphorylation. Furthermore, the NtrB–GlnB complex mediates rapid dephosphorylation of NtrC (Fig. 4.1; Pioszak and Ninfa 2004). The dephosphorylated form of NtrC is no longer able to activate *nifA* transcription, and consequently, expression of the nitrogen fixation machinery fails to occur.

As outlined below, GlnB and GlnK substitute for each other in post-translational control of NifA and nitrogenase activities (Drepper et al. 2003). However, two lines of evidence suggest that GlnK cannot replace GlnB regarding control of NtrB activity. First, in contrast to GlnB, GlnK failed to interact with NtrB in yeast two-hybrid studies (Pawlowski et al. 2003). Second, deletion of the *glnB* gene is sufficient to abolish ammonium inhibition of *nifA1* and *nifA2* transcription (Drepper et al. 2003).

4.2.4 Post-translational Control of NifA Activity

NifA proteins function as transcriptional activators of *nif* genes in proteobacteria (Fischer 1994; Martinez-Argudo et al. 2004). They bind to conserved enhancer-binding sites upstream of *nif* promoters and activate *nif* gene transcription in concert with RNA polymerase (RNAP) containing the nitrogen-specific sigma factor RpoN. Activity of NifA proteins is controlled by the nitrogen status and by oxygen, but the mechanisms underlying sensing of nitrogen and oxygen differ between species.

NifA proteins, including *R. capsulatus* NifA1 and NifA2, typically consist of three domains. The N-terminal GAF domain is involved in the nitrogen control of NifA activity (Paschen et al. 2001). The central AAA domain is required for ATP hydrolysis, and for interaction with RNAP–RpoN holoenzyme. The C-terminal helix-turn-helix domain binds to enhancer-binding sites of the form TGT-N_{10}-ACA.

Rhodobacter capsulatus NifA1 and NifA2 belong to the oxygen-sensitive "rhizobial" class of NifA proteins, which is characterized by a cysteine-rich linker between the central and the C-terminal domains (Fischer et al. 1988; Fischer 1994). This interdomain linker is thought to respond to changes in the redox status, thus preventing *nif* gene expression under conditions unfavorable for the extremely oxygen-sensitive nitrogenase enzyme.

Either GlnB or GlnK is sufficient to inhibit NifA1 and NifA2 activity upon ammonium addition to a nitrogen-fixing culture (Drepper et al. 2003). Both PII proteins interact directly with NifA1 and NifA2 (Pawlowski et al. 2003). Since ammonium addition typically leads to deuridylylation of PII proteins, it seems likely that the unmodified forms of GlnB and GlnK interact with NifA1 and NifA2. Concurrent deletion of both PII genes relieves ammonium inhibition of NifA1 and NifA2, suggesting that neither GlnB nor GlnK alone is essential for NifA1 or NifA2 activity (Drepper et al. 2003).

4.2.5 Post-translational Control of Mo-nitrogenase Activity

Given the relatively high energy demands of nitrogenase it is remarkable that relatively few diazotrophs are capable of modulating its activity in response to sudden environmental changes. However, many purple nonsulfur photosynthetic bacteria, including *R. capsulatus*, as well as a number of other proteobacteria, possess a nitrogenase Fe protein (NifH) modification/demodification system, DraT (dinitrogenase reductase ADP-ribosyltransferase) and DraG (dinitrogenase reductase glycohydrolase), that reversibly inhibits its interaction with the nitrogenase MoFe protein through the addition of an ADP-ribose group to Arg101 of one of the two subunits of the homodimer. In fact, DraT is capable of modifying both NifH and AnfH (Masepohl et al. 1993). A simple view is that modification controls nitrogenase activity. Modification is triggered by the perception of environmental signals, either an increase in fixed nitrogen or the sudden absence of light (which presumably leads to energy deprivation). Subsequent reversal of the signal brings about demodification. However, analysis shows that rather than being an all or nothing phenomenon, cells growing under some nitrogen regimes (partial nitrogen limitation) contain a fraction of their NifH in its ADP-ribosylated form (Yakunin and Hallenbeck 1999). As well, the nitrogen status of the cells, which obviously can be affected by particular growth conditions, affects the manner in which the modification reaction occurs (Yakunin and Hallenbeck 1998). Highly nitrogen-limited cultures show classical switch-off, with almost immediate and complete cessation of nitrogenase activity upon ammonium addition followed by complete recovery after consummation of the added ammonium. On the other hand, moderately nitrogen-limited cultures show a different kinetic response which has been termed the magnitude response. Similar effects can be seen with photoheterotrophically grown cells that are assayed in the dark in the presence of small amounts of oxygen (Yakunin and Hallenbeck 2000).

However, the situation appears to be more complex in *R. capsulatus* which possesses a second system capable of the post-translational control of nitrogenase activity. This can be readily demonstrated with mutant strains that lack DraT/DraG activity (Yakunin and Hallenbeck 1998) or in strains carrying mutant NiFH alleles that are unmodifiable due to alteration of Arg101 (Pierrard et al. 1993). How this second system, independent of ADP-ribosylation, functions is unclear, but it shares common regulatory elements with the better characterized DraT/DraG system, including the PII proteins GlnB/GlnK and AmtB (see below).

There is a growing consensus that PII proteins play a direct role in regulating DraT/DraG activity based on detailed studies with *R. capsulatus, R. rubrum* (see later), and *Azospirillum brasilense*. Mo-nitrogenase is synthesized in a *glnB–glnK* double mutant and, moreover, it is active and unmodified (in the presence of ammonia), indicating that one or both of the PII proteins are absolutely required for DraT/DraG regulation (Drepper et al. 2003). In addition, yeast two-hybrid studies have shown that GlnB and GlnK interact with DraT (Pawlowski et al. 2003). This might lead one to think that one PII protein could substitute for the other and that therefore single mutants would show normal nitrogenase regulation. However, this is not the case as shown by a recent study in which individual GlnB and GlnK knockout mutants were generated and studied (Tremblay et al. 2007). For reasons that are not completely clear both GlnB and GlnK appear necessary to properly regulate nitrogenase. GlnB and GlnK are themselves subject to covalent modification, addition/removal of an UMP group by GlnD, and their modification status is responsive to the nitrogen status of the cell as well as regulating their activity in controlling nitrogen metabolism. Thus, absence of one of the two PII proteins could lead to aberrant levels of nitrogen metabolites thereby affecting the activity of the PII protein that is present and indirectly abrogating nitrogenase regulation. Indirect evidence for this is the fact that the modification state of the PII protein present in mutant cells is different than that found with wild-type cells under the same conditions (Tremblay et al. 2007). Future analysis of intracellular metabolite levels could be used to directly verify this hypothesis. Possible roles of the PII proteins are shown in Fig. 4.2 and further discussed below.

Fig. 4.2 (continued) modification with UMP, which itself is sensitive to changes in these metabolite pools. Under limiting nitrogen conditions, the 2-OG/Gln ratio is high and both enzymes are largely inactive. Upon addition of ammonium, DraT becomes active, modifying the Fe protein (NifH) of nitrogenase with an ADP-ribose and inactivating it. At the same time, GlnK is sequestered by the membrane-bound AmtB and probably forms a ternary complex with the nitrogenase–demodifying enzyme DraG. Membrane-bound DraG is thought to be inactive. GlnK binding to AmtB is thought to inactivate ammonium transport. AmtB must bind NH_3, or perhaps even partially transport it, for the complex to be productive. A possible mechanism for switch-off independent of ADP-ribosylation is shown in (**b**). Studies have shown that DraT and DraG are dispensable for this process but that both GlnB and GlnK, as well as AmtB, are required. Here the hypothesis is that nitrogenase activity is controlled via effects on the Rnf complex, a specific membrane-bound complex essential for electron transport to nitrogenase under photoheterotrophic conditions. Changes in metabolite pools caused by ammonium addition would again cause membrane sequestration of GlnK by AmtB. In this case, it is suggested that the GlnK–AmtB complex could recruit one or more components of the Rnf complex, thereby inactivating its ability to supply the necessary electrons for nitrogenase activity. Experimental evidence in support of this has not yet been obtained

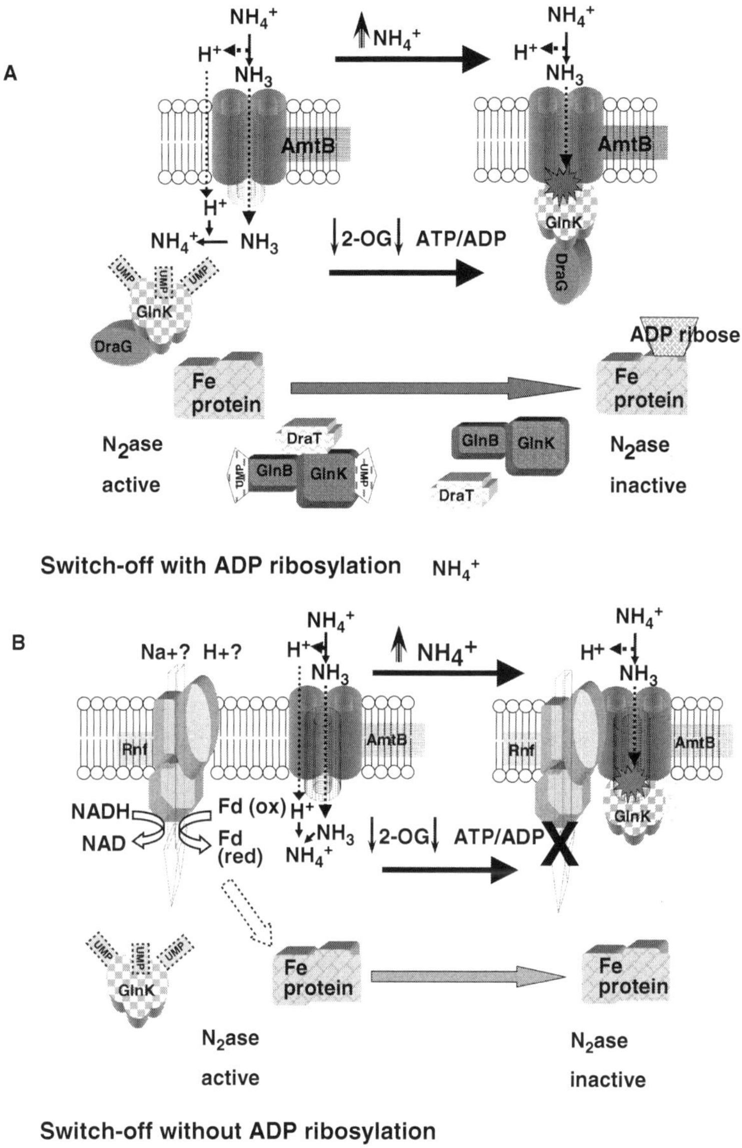

Fig. 4.2 Post-translational control of nitrogenase activity and membrane sequestration of PII proteins in *R. capsulatus*. GlnB, GlnK, and AmtB have all been implicated in both mechanisms of the post-translational control of nitrogenase activity, often called nitrogenase switch-off: the ADP-ribosylation process (**a**) and an ADP-ribosylation-independent process (**b**). However, not all the molecular details of these processes are known. Here we present hypotheses as to how these may function based on analogy with other systems or speculation, given some of the factors that are specific to *R. capsulatus*. Switch-off with ADP-ribosylation is shown in (**a**). The PII proteins, GlnB and GlnK, control the activity of the two modifying enzymes DraT and DraG, most likely through direct protein–protein interactions, as shown for DraT. Association can be driven either directly by changes in metabolite pools, 2-oxoglutarate (2-OG) and glutamine (Gln), or ATP/ADP, or by

4.2.6 Nitrogen Control of the Alternative Fe-nitrogenase

Expression of *anfA* (coding for the central activator of Fe-nitrogenase genes) is under dual control. First, as described above for *nifA1* and *nifA2*, NtrC is strictly required for activation of *anfA* transcription upon depletion of fixed nitrogen (Kutsche et al. 1996). Second, transcription of *anfA* is repressed by molybdenum (see below; Kutsche et al. 1996; Wiethaus et al. 2006). Thus, *anfA* is exclusively expressed upon dual limitation for fixed nitrogen and molybdenum. In turn, AnfA activates transcription of the structural genes of Fe-nitrogenase, *anfHDGK*, and its other target genes in concert with RNAP–RpoN (Schüddekopf et al. 1993).

Similar to NifA1 and NifA2, AnfA activity does not depend on GlnB or GlnK (Drepper et al. 2003). As is the case for the NifA proteins, AnfA activity is inhibited by ammonium (Drepper et al. 2003). However, the mechanisms underlying ammonium inhibition of the NifA proteins and AnfA differ. While ammonium inhibition of NifA1 and NifA2 is mediated by direct interaction with GlnB or GlnK, ammonium inhibition of AnfA does not depend on the PII proteins. Like Mo-nitrogenase, Fe-nitrogenase is post-translationally controlled by the DraT/DraG system (Drepper et al. 2003; Masepohl et al. 1993).

4.3 Nitrogen Control of Nitrogen Fixation by AmtB

4.3.1 Transport and Sensing of Ammonium by Bacterial AmtB Proteins

Members of the Amt/Rh family of transporters are found almost ubiquitously in all forms of life. For a long time it was thought that their primary function was to transport ammonium (NH_4^+), but recent studies have shown that some members of this family may function as sensors, regulating cellular metabolism in response to changes in external ammonium concentrations (Tremblay and Hallenbeck 2009). Indeed, since biological membranes are permeable to ammonia (NH_3), which exists in equilibrium with ammonium, under most cases there is little need for a dedicated ammonium transporter. However, Amt is necessary for the optimal growth of some microorganisms at acidic pH when the extracellular concentration of NH_3 is quite low (Marini et al. 1997; Soupene et al. 1998). Another important transport-related activity of Amt is the cyclic retention of NH_3 lost from the cell by diffusion, particularly relevant in the case of diazotrophs, such as *R. capsulatus*, where AmtB recovers NH_3 molecules produced by nitrogenase at high energetic cost (Yakunin and Hallenbeck 2002). Bacterial AmtB, although acting on NH_4^+, appears to actually transport the uncharged species, NH_3. NH_4^+ is deprotonated at the AmtB periplasmic interface and reprotonated at the cytoplasmic interface with a net transfer (co-transport) of a proton (Fong et al. 2007; Ludewig et al. 2007).

Recent evidence indicates that Amt proteins can also serve as important sensor functions, at least in some organisms (Tremblay and Hallenbeck 2009). In a variety of fungi Amt senses the extracellular NH_4^+ concentration leading to diploid pseudohyphal growth or haploid invasive growth (Marini et al. 2006; Rutherford et al. 2008). In *Saccharomyces cerevisiae*, NH_4^+ transport appears to be necessary but not sufficient to trigger cell differentiation (Boeckstaens et al. 2008; Rutherford et al. 2008). The first evidence that prokaryotic Amt proteins (AmtB) can function as a sensor was obtained in *R. capsulatus* when it was found that an AmtB null mutant was totally defective in the post-translational regulation of nitrogenase with both the ADP-ribosylation-dependent and ADP-ribosylation-independent pathways being affected (Yakunin and Hallenbeck 2002). This signaling mechanism, also being studied in *R. rubrum* (Wang et al. 2005; Wolfe et al. 2007) and *A. brasilense* (Huergo et al. 2007), acts through the PII proteins (Tremblay et al. 2007; see also below). Moreover, recent evidence has been presented that transport by AmtB is essential for its sensing function and the consequent regulation (Tremblay and Hallenbeck 2008).

4.3.2 Membrane Sequestration of PII Proteins by AmtB

Amt proteins are often encoded by *glnK–amtB* or *amtB–glnK* operons, strongly suggesting a possible physical interaction between the *glnK*-encoded PII proteins and AmtB of potential regulatory significance (Javelle and Merrick 2005). Indeed, a number of studies have shown that ammonium treatment induces AmtB-dependent membrane sequestration of PII proteins, and this has been shown to be the case for *R. capsulatus* (Tremblay et al. 2007). *Rhodobacter capsulatus* possesses two PII proteins, GlnB and GlnK, and two potential Amt proteins, AmtB and AmtY. The function of AmtY, which should probably be renamed AmtB2, is unclear since, even though it is differentially expressed under nitrogen-limiting conditions, it does not transport ammonium nor does it seem to be involved in nitrogenase regulation. AmtB (AmtB1) is capable of binding both GlnB and GlnK but appears to preferentially bind GlnK in response to an ammonium addition (Tremblay et al. 2007), potentially allowing for differential regulation of the different targets of the two PII proteins. As mentioned before, binding of GlnK by AmtB in response to an ammonium shock is a prerequisite for nitrogenase regulation. Under the culture conditions used, nitrogenase regulation in response to energy deprivation, i.e., in response to darkness, appears to be independent of the sequestration of GlnK by AmtB (Tremblay et al. 2007; Yakunin and Hallenbeck 2002).

Understanding the details of nitrogenase regulation in response to ammonium therefore requires knowledge of the factors that drive the GlnK–AmtB association. X-ray analysis of an *E. coli* AmtB–GlnK complex suggests that only the deuridylylated form of GlnK could bind as modification of the GlnK T-loop should sterically prevent its interaction with AmtB. However, several lines of evidence suggest that the modification status of *R. capsulatus* GlnK is not the factor driving the interaction (Tremblay et al. 2007). First, the sequestration of an unmodifiable (Y51F) GlnK variant is nevertheless appropriately regulated by the absence and the presence of ammonium. Second, wild-type GlnK is apparently largely unmodified, even under nitrogen-limiting conditions, in a GlnB-deficient strain and in this case too, sequestration proceeds normally. Therefore, membrane association must be responding to changes in small metabolite pools, probably 2-oxoglutarate and/or ATP–ADP, effectors known to bind to PII proteins and alter their properties. In fact, a recent study in the archaeon *Methanococcus jannaschii* showed that complex formation is specifically prevented by changes in the structure of GlnK induced by the effector molecules Mg-ATP and 2-oxoglutarate (Yildiz et al. 2007). Finally, it should be noted that even though binding of GlnK by AmtB is required for nitrogenase activity regulation, by itself complex formation is not sufficient (Tremblay and Hallenbeck 2008). AmtB transport function also appears to be required since some transport-incompetent AmtB variants, i.e., F131A, H193A, and H342A, form ammonium-induced complexes with GlnK but fail to properly regulate nitrogenase. Ultimately, the function of an effective membrane complex of GlnK and AmtB is probably to sequester DraG, as has been shown in *A. brasilense* and *R. rubrum* (see below), thus physically separating it from its substrate, NifH, and allowing the inactive ADP-ribose modified form to accumulate. Formation of a ternary complex involving DraG has yet to be experimentally demonstrated in *R. capsulatus*. A model summarizing these results and hypotheses is shown in Fig. 4.2.

4.4 Molybdenum Control of Nitrogen Fixation

4.4.1 Molybdenum Regulation of Mo Uptake and Fe-Nitrogenase Expression

In general, Mo-nitrogenases exhibit higher N_2 reduction rates than do Mo-free nitrogenases, and therefore, bacteria preferentially synthesize Mo-nitrogenase as long as sufficient amounts of molybdenum are available (Masepohl et al. 2002). Naturally, when molybdenum becomes limiting under nitrogen-limiting conditions, synthesis of Mo-free nitrogenases becomes essential for nitrogen fixation-dependent growth.

Like many bacteria, *R. capsulatus* synthesizes a high-affinity molybdenum uptake system, ModABC (Wang et al. 1993). This ABC-type transporter consists of a periplasmic molybdate-binding protein (ModA), a membrane-spanning protein (ModB), and a cytoplasmic ATP-binding protein (ModC). The ModABC uptake system enables *R. capsulatus* to fix N_2 by Mo-nitrogenase down to nanomolar concentrations of molybdenum in the medium (Wang et al. 1993; Wiethaus et al. 2006). Mutants defective for the transporter require about 500 times higher molybdenum concentrations than the wild type to reach comparable Mo-nitrogenase activity. Accordingly, *modABC* mutants express Fe-nitrogenase at molybdenum concentrations of up to 1 μM, while the wild type represses Fe-nitrogenase at much lower concentrations (see below; Wang, Angermüller and Klipp 1993).

The *R. capsulatus* ModABC transporter genes form part of the *mopA–modABC* operon (Wang et al. 1993). Transcription starting at the *mopA* promoter is under dual control (Kutsche et al. 1996). First, NtrC activates *mopA* transcription upon nitrogen depletion. Second, MopA and MopB repress *mopA* transcription in the presence of molybdenum. Thus, MopA negatively autoregulates its own expression (Fig. 4.3a). The monocistronic *mopB* gene is located upstream of *mopA* but reads in the opposite direction. In contrast to *mopA*, expression of *mopB* is not controlled by fixed nitrogen or molybdenum (Wiethaus et al. 2006). Consequently, the MopA/MopB ratio is thought to vary significantly in response to the availability of fixed nitrogen and molybdenum, thus providing a mechanism for fine-tuning of target gene expression (see below; Wiethaus et al. 2009). In the end, repression of the *mopA–modABC* operon by MopA and MopB limits the amount of the molybdenum uptake system at high molybdenum concentrations. Uptake of molybdenum from the medium at concentrations in the micromolar range and above does not depend on ModABC but involves an as yet uncharacterized transport system (Wang et al. 1993; Wiethaus et al. 2006).

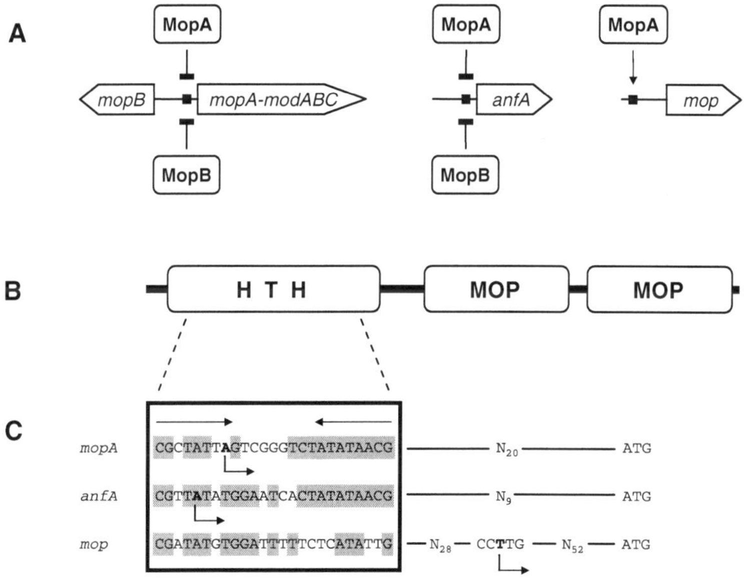

Fig. 4.3 Gene regulation by MopA and MopB in *R. capsulatus*. (**a**) Either MopA or MopB is sufficient to repress transcription of the *mopA–modABC* operon and the *anfA* gene at high molybdenum concentrations. Exclusively MopA is able to activate *mop* transcription. (**b**) MopA and MopB consist of an N-terminal DNA-binding domain (encompassing a helix-turn-helix motif) and a C-terminal molybdate-binding bi-MOP domain. (**c**) MopA and MopB bind to conserved palindromic sequences (Mo-boxes) of their respective target genes. Conserved nucleotides are highlighted in *grey*. Transcription start sites are marked by folded arrows below the nucleotide sequences. Distances of the Mo-boxes to the translation start sites (ATG) are indicated

Like the *mopA–modABC* operon, the *anfA* gene is repressed by MopA and MopB in response to increasing molybdenum concentrations (Fig. 4.3a; Kutsche et al. 1996; Wiethaus et al. 2006). AnfA is essential for activation of Fe-nitrogenase genes and consequently, *anfA* repression prevents expression of Fe-nitrogenase genes. In vivo activity of Fe-nitrogenase is maximal at molybdenum concentrations below 1 nM (Wang et al. 1993). At concentrations above 100 nM, Fe-nitrogenase activity is no longer detectable. At molybdenum concentrations around 10 nM both Fe- and Mo-nitrogenases exhibit half-maximal activities.

4.4.2 Gene Regulation by the Molybdenum-Responsive Regulators MopA and MopB

The molybdenum-binding regulators MopA and MopB of *R. capsulatus* encompass 265 and 261 amino acid residues, respectively. Both exhibit clear similarity to their *E. coli* counterpart, ModE, over the entire length of the proteins. ModE-like regulators consist of an N-terminal DNA-binding domain with a helix-turn-helix motif and a C-terminal molybdate-binding domain (Fig. 4.3b; Hall et al. 1999; Schüttelkopf et al. 2003). The latter domain consists of two MOP domains each encompassing about 70 amino acid residues. MOP domains typically include a SARN motif and other highly conserved amino acids involved in molybdate coordination (Pau 2004; Wiethaus et al. 2009). Binding of molybdate to the bi-MOP domain leads to conformational changes which increase the affinity of the regulators for their target promoters. These target promoters contain highly conserved palindromic sequences (Mo-boxes), which are recognized and bound by the regulators (Fig. 4.3c; Kutsche et al. 1996; Wiethaus et al. 2006).

MopA and MopB form homodimers as apo-proteins and in the molybdate-bound state (Wiethaus et al. 2009). Both the DNA-binding and the molybdate-binding domains contribute to dimer formation. MopA and MopB bind four molybdate oxyanions per dimer, while ModE binds only two molybdate oxyanions per dimer (Gourley et al. 2001). This difference might be explained by a low conservation of the molybdate-coordinating amino acids in the second MOP domain of ModE. In addition to homodimers, MopA and MopB form heteromers thought to be involved in fine-tuning of target gene expression (Wiethaus et al. 2009).

MopA or MopB exhibits overlapping functions as either of these regulators is sufficient to repress transcription of the *mopA–modABC* operon and the *anfA* gene in response to increasing molybdenum concentrations (Fig. 4.3a; Kutsche et al. 1996; Wiethaus et al. 2006). In addition to its role as a repressor, MopA acts as an activator of *mop* gene transcription (Wiethaus et al. 2006). This role of MopA is specific, since MopB cannot substitute for MopA regarding *mop* activation (Fig. 4.3a).

The *mop* gene codes for a small molbindin-like protein consisting of a single MOP domain (Wiethaus et al. 2006). Molbindins have been hypothesized to play roles in molybdenum homeostasis and storage, but to date, experimental evidence for this assumption is lacking (Grunden and Shanmugam 1997; Pau 2004). The *R. capsulatus* Mop protein forms hexamers binding six molybdate oxyanions (Wiethaus et al. 2009). Mop specifically interacts with MopB (but not with MopA) suggesting that Mop might serve as a specific molybdenum buffer for MopB. It is worth noting, however, that *R. capsulatus* strains lacking Mop are not impaired in diazotrophic growth via Mo-nitrogenase or in molybdenum repression of *anfA*.

MopA and MopB bind to the promoters of the *anfA* gene and the *mopA–modABC* operon (Wiethaus et al. 2006). The *anfA*–Mo-box is both essential and sufficient for binding of both regulators and *anfA* repression (Kutsche et al. 1996; Müller and Masepohl, unpublished results). As typical repressor binding sites, the *anfA*–Mo-box and the *mopA*–Mo-box overlap the respective transcription start sites (Fig. 4.3c; Kutsche et al. 1996). Thus, binding of MopA or MopB prevents binding of RNA polymerase, and consequently, transcription is not initiated.

MopA (but not MopB) binds to the *mop* promoter region (Wiethaus et al. 1996). The *mop*–Mo-box is essential for MopA binding and *mop* gene activation (Müller and Masepohl, unpublished results). As a typical activator-binding site, the *mop*–Mo-box does not overlap the transcription start site but instead precedes the putative promoter (Fig. 4.3c; Wiethaus et al. 2006). Close proximity of the *mop*–Mo-box and the putative promoter suggests that MopA and RNA polymerase directly interact without the necessity of DNA bending.

4.5 Nitrogen Fixation in Other Purple Nonsulfur Bacteria

4.5.1 Nitrogen Fixation in R. sphaeroides

Rhodobacter sphaeroides synthesizes a molybdenum-dependent nitrogenase. Most studies on nitrogen fixation in this species were done with strain 2.4.1, which does not synthesize an alternative nitrogenase. In contrast, the halophilic *R. sphaeroides* strain KD131 possibly synthesizes Fe-nitrogenase as predicted from genome sequence analysis (Lim et al. 2009).

As is the case for *R. capsulatus* and other free-living diazotrophs, expression of *R. sphaeroides nif* genes is inhibited by ammonium (Zinchenko et al. 1997). However, strains defective for carbon dioxide fixation accumulate mutations, leading to synthesis of active Mo-nitrogenase even in the presence of ammonium (Joshi and Tabita 1996). In these mutants, Mo-nitrogenase is thought to substitute for ribulose bisphosphate carboxylase as an electron sink. At least some of the "gain-of-function" mutations involved appear to be in *nifA* (R. Tabita, personal communication). *Rhodobacter sphaeroides* is devoid of a DraT/DraG system controlling the activity of nitrogenase reductase (Yakunin et al. 2001). Transfer of the *R. capsulatus draTG* genes into *R. sphaeroides*, however, enables this strain to reversibly modify nitrogenase reductase upon ammonium addition.

4.5.2 Nitrogen Fixation in R. palustris

Rhodopseudomonas palustris is one of the only few diazotrophs able to synthesize three nitrogenases: Mo-nitrogenase, V-nitrogenase, and Fe-nitrogenase (Oda et al. 2005). As described above for *R. capsulatus*, Mo-nitrogenase is the preferred enzyme in *R. palustris*, but regulation of alternative nitrogenases clearly differs between the two species. In the presence of molybdenum, synthesis of *R. capsulatus* Fe-nitrogenase is inhibited via *anfA* repression by the ModE-like regulators MopA and MopB (Kutsche et al. 1996; Wiethaus et al. 2006). *Rhodopseudomonas palustris* harbors a *modE*-like gene (RPA0147), suggesting that this bacterium in principle regulates gene expression in response to molybdenum. In contrast to *R. capsulatus*, however, expression of *vnf* and *anf* genes is not repressed by molybdenum in *R. palustris* (Oda et al. 2005). *Rhodopseudomonas palustris* mutants defective for Mo-nitrogenase synthesize functional V- and Fe-nitrogenases even at high molybdenum or vanadium concentrations. Thus, *R. palustris* might control expression of both alternative nitrogenases primarily in response to the fixed nitrogen status indirectly reflecting molybdenum availability by Mo-nitrogenase activity.

Rhodopseudomonas palustris mutants constitutively producing molecular hydrogen have been isolated which synthesize NifA variants carrying substitutions between the N-terminal regulatory GAF domain and the central AAA domain (Rey, Heiniger and Harwood 2007). These NifA variants are sufficient to activate *nif* gene expression even in the presence of ammonium. Interestingly, during growth in the presence of ammonium, Mo-nitrogenase is not significantly inactivated by DraT, which is constitutively expressed in *R. palustris*.

4.5.3 Nitrogen Fixation in R. rubrum

In contrast to *R. capsulatus*, expression of the *R. rubrum nifA* gene is independent of NtrC and occurs even in the presence of ammonium (Zhang et al. 1995; Zhang et al. 2000). In the presence of ammonium, NifA exists in an inactive form which does not promote *nif* gene transcription. Upon nitrogen limitation, NifA is converted into its active form. NifA activation requires interaction with the uridylylated form of GlnB (Zhang et al. 2005). Several GlnB variants showing improved interaction with NifA were identified, which activate NifA in the presence of ammonium as well as in a *glnD* mutant background (Zhu et al. 2006). GlnD is required for GlnB uridylylation suggesting that these GlnB variants mimic the uridylylated form of the signal transduction protein.

Conversion of *R. rubrum* NifA from its inactive into its active form involves a conformational change. Several NifA variants with substitutions in the N-terminal GAF domain were identified which no longer require GlnB for activation (Zou et al. 2008). These substitutions most likely keep these NifA variants in an active conformation.

In comparison with the *R. capsulatus* system for the post-translational regulation of nitrogenase, the *R. rubrum* system presents some interesting differences and complexities. For one thing, *R. rubrum* possesses three PII proteins: GlnB, two GlnK homologues, GlnK and GlnJ, with GlnB being absolutely required for activation of *nif* gene transcription, and either GlnB or GlnJ required for post-translational regulation of nitrogenase (Zhang et al. 2001). Uridylylation of either GlnB or GlnJ is necessary to activate DraG since *glnD* (uridylyl transferase) mutants fail to reactivate nitrogenase after darkness-induced nitrogenase switch-off (Zhang et al. 2005).

Rhodospirillum rubrum possesses two AmtB homologues, AmtB1 and AmtB2, but unlike *R. capsulatus* where the second homologue, AmtY, is not associated with a PII protein, both homologues are in operons with PII proteins: *glnJ–amtB1* and *glnK–amtB2* (Zhang et al. 2006). As well, it appears that only AmtB1, in concert with either GlnB or GlnJ, is effective in carrying out nitrogenase regulation. The possible cellular function of AmtB2 remains unknown.

In contrast to what has been observed with *R. capsulatus* (Yakunin and Hallenbeck 2002), *R. rubrum* cultures in minimal medium without fixed nitrogen show a dependence on AmtB1 for darkness-induced nitrogenase regulation. As might be expected, nitrogenase regulation depends upon the formation of an AmtB1–GlnJ membrane complex (Wang et al. 2005; Zhang et al. 2006). As suggested above for *R. capsulatus*, in vitro studies have shown that AmtB–PII complex formation is disrupted by high levels of 2-oxoglutarate in the absence of ATP or low levels of 2-oxoglutarate in the presence of ATP (Wolfe et al. 2007). A role for PII as an energy sensor responding to the ratio of ATP to ADP is suggested by the fact that ADP inhibits the destabilization of the GlnJ–AmtB1 complex in the presence of ATP and 2-oxoglutarate. The importance of complex formation in nitrogenase regulation has been directly shown by demonstrating the association of DraG with AmtB1 in response to an ammonium shock (Wang et al. 2005). Thus, the major steps in the ammonium-induced post-translational regulation of nitrogenase are similar between *R. rubrum*, *A. brasilense* (Huergo et al. 2007), and *R. capsulatus*.

Like *R. capsulatus*, *R. rubrum* synthesizes an alternative Fe-nitrogenase (Lehman and Roberts 1991). As described above for the alternative nitrogenases of *R. palustris*, synthesis of *R. rubrum* Fe-nitrogenase is not inhibited by molybdenum, but instead Fe-nitrogenase is expressed in strains lacking active Mo-nitrogenase.

4.6 Conclusions

The control of nitrogen fixation appears to employ similar principles in many different diazotrophic bacteria. For example, nitrogen fixation is almost universally inhibited by ammonium, and Mo-nitrogenases are preferentially made over alternative nitrogenases as long as molybdenum is available. However, "the devil is in the details"; there is great variety in the underlying mechanisms by which these principles are put into practice even among relatively closely related bacteria. In many proteobacteria, including *R. capsulatus*, *nifA* transcription is activated by NtrC upon nitrogen limitation, while *nifA* expression in some other species, like the related *R. rubrum*, does not depend on NtrC. Only relatively few diazotrophs control nitrogenase activity by a switch-off/on mechanism, while most other species "rely" upon control of *nifA* transcription and/or control of NifA activity for regulation of nitrogenase activity over a longer timescale. *R. capsulatus* controls expression of Fe nitrogenase via Mo repression of *anfA*. In contrast, expression of alternative nitrogenases is not repressed by molybdenum in *R. palustris* and *R. rubrum*. This diversity is also seen by the fact that although different bacteria use similar modules to control nitrogen fixation, the mechanisms of these modules differ in various aspects between different species. For example, in *R. rubrum*, NifA requires GlnB in order to be active, while in *R. capsulatus*, NifA is fully active in the absence of PII proteins, which, on the contrary, are capable of the negative modulation of NifA activity in response to fixed nitrogen.

It is also interesting to note that several regulatory proteins controlling nitrogen fixation are duplicated in *R. capsulatus* (GlnB/GlnK, NifA1/NifA2, MopA/MopB). In some cases, deletion of individual genes does not produce a clear phenotype, at least under the growth conditions as tested in the laboratory, suggesting that the respective proteins (at least partially) substitute for each other. However, recent findings clearly show that deletion of either *glnB* or *glnK* clearly affects the remaining PII protein. Furthermore, the possible formation of heteromeric forms (GlnB–GlnK and MopA–MopB) makes the interdependence of homologous proteins even more difficult to clarify. Such heteromeric forms may indeed play a role in the fine-tuning of nitrogen control, but, if so, this largely remains to be elucidated.

Finally, laboratory investigations often compare extreme environments, e.g., nitrogen-replete versus nitrogen-sufficient conditions or very low versus relatively high concentrations of molybdenum. While this provides conditions suitable for the clear-cut demonstration of underlying mechanisms, these situations, for example, the addition of large amounts of ammonium to a nitrogen-fixing culture, are unlikely to occur in nature. More often, an intermediate state is more likely to pertain, and thus a graded response probably occurs. For example, the magnitude response described for post-translational control of nitrogenase activity might more accurately reflect the natural situation.

Acknowledgments Research in the authors' laboratories is supported by Deutsche Forschungsgemeinschaft (BM) and the Natural Sciences and Engineering Research Council of Canada Discovery Grants Program (PCH).

References

Andrade SL, Einsle O (2007) The Amt/Mep/Rh family of ammonium transport proteins. Mol Membr Biol 24:357–365

Boeckstaens M, André B, Marini AM (2008) Distinct transport mechanisms in yeast ammonium transport/sensor proteins of the Mep/Amt/Rh family and impact on filamentation. J Biol Chem 283:21362–21370

Bowman WC, Kranz RG (1998) A bacterial ATP-dependent, enhancer binding protein that activates the housekeeping RNA polymerase. Genes Dev 12:1884–1893

Cullen PJ, Bowman WC, Hartnett DF, Reilly SC, Kranz RG (1998) Translational activation by an NtrC enhancer-binding protein. J Mol Biol 278:903–914

Cullen PJ, Bowman WC, Kranz RG (1996) *In vitro* reconstitution and characterization of the *Rhodobacter capsulatus* NtrB and NtrC two-component system. J Biol Chem 271:6530–6536

Cullen PJ, Foster-Hartnett D, Gabbert KK, Kranz RG (1994) Structure and expression of the alternative sigma factor, RpoN, in *Rhodobacter capsulatus*; Physiological relevance of an autoactivated *nifU2-rpoN* superoperon. Mol Microbiol 11:51–65

Dixon R, Kahn D (2004) Genetic regulation of biological nitrogen fixation. Nat Rev 2:621–631

Drepper T, Groß S, Yakunin AF, Hallenbeck PC, Masepohl B, Klipp W (2003) Role of GlnB and GlnK in ammonium control of both nitrogenase systems in the phototrophic bacterium *Rhodobacter capsulatus*. Microbiol 149:2203–2212

Fischer H-M (1994) Genetic regulation of nitrogen fixation in rhizobia. Microbiol Rev 58:352–386

Fischer H-M, Bruderer T, Hennecke H (1988) Essential and non-essential domains in the *Bradyrhizobium japonicum* NifA protein: Identification of indispensable cysteine residues potentially involved in redox reactivity and/or metal binding. Nucleic Acids Res 16: 2207–2224

Fong RN, Kim KS, Yoshihara C, Inwood WB, Kustu S (2007) The W148L substitution in the *Escherichia coli* ammonium channel AmtB increases flux and indicates that the substrate is an ion. Proc Natl Acad Sci USA 104:18706–18711

Forchhammer K (2008) P_{II} signal transducers: Novel functional and structural insights. Trends Microbiol 16:65–72

Foster-Hartnett D, Cullen PJ, Monika EM, Kranz RG (1994) A new type of NtrC transcriptional activator. J Bacteriol 176:6175–6187

Foster-Hartnett D, Kranz RG (1992) Analysis of the promoters and upstream sequences of *nifA1* and *nifA2* in *Rhodobacter capsulatus*; activation requires *ntrC* but not *rpoN*. Mol Microbiol 6:1049–1060

Glazer AN, Kechris KJ (2009) Conserved amino acid sequence features in the α subunits of MoFe, VFe, and FeFe nitrogenases. PLoS ONE 4:e6136

Gourley DG, Schüttelkopf AW, Anderson LA, Price NC, Boxer DH, Hunter WN (2001) Oxyanion binding alters conformation and quaternary structure of the C-terminal domain of the transcriptional regulator ModE. Implications for molybdate-dependent regulation, signaling, storage, and transport. J Biol Chem 276:20641–20647

Grunden AM, Shanmugam KT (1997) Molybdate transport and regulation in bacteria. Arch Microbiol 168:345–354

Hall DR, Gourley DG, Leonard GA, Duke EM, Anderson LA, Boxer DH, Hunter WN (1999) The high-resolution crystal structure of the molybdate-dependent transcriptional regulator (ModE) from *Escherichia coli*: A novel combination of domain folds. EMBO J 18:1435–1446

Hu Y, Fay AW, Lee CC, Yoshizawa J, Ribbe MW (2008) Assembly of nitrogenase MoFe protein. Biochemistry 47:3973–3981

Hübner P, Masepohl B, Klipp W, Bickle TA (1993) *nif* gene expression studies in *Rhodobacter capsulatus*: *ntrC*-independent repression by high ammonium concentrations. Mol Microbiol 10:123–132

Huergo LF, Merrick M, Pedrosa FO, Chubatsu LS, Araujo LM, Souza EM (2007) Ternary complex formation between AmtB, GlnZ and the nitrogenase regulatory enzyme DraG reveals a novel facet of nitrogen regulation in bacteria. Mol Microbiol 66:1523–1535

Javelle A, Merrick M (2005) Complex formation between AmtB and GlnK: An ancestral role in prokaryotic nitrogen control. Biochem Soc Trans 33:170–172

Jiang P, Peliska JA, Ninfa AJ (1998a) Enzymological characterization of the signal-transducing uridylyltransferase/uridylyl-removing enzyme (EC 2.7.7.59) of *Escherichia coli* and its interaction with PII protein. Biochemistry 37:12782–12794

Jiang P, Peliska JA, Ninfa AJ (1998b) Reconstitution of the signal-transduction bicyclic cascade responsible for the regulation of Ntr gene transcription in *Escherichia coli*. Biochemistry 37:12795–12801

Jiang P, Peliska JA, Ninfa AJ (1998c) The regulation of *Escherichia coli* glutamine synthetase revisited: Role of 2-ketoglutarate in the regulation of glutamine synthetase adenylylation state. Biochemistry 37:12802–12810

Joshi HM, Tabita FR (1996) A global two component signal transduction system that integrates the control of photosynthesis, carbon dioxide assimilation, and nitrogen fixation. Proc Natl Acad Sci USA 93:14515–14520

Kutsche M, Leimkühler S, Angermüller S, Klipp W (1996) Promoters controlling expression of the alternative nitrogenase and the molybdenum uptake system in *Rhodobacter capsulatus* are activated by NtrC, independent of σ^{54}, and repressed by molybdenum. J Bacteriol 178: 2010–2017

Lehman LJ, Roberts GP (1991) Identification of an alternative nitrogenase system in *Rhodospirillum rubrum*. J Bacteriol 173:5705–5711

Lim S-K, Kim SJ, Cha SH, Oh Y-K, Rhee H-J, Kim M-S, Lee JK (2009) Complete genome sequence of *Rhodobacter sphaeroides* KD131. J Bacteriol 191:1118–1119

Ludewig U, Neuhäuser B, Dynowsky M (2007) Molecular mechanisms of ammonium transport and accumulation in plants. FEBS Lett 581:2301–2308

Madigan M, Cox SS, Stegeman RA (1984) Nitrogen fixation and nitrogenase activities in members of the family *Rhodospirillaceae*. J Bacteriol 157:73–78

Marini AM, Soussi-Boudekou S, Vissers S, André B (1997) A family of ammonium transporters in *Saccharomyces cerevisiae*. Mol Cell Biol 17:4282–4293

Marini AM, Boeckstaens M, André B (2006) From yeast ammonium transporters to Rhesus proteins, isolation and functional characterization. Transfus Clin Biol 13:95–96

Martinez-Argudo I, Little R, Shearer N, Johnson P, Dixon R (2004) The NifL–NifA system: A multidomain transcriptional regulatory complex that integrates environmental signals. J Bacteriol 184:601–610

Masepohl B, Forchhammer K (2007) Regulatory cascades to express nitrogenase. In: Bothe H, Ferguson SJ, Newton WE (eds) Biology of the nitrogen cycle. Elsevier, Amsterdam pp. 131–145

Masepohl B, Kaiser B, Isakovic N, Richard CL, Kranz RG, Klipp W (2001) Urea utilization in the phototrophic bacterium *Rhodobacter capsulatus* is regulated by the transcriptional activator NtrC. J Bacteriol 183:637–643

Masepohl B, Klipp W (1996) Organization and regulation of genes encoding the molybdenum nitrogenase and the alternative nitrogenase in *Rhodobacter capsulatus*. Arch Microbiol 165: 80–90

Masepohl B, Klipp W, Pühler A (1988) Genetic characterization and sequence analysis of the duplicated *nifA*/*nifB* gene region of *Rhodobacter capsulatus*. Mol Gen Genet 212:27–37

Masepohl B, Kranz RG (2009) Regulation of nitrogen fixation. In: Hunter CN, Daldal F, Thurnauer MC, Beatty JT (eds) The purple phototrophic bacteria. Springer, Dordrecht pp. 759–775

Masepohl B, Krey R, Klipp W (1993) The *draTG* gene region of *Rhodobacter capsulatus* is required for post-translational regulation of both the molybdenum and the alternative nitrogenase. J Gen Microbiol 139:2667–2675

Masepohl B, Schneider K, Drepper T, Müller A, Klipp W (2002) Alternative nitrogenases. In: Leigh GJ (ed) Nitrogen fixation at the millennium. Elsevier, Amsterdam pp. 191–222

Maynard RH, Premakumar R, Bishop PE (1994) Mo-independent nitrogenase 3 is advantageous for diazotrophic growth of *Azotobacter vinelandii* on solid medium containing molybdenum. J Bacteriol 176:5583–5586

Oda Y, Samanta SK, Rey FE, Wu L, Liu X, Yan T, Zhou J, Harwood CS (2005) Functional genomic analysis of three nitrogenase isozymes in the photosynthetic bacterium *Rhodopseudomonas palustris*. J Bacteriol 187:7784–7794

Paschen A, Drepper T, Masepohl B, Klipp W (2001) *Rhodobacter capsulatus nifA* mutants mediating *nif* gene expression in the presence of ammonium. FEMS Microbiol Lett 200:207–213

Pau RN (2004) Molybdenum uptake and homeostasis. In: Klipp W, Masepohl B, Gallon JR, Newton WE (eds) Genetics and regulation of nitrogen fixation in free-living bacteria. Kluwer, Dordrecht pp. 225–256

Pawlowski A, Riedel K-U, Klipp W, Dreiskemper P, Groß S, Bierhoff H, Drepper T, Masepohl B (2003) Yeast two-hybrid studies on interaction of proteins involved in regulation of nitrogen fixation in the phototrophic bacterium *Rhodobacter capsulatus*. J Bacteriol 185: 5240–5247

Pierrard JP, Ludden PW, Roberts GP (1993) Posttranslational regulation of nitrogenase in *Rhodobacter capsulatus*: Existence of two independent regulatory effects of ammonium. J Bacteriol 175:1358–1366

Pioszak AA, Ninfa AJ (2004) Mutations altering the N-terminal receiver domain of NRI (NtrC) that prevent dephosphorylation by the NRII-PII complex in *Escherichia coli*. J Bacteriol 186: 5730–5740

Preker P, Hübner P, Schmehl M, Klipp W, Bickle TA (1992) Mapping and characterization of the promoter elements of the regulatory *nif* genes *rpoN*, *nifA1* and *nifA2* in *Rhodobacter capsulatus*. Mol Microbiol 6:1035–1047

Rey FE, Heiniger EK, Harwood CS (2007) Redirection of metabolism for biological hydrogen production. Appl Environ Microbiol 73:1665–1671

Rutherford JC, Chua G, Hughes T, Cardenas ME, Heitman J (2008) A Mep2-dependent transcriptional profile links permease function to gene expression during pseudohyphal growth in *Saccharomyces cerevisiae*. Mol Biol Cell 19:3028–3039

Schmehl M, Jahn A, Meyer zu Vilsendorf A, Hennecke S, Masepohl B, Schuppler M, Marxer M, Oelze J, Klipp W (1993) Identification of a new class of nitrogen fixation genes in *Rhodobacter capsulatus*: A putative membrane complex involved in electron transport to nitrogenase. Mol Gen Genet 241:602–615

Schneider K, Gollan U, Selsemeier-Voigt S, Plass W, Müller A (1994) Rapid purification of the protein components of a highly active "iron only" nitrogenase. Naturwissenschaften 81: 405–408

Schüddekopf K, Hennecke S, Liese U, Kutsche M, Klipp W (1993) Characterization of *anf* genes specific for the alternative nitrogenase and identification of *nif* genes required for both nitrogenases in *Rhodobacter capsulatus*. Mol Microbiol 8:673–684

Schüttelkopf AW, Boxer DH, Hunter WN (2003) Crystal structure of activated ModE reveals conformational changes involving both oxyanion and DNA-binding domains. J Mol Biol 326:761–767

Seefeldt LC, Hoffman BM, Dean DR (2009) Mechanism of Mo-dependent nitrogenase. Annu Rev Biochem 78:701–722

Self WT, Grunden AM, Hasona A, Shanmugam KT (2001) Molybdate transport. Res Microbiol 152:311–321

Soupene E, He L, Yan D, Kustu S (1998) Ammonia acquisition in enteric bacteria: physiological role of the ammonium/methyl ammonium transport B (AmtB) protein. Proc Natl Acad Sci USA 95:7030–7034

Tremblay P-L, Drepper T, Masepohl B, Hallenbeck PC (2007) Membrane sequestration of PII proteins and nitrogenase regulation in the photosynthetic bacterium *Rhodobacter capsulatus*. J Bacteriol 189:5850–5859

Tremblay P-L, Hallenbeck PC (2008) Ammonia-induced formation of an AmtB-GlnK complex is not sufficient for nitrogenase regulation in the photosynthetic bacterium *Rhodobacter capsulatus*. J Bacteriol 190:1588–1594

Tremblay P-L, Hallenbeck PC (2009) Of blood, brains and bacteria, the Amt/Rh transporter family: Emerging role of Amt as a unique microbial sensor. Mol Microbiol 71:12–22

Wang G, Angermüller S, Klipp W (1993) Characterization of *Rhodobacter capsulatus* genes encoding a molybdenum transport system and putative molybdenum–pterin-binding proteins. J Bacteriol 175:3031–3042

Wang H, Franke CC, Nordlund S, Norén A (2005) Reversible membrane association of dinitrogenase reductase activating glycohydrolase in the regulation of nitrogenase activity in *Rhodospirillum rubrum*; dependence on GlnJ and AmtB1. FEMS Microbiol Lett 253: 273–279

Wiethaus J, Müller A, Neumann M, Neumann S, Leimkühler S, Narberhaus F, Masepohl B (2009) Specific interactions between four molybdenum-binding proteins contribute to Mo-dependent gene regulation in *Rhodobacter capsulatus*. J Bacteriol 191:5205–5215

Wiethaus J, Wirsing A, Narberhaus F, Masepohl B (2006) Overlapping and specialized functions of the molybdenum-dependent regulators MopA and MopB in *Rhodobacter capsulatus*. J Bacteriol 188:8441–8451

Wolfe DM, Zhang Y, Roberts GP (2007) Specificity and regulation of interaction between the PII and AmtB1 proteins in *Rhodospirillum rubrum*. J Bacteriol 189:6861–6869

Yakunin AF, Fedorov AS, Laurinavichene TV, Glaser VM, Egorov NS, Tsygankov A, Zinchenko VS, Hallenbeck PC (2001) Regulation of nitrogenase in the photosynthetic bacterium *Rhodobacter sphaeroides* containing *draTG* and *nifHDK* genes from *Rhodobacter capsulatus*. Can J Microbiol 47:206–212

Yakunin AF, Hallenbeck PC (1998) Short-term nitrogenase regulation in *Rhodobacter capsulatus:* Multiple in vivo nitrogenase responses to NH_4^+ addition. J Bacteriol 180:6392–6395

Yakunin AF, Hallenbeck PC (1999) The presence of ADP-ribosylated Fe protein of nitrogenase in *Rhodobacter capsulatus* is correlated with the cellular nitrogen status. J Bacteriol 181: 1994–2000

Yakunin AF, Hallenbeck PC (2000) Regulation of nitrogenase activity in *Rhodobacter capsulatus* under dark microoxic conditions. Arch Microbiol 173:366–372

Yakunin AF, Hallenbeck PC (2002) AmtB is necessary for NH_4^+-induced nitrogenase switch-off and ADP-ribosylation in *Rhodobacter capsulatus*. J Bacteriol 184:4081–4088

Yildiz O, Kalthoff C, Raunser, Kuhlbrandt W (2007) Structure of GlnK1 with bound effectors indicates regulatory mechanism for ammonia uptake. EMBO J 26:589–599

Zhang Y, Cummings AD, Burris RH, Ludden PW, Roberts GP (1995) Effect of an *ntrBC* mutation on the posttranslational regulation of nitrogenase activity in *Rhodospirillum rubrum*. J Bacteriol 177:5322–5326

Zhang Y, Pohlmann EL, Ludden PW, Roberts GP (2000) Mutagenesis and functional characterization of the *glnB*, *glnA*, and *nifA* genes from the photosynthetic bacterium *Rhodospirillum rubrum*. J Bacteriol 182:983–992

Zhang Y, Pohlmann EL, Ludden PW, Roberts GP (2001) Functional characterization of three GlnB homologs in the photosynthetic bacterium *Rhodospirillum rubrum*: Roles in sensing ammonium and energy status. J Bacteriol 183:6159–6168

Zhang Y, Pohlmann EL, Roberts GP (2005) GlnD is essential for NifA activation, NtrB/NtrC-regulated gene expression, and posttranslational regulation of nitrogenase activity in the photosynthetic, nitrogen-fixing bacterium *Rhodospirillum rubrum*. J Bacteriol 187:1254–1265

Zhang Y, Wolfe DM, Pohlmann EL, Conrad MC, Roberts GP (2006) Effect of AmtB homologues on the post-translational regulation of nitrogenase activity in response to ammonium and energy signals in *Rhodospirillum rubrum*. Microbiology 152:2075–2089

Zhu Y, Conrad MC, Zhang Y, Roberts GP (2006) Identification of *Rhodospirillum rubrum* GlnB variants that are altered in their ability to interact with different targets in response to nitrogen status signals. J Bacteriol 188:1866–1874

Zinchenko V, Babykin M, Glaser S, Mekhedov S, Shestakov S (1997) Mutation in *ntrC* gene leading to the derepression of nitrogenase synthesis in *Rhodobacter sphaeroides*. FEMS Microbiol Lett 147:57–61

Zou X, Zhu Y, Pohlmann EL, Li J, Zhang Y, Roberts GP (2008) Identification and functional characterization of NifA variants that are independent of GlnB activation in the photosynthetic bacterium *Rhodospirillum rubrum*. Microbiology 154:2689–2699

Chapter 5
The Network of P_{II} Signalling Protein Interactions in Unicellular Cyanobacteria

Karl Forchhammer

Abstract P_{II} signalling proteins constitute a large superfamily of signal perception and transduction proteins, which is represented in all domains of life and whose members play central roles in coordinating nitrogen assimilation. Generally, P_{II} proteins act as sensors of the cellular adenylylate energy charge and 2-oxoglutarate level, and in response to these signals, they regulate central nitrogen assimilatory processes at various levels of control (from nutrient transport to gene expression) through protein–protein interactions with P_{II} receptor proteins. An examination of the phylogeny of cyanobacteria reveals that specific functions of P_{II} signalling evolved in this microbial lineage, which are not found in other prokaryotes. At least one of these functions, regulation of arginine biosynthesis by controlling the key enzyme *N*-acetyl-L-glutamate kinase (NAGK), was transmitted by the ancestral cyanobacterium, which gave rise to chloroplasts, into the eukaryotic domain and was conserved during the evolution of planta. We have investigated in some detail the P_{II} signalling protein, its signal perception and its interactions with receptors in the unicellular cyanobacteria *Synechococcus elongatus* PCC 7942 and *Synechocystis* PCC 6803 and have performed comparative analysis with *Arabidopsis thaliana* P_{II}–NAGK interaction. This chapter will summarize these studies and will describe the emerging picture of a complex network of P_{II} protein interactions in the unicellular cyanobacteria.

5.1 Introduction to the P_{II} Signalling System

5.1.1 General Purpose and Evolution of P_{II} Signalling

P_{II} signalling proteins constitute a large superfamily of signal perception and transduction proteins, which is represented in all domains of life (Sant' Anna et al.

K. Forchhammer (✉)
Institut für Mikrobiologie der Eberhard-Karls-Universität Tübingen, Auf der Morgenstelle 28, D-72076, Tübingen, Germany
e-mail: karl.forchhammer@uni-tuebingen.de

P.C. Hallenbeck (ed.), *Recent Advances in Phototrophic Prokaryotes*, Advances in Experimental Medicine and Biology 675, DOI 10.1007/978-1-4419-1528-3_5,

2009). In all cases studied so far, the P_{II} signalling proteins are central in regulating processes related to nitrogen assimilation (Ninfa and Jiang 2005; Leigh and Dodsworth 2007; Forchhammer 2008). An examination of the various phylogenetic lineages of bacteria and archaea reveals a multitude of P_{II}-dependent processes and mechanisms of P_{II} signalling. To our knowledge, P_{II} signalling in eukaryotes is limited to chloroplast-bearing organisms, where it is localized in the plastids and descends from cyanobacterial P_{II} signalling (Moorhead et al. 2007; Osanai and Tanaka 2007). In spite of the general variability of regulatory roles played by P_{II}, there appears to be a common mode of function: P_{II} proteins act as sensors of the cellular adenylylate energy charge (by binding to the same ATP and ADP sites) and of the 2-oxoglutarate level. These metabolites are sensed by P_{II} through binding to interdependent binding sites, on the one hand three sites for ATP or ADP (binding competitively to the same sites) and on the other hand three sites for 2-oxoglutarate. The occupation of these sites appears to result in various P_{II} conformations, leading to signal output by differential binding of P_{II} to its receptors, which may be signal transduction proteins, metabolic enzymes or transporters (Forchhammer 2008).

The wide phylogenetic distribution of P_{II} with its high degree of structural conservation on one side, but functional variability on the other side, suggests that it represents an ancient regulatory module, which was used in the course of evolution to regulate a large variety of nitrogen-controlled processes, making use of the unique signal perception mode of P_{II} (Arcondeguy et al. 2001; Ninfa and Jiang 2005; Forchhammer 2008). The phylogeny of cyanobacteria reveals that specific functions of the P_{II} signal transduction pathway evolved which are not found in other prokaryotes (Forchhammer 2004, 2008; Osanai and Tanaka 2007). At least one of these functions, the regulation of arginine biosynthesis by control of the key enzyme *N*-acetyl-L-glutamate kinase (NAGK), was transmitted by the ancestral cyanobacterium, which gave rise to chloroplasts, into the eukaryotic domain and was conserved during the evolution of planta. Other functions, such as the modulation of nitrogen-dependent gene expression through interaction with a transcriptional co-activator, seem to be restricted to cyanobacteria.

5.1.2 Nomenclature and Phylogenetic Distribution of P_{II} Signalling Proteins

With a few exceptions, P_{II} proteins are encoded by genes termed *glnB*, *glnK* or *nifI* (Leigh and Dodsworth 2007). Since P_{II} proteins per se have no own enzymatic function, but are functionally versatile interacting proteins, there are no simple functional criteria defining P_{II} proteins. I propose that the term P_{II} protein should be used for proteins which display the structural and functional hallmarks of the hitherto characterized P_{II} proteins, namely the typical trimeric architecture and 3D structure, with surface-exposed T-loops and effector molecule binding sites in the intersubunit clefts (see Fig. 5.1). Historically, the term P_{II} designated an activity from an *Escherichia coli* extract, regulating glutamine synthetase activity, that was found in peak 2 of a gel filtration eluate (hence P_{II}) (Stadtman 1990; Arcondeguy et al. 2001).

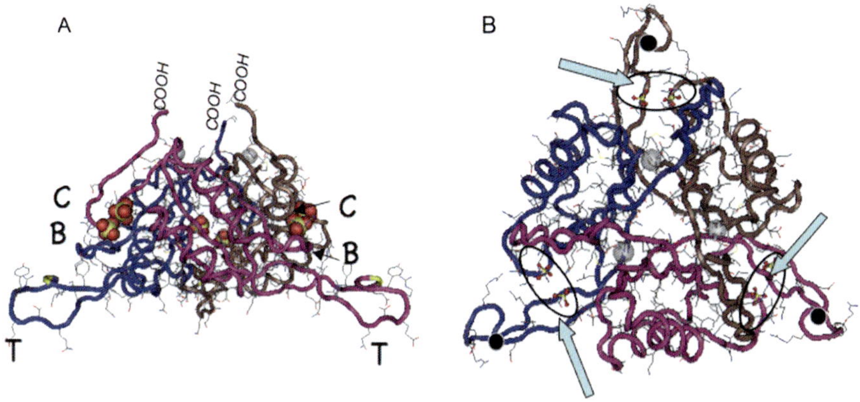

Fig. 5.1 Structure of the non-liganded P_{II} protein from *S. elongatus* PCC 7942. In this structure (Xu et al. 2003; pdb 1qy7), the T-loops are in an extended conformation. Other T-loop conformations may appear upon ligand or receptor binding (see Forchhammer 2008 for review). The three subunits are represented in *magenta*, *blue* and *brown*. (**a**) Side view of the trimeric P_{II} protein. The T-, B- and C-loops of two subunits are indicated; sulphur ions, occupying the adenylylate binding sites in the crystal, are shown. (**b**) View on the P_{II} trimer from the bottom orientation of part (**a**). The three adenylylate binding sites are indicated as *ovals* and are highlighted by the *arrows*; residue 49, the site of phosphorylation, is indicated as *black dot*

After the corresponding protein was identified, the term P_{II} was maintained and the gene encoding this protein was termed *glnB*, since it was regulating the product of the *glnA* gene, glutamine synthetase.

In proteobacteria, several P_{II} paralogues with almost identical 3D structure and high sequence conservation were subsequently discovered (Xu et al. 1998, 2001; Benelli et al. 2002; for review, see Forchhammer 2008), and very often, the gene for one or two of these paralogues is co-localized with an *amtB* gene encoding an ammonium transporter (reviewed in Leigh and Dodsworth 2007). Those P_{II} encoding genes are now termed, with a few exceptions, *glnK* (Arcondeguy et al. 2001; Sant' Anna et al. 2009). The co-localization of *glnK* and *amtB* reflects their tight functional interaction as a GlnK–AmtB complex, whose structure has been resolved (Gruswith et al. 2007; Conroy et al. 2007). Molecular genetic analysis and genome sequencing have revealed the presence of numerous other P_{II} homologues: In nitrogen-fixing archaea and in some anaerobic bacteria, P_{II}-like proteins, termed NifI, have been found that are genetically linked to nitrogenase genes and are phylogenetically related to each other (Leigh and Dodsworth 2007; Sant' Anna et al. 2009).

A recent bioinformatics analysis suggested on the basis of sequence similarity and by improving the PROSITE signature of P_{II} a novel group of P_{II}-like proteins. These are genetically linked to heavy metal resistance transporters (Sant' Anna et al. 2009). Members of this novel group of potential P_{II} proteins, which was termed P_{II}-NG (new group), have not been functionally characterized so far, and it remains to be shown if these proteins really display P_{II} features as defined above.

P_{II} proteins from cyanobacteria and plants are likely to be homophyletic (Osanai and Tanaka 2007; Sant' Anna et al. 2009). At the time of the first discovery of a cyanobacterial P_{II}-encoding gene, P_{II} paralogues were not yet known and thus, the gene was named according to the only known P_{II} gene at that time, *glnB* (Tsinoremas et al. 1991). In most cases, cyanobacterial *glnB* genes are monocistronic, and in most cyanobacteria only one clearly evident P_{II} paralogue is present. However, recent genome sequences have revealed cyanobacterial species which encode unusual P_{II} paralogues, such as *Gloeobacter* sp., *Acaryochloris marina* or the picocyanobacterium *Synechococcus* WH5701 (Scanlan et al. 2009). These gene products have unusual non-conserved residues in the T-loop region and, so far, they have not been functionally investigated. The P_{II} proteins from the unicellular freshwater strains *S. elongatus* PCC 7942 and *Synechocystis* PCC 6803 are the model P_{II} proteins for cyanobacteria (reviewed in Forchhammer 2004) and they will be presented in the following.

5.2 Properties of the P_{II} Proteins from *S. elongatus* and *Synechocystis* PCC 6803

5.2.1 Structure and Metabolite Binding Properties

The structures of the P_{II} proteins from *S. elongatus* and *Synechocystis* PCC 6803 have been elucidated (Xu et al. 2003). As typical P_{II} proteins, they are homotrimeric proteins composed of subunits of 112 amino acids (12.3 kDa) with a highly conserved 3D structure (Fig. 5.1). The body of the P_{II} trimer resembles a flat barrel, from which three prominent loops (T-loop; one per subunit) protrude like legs into the solvent. The T-loops are flexible in solution; upon crystal packing they may adopt different conformations, depending on whether P_{II} ligands are present or not (Llacer et al. 2007; Forchhammer 2008); T-loops of all P_{II} proteins investigated so far are of paramount importance for protein–protein interactions (see below).

The binding sites for ATP/ADP are located in the three intersubunit clefts, involving residues from the bases of the T-loop, the B-loop and the juxtaposed C-loop from the opposing subunit. The three ATP binding sites exhibit negative cooperativity and they can be occupied by ADP in a competitive manner. The affinity towards ATP is enhanced in the presence of 2-oxoglutarate (Forchhammer and Hedler 1997), whereas this is apparently not the case for the response towards ADP (Ruppert et al. 2002). 2-Oxoglutarate binds to cyanobacterial P_{II} protein only in the presence of ATP (Forchhammer and Hedler 1997). The 2-oxoglutarate binding sites have not been clearly resolved until now. The fact that ATP is required to bind 2-oxoglutarate, whereas ADP is not sufficient, indicates an important contribution of the γ-phosphate of ATP for 2-oxoglutarate binding, resulting in a tight coupling of 2-oxoglutarate and adenylylate charge measurement.

5.2.2 Phosphorylation of P_{II} Proteins in Cyanobacteria

In addition to binding and responding to the low molecular effector molecules, there is a second signal perception mode of P_{II}, which is not universally conserved. In *S. elongatus*, it was shown that the seryl residue 49 at the apex of the T-loop is phosphorylated in an ATP-dependent reaction (Forchhammer and Tandeau de Marsac 1995b). The same type of phosphorylation occurs in *Synechocystis* PCC 6803 (Kloft et al. 2005) and some other unicellular cyanobacteria. However, it appears to be absent in *Prochlorococcus* (Palinska et al. 2002) or in *Anabaena* PCC 7120 (Zhang et al. 2007). In the latter strain, an unusual type of modification at the T-loop was reported, nitration of tyrosyl residue 51, whose physiological relevance remains unclear. Intriguingly, the homologous position is the site of P_{II} uridylylation found in proteobacteria.

The in vivo phosphorylation state of the P_{II} protein in *S. elongatus* depends on growth conditions: The more the cells are starved for combined nitrogen sources, the more the P_{II} protein is found in phosphorylated states (Forchhammer and Tandeau de Marsac 1994; Forchhammer 1999). Each subunit of the trimeric P_{II} protein can be phosphorylated individually at the S49 residue, resulting in three different phosphorylated states, with one, two or three phosphorylated subunits (designated P_{II}^1, P_{II}^2 and P_{II}^3), and one non-modified form (designated P_{II}^0) (see Fig. 5.2). When grown in the presence of nitrate, intermediate phosphorylation states of P_{II}

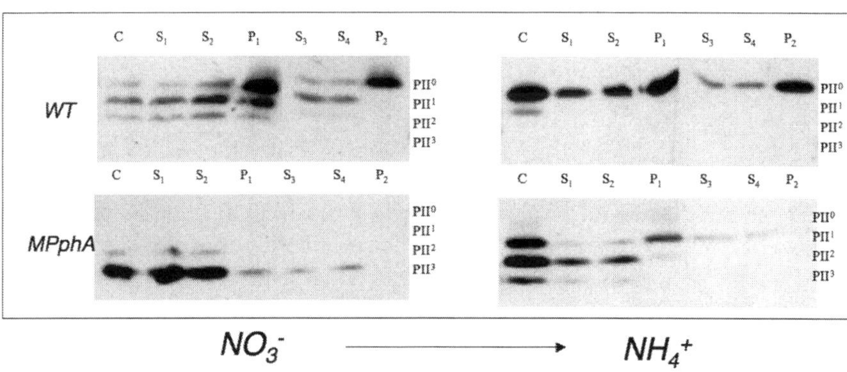

Fig. 5.2 Phosphorylation state and cellular localization of P_{II} from *Synechocystis* PCC 6803 wild-type cells (WT) and the PphA-deficient mutant MPphA. Extracts were prepared from nitrate-grown cells (*left part*) and after 15 min following the addition of 5 mM ammonium chloride (*right part*). The crude extracts (c) were fractionated by ultracentrifugation (120,000×*g* for 1 h) in an upper and lower part of the supernatant (S1 and S2) and the first sediment (P1). The P1 fraction was washed and centrifuged again at 120,000×*g* to yield supernatants S3 and S4, as well as the washed pellet (P2), representing the final membrane-associated fraction. The proteins were separated by non-denaturing PAGE to resolve the different phosphorylated forms of P_{II}, which display different electrophoretic mobilities (Forchhammer and Tandeau de Marsac 1994). P_{II}^0, P_{II}^1, P_{II}^2 and P_{II}^3 designate the trimeric P_{II} protein with zero, one, two and three phosphorylated subunits

are present, with the level of P_{II} phosphorylation increasing with higher inorganic C-supply (Forchhammer 1999). In the presence of ammonium, P_{II} is almost completely dephosphorylated. The kinase, which phosphorylates P_{II}, requires ATP and high 2-oxoglutarate concentrations (Forchhammer and Tandeau de Marsac 1995).

This suggests that P_{II} has to have all three 2-oxoglutarate sites occupied in order to react with the kinase, since the dissociation constant of the first two 2-oxoglutarate binding sites is too low, and only the third binding site needs concentrations near the millimolar range, which resembles the high 2-oxoglutarate needed for the phosphorylation reaction (Forchhammer and Hedler 1997; Forchhammer and Fokina unpublished results). Despite several attempts to identify the P_{II} kinase, this enzyme is still enigmatic. The activity is highly unstable and disappears upon purification. Nevertheless, two activity peaks are seen during gel filtration, a smaller peak eluting in the range of approx. 30–40 kDa proteins and a larger peak comigrating with the high molecular mass exclusion volume (Irmler et al. 1997). Attempts to identify the P_{II} kinase by genetic means, by analysing knockout mutants of presumptive protein kinase genes, have also failed, since all mutants analysed so far possess P_{II} kinase activity (unpublished results). In contrast, the search for the P_{II} phosphatase by analysing mutants of predicted protein phosphatases in *Synechocystis* PCC 6803 led to the identification of the P_{II} phosphatase PphA (Irmler and Forchammer 2001).

5.2.3 P_{II} Interaction with Protein Phosphatase PphA

5.2.3.1 P_{II}-P Phosphatase PphA from *Synechocystis* PCC 6803

The phospho-P_{II} (P_{II}-P) phosphatase PphA was discovered in *Synechocystis* PCC 6803 by systematically inactivating genes encoding potential protein phosphatases. Mutation of the open reading frame sll1771 yielded a mutant that was unable to dephosphorylate P_{II}-P, and the phosphorylation status of P_{II} was markedly increased compared to the wild type under all growth conditions (Irmler and Forchhammer 2001). All other mutants were not impaired in dephosphorylating P_{II}-P. Moreover, whereas phospho-P_{II} (P_{II}-P) was readily dephosphorylated with *Synechocystis* wild-type extract, almost no dephosphorylation occurred with extracts of the sll1771 mutant in spite of the fact that the encoded phosphatase is only one out of nine potential protein phosphatase homologues (Kloft et al. 2005). This strong phenotype implied that the sll1771 gene, which was designated *pphA*, encodes the only specific P_{II}-P phosphatase in *Synechocystis*. PphA belongs to the PPM family of protein phosphatases, which is defined by 11 conserved sequence motifs comprising the residues of the catalytic core. Since human protein phosphatase 2C (PP2C) is the defining enzyme of this family, the PPM members are also referred as PP2C-like phosphatases (Barford et al. 1998). Although these enzymes have long been regarded as typical eukaryotic-like phosphatases, genomic sequences reveal that PPM family members are particularly widespread in bacteria (Zhang and Shi 2004).

PphA has a molecular mass of 28.4 kDa and consists of 254 amino acids, which corresponds to the catalytic core of PP2C enzymes. With this property it belongs to a group of bacterial PP2C homologues, which are characterized by their small size, corresponding only to the PPM catalytic core and lacking putative regulatory domains (Irmler and Forchhammer 2001). This raises questions regarding the specificity of PphA towards P_{II}-P and the regulation of its activity. The biochemical properties of purified PphA are typical for PP2C enzymes with respect to Mn^{2+}/Mg^{2+}-dependent activity and sensitivity towards protein phosphatase inhibitors (Ruppert et al. 2002).

Depending on the assay conditions, PphA dephosphorylates a wide range of phosphorylated substrates, among them artificial substrates such as *p*-nitrophenyl phosphate or synthetic peptides containing phosphoseryl, phosphothreonyl or phosphotyrosyl residues, albeit with a catalytic activity that is orders of magnitude lower than that towards serine/threonine-phosphorylated proteins. The reactivity towards these low molecular weight substrates is stimulated by Mn^{2+} ions and alkaline pH. At physiological pH and in the presence of Mg^{2+} ions, PphA readily reacts only with serine/threonine-phosphorylated proteins such as phosphocasein or phosphohistones and with its physiological substrate, P_{II}-P. In the absence of effector molecules, the reactivity towards the non-physiological phosphoproteins was estimated to be similar to that towards P_{II}-P. In the presence of the P_{II} effector molecules, notably ADP or ATP and 2-oxoglutarate, the reactivity of PphA towards P_{II}-P was strongly inhibited, whereas the reactivity towards phosphocasein was unaffected.

In agreement with the lack of regulatory domains in PphA and with the biochemical properties of *S. elongatus* P_{II}, this suggests that the effector molecules specifically affect the molecular recognition between PphA and P_{II}-P, presumably by causing conformational changes of the T-loop of P_{II}, impairing its recognition by PphA. In the presence of physiological concentrations of ATP (1–2.5 mM), P_{II}-P dephosphorylation was very sensitive to low 2-oxoglutarate concentrations: 10 μM of this metabolite already caused significant inhibition and at 100 μM, P_{II}-P dephosphorylation was almost completely inhibited. In addition to 2-oxoglutarate, oxaloacetate is also able to inhibit P_{II}-P dephosphorylation in synergy with ATP, although 20-fold higher concentrations are required to yield similar effects. Other glycolytic and Krebs cycle intermediates, as well as various amino acids, did not affect PphA-catalysed P_{II}-P dephosphorylation (Ruppert et al. 2002). At 2-oxoglutarate concentrations in the low micromolar range, inhibition of P_{II}-P dephosphorylation in vitro could be enhanced by raising the ATP concentration.

This suggests that under physiological conditions of low cellular 2-oxoglutarate levels, P_{II}-P dephosphorylation may be sensitive to the cellular concentration of ATP. However, at various ATP:ADP ratios in the presence of 10 μM 2-oxoglutarate, keeping the total concentration of ATP plus ADP at constant 3 mM, the level of P_{II}-P dephosphorylation did not markedly respond to various ATP:ADP ratios (Forchhammer et al. 2004). Therefore, these in vitro results suggest that in vivo P_{II}-P dephosphorylation is activated by a drop in cellular 2-oxoglutarate levels.

5.2.3.2 Structure of the PphA Homologue from *Thermosynechococcus elongatus*

In order to address the question of substrate recognition and turnover by PphA, the X-ray structure of the homologous enzyme tPphA from the thermophilic unicellular cyanobacterium *T. elongatus* was resolved (see Fig. 5.3, Schlicker et al. 2008). The structures of the PPM members resolved so far display a highly conserved catalytic core. An invariant binuclear metal centre, occupied by Mg^{2+} or Mn^{2+} ions, co-ordinates a catalytically active nucleophilic water molecule. In agreement with the structures of three other bacterial PPM members, tPphA binds a third metal ion (M3) in the active site. M3 has not been observed in human PP2C, the defining member of this class of enzymes (Das et al. 1996). The function of the third metal in catalysis is debated. In general, the protein structure of tPphA is highly similar to the catalytic core of other PPM phosphatases, with the exception of a segment of the protein termed the flap subdomain, which is located near the catalytic cleft (see Fig. 5.3). The flap subdomain of tPphA has been shown to be flexible and may adopt various conformations during catalysis, and the flap of tPphA has been proposed to control access to the catalytic site (Schlicker et al. 2008). In order to further identify amino acids involved in the substrate specificity of tPphA, a targeted mutagenesis study was performed and the resulting mutants were characterized with respect to catalytic activity towards various substrates and with respect to the binding of the physiological substrate, phosphorylated P_{II} protein (Su Jiyong and Forchhammer unpublished). In addition to flap domain residues, a histidine residue near the entry to the catalytic cleft was shown to greatly modulate substrate specificity. Further, mutations affecting the M3 binding site cause a loss in catalytic activity, and catalysis is inhibited by elevated Ca^{2+} concentrations, which were shown to bind to the M3 site. In the presence of Ca^{2+}, reaction intermediates between P_{II}-P and tPphA could be isolated following stabilization of the interaction complex with

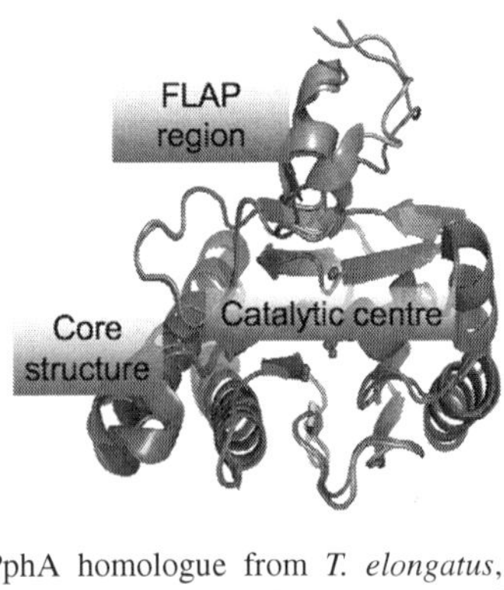

Fig. 5.3 Structure of the PphA homologue from *T. elongatus*, tPphA. The catalytic centre with a trinuclear metal centre is embedded in a highly conserved core structure, from which a non-conserved and flexible flap subdomain protrudes (Schlicker et al. 2008; pdb 2J86). In the asymmetric unit of the crystal, two states of tPphA could be resolved (molecule A and molecule B), which are shown as an overlay. From the overlay the flexibility of the flap subdomain is visible, and due to this flexibility part of the apex of the flap is not resolved

glutaraldehyde. The fact that 2-oxoglutarate and ATP as well as ADP impair this in vitro P_{II}-P–tPphA interaction emphasizes that this complex represents the physiologically relevant reaction intermediate. We propose that the effector molecules bound to P_{II} control the entry of the phopshorylated T-loop into the catalytic cavity of the phosphatase.

5.2.3.3 The Physiological Significance of P_{II} Dephosphorylation

Physiological analysis of a PphA-deficient *Synechocystis* mutant revealed a close relation between the P_{II} phosphorylation state and the regulation of nitrate utilization. PphA-deficient mutants were overgrown by wild-type cells only when nitrate was used as nitrogen source, but not with ammonia or urea. When PphA mutant cells are grown with nitrate as nitrogen source, they excrete nitrite into the medium, especially at low photon flux densities, whereas increasing photon fluence densities override this effect (Kloft and Forchhammer 2005). A similar phenotype was found in the P_{II}-deficient mutant MP2. PphA and MP2 have in common that they do not contain non-phosphorylated P_{II}. By analysing the activities of nitrate and nitrite reductases and nitrate/nitrite uptake, we concluded that nitrite excretion was caused by uncontrolled nitrate uptake and reduction, overloading the capacity of nitrite reductase. The dephosphorylated form of P_{II} (P^0_{II}), which is generated by PphA activity, appears to be responsible for adjusting nitrate uptake to the capacity of nitrite reductase. This fine-tuning of nitrate utilization, which requires PphA, may be related to an effect of nitrite on the cellular accumulation of PphA: Kloft et al. (2005) showed that the cellular PphA level increases significantly in the presence of nitrite and the elevated PphA levels may thus enhance P_{II}-P dephosphorylation. This could result in a regulatory feedback loop for nitrate utilization such that limiting reductant from photosynthesis causes a limitation of nitrite reductase capacity, creating a transient accumulation of nitrite. This increases PphA levels and favours P_{II} dephopshorylation, which in turn will tune down nitrate uptake, until excess nitrite is completely reduced.

In contrast to the fine-tuning of nitrate utilization, regulation of nitrate uptake in response to ammonium treatment was still functional in the PphA mutant, but was defective in a P_{II}-deficient mutant (Kloft and Forchhammer 2005). This indicates that the P_{II} protein, but not its dephosphorylation, is required to inhibit NRT (nitrate transport system) in response to ammonium. A possible explanation could be that ammonium strongly depletes the cellular 2-oxoglutarate levels, which may promote a direct, modification-independent P_{II} response that is only based on the effector molecule-dependent P_{II} conformation.

5.3 Receptor Interactions of P_{II} Proteins in *S. elongatus* and *Synechocystis* PCC 6803

To identify regulatory targets of P_{II} signalling, two major strategies were employed: analysis of the phenotype of P_{II}-deficient mutants from *S. elongatus* (Forchhammer and Tandeau de Marsac 1995a; Lee et al. 2000; Aldehni et al. 2003) and the straightforward screening for P_{II} interaction partners by the yeast two-hybrid method using *glnB* genes from *S. elongatus* or *Synechocystis* PCC 6803 as bait (Burillo et al. 2004; Heinrich et al. 2004, Osanai et al. 2005a). Whereas the first approach revealed several P_{II}-regulated functions without identifying P_{II} interaction partners, the second approach led to the discovery of the three presently known downstream signalling receptors of P_{II}, namely *N*-acetyl-L-glutamate kinase (NAGK), a key enzyme of arginine synthesis, PipX, a co-factor of transcription factor NtcA, and PamA, a membrane protein of unknown activity. In the following section, the various interactions of P_{II} with these target proteins will be described.

5.3.1 P_{II} Interaction with N-acetyl-L-glutamate Kinase

5.3.1.1 Enzymatic Activation of NAGK by P_{II} Complex Formation

N-acetyl-L-glutamate kinase is the controlling enzyme of the cyclic pathway of arginine biosynthesis. Glutamate enters the pathway by reaction with a pathway intermediate *N*-acetyl-L-ornithine to yield ornithine and *N*-acetyl-L-glutamate (NAG). The next step, phosphorylation of NAG to NAG phosphate by *N*-acetyl-L-glutamate kinase (NAGK), is the rate-limiting step in arginine synthesis, and this enzyme is strictly feedback regulated by the pathway end-product arginine (Llacer and Rubio 2009). In the search for P_{II} receptors, yeast two-hybrid screenings using the *S. elongatus glnB* gene as bait retrieved the *argB* gene (encoding NAGK) as a potential P_{II} interaction partner (Burillo et al. 2004; Heinrich et al. 2004). Biochemical investigation showed that this interaction was specific and was highly susceptible towards modifications at the seryl 49 residue, the site of P_{II} phosphorylation, with only the non-modified form of P_{II} able to bind NAGK (Heinrich et al. 2004).

Binding of P_{II} to NAGK dramatically alters the catalytic properties of the enzyme; catalytic efficiency was originally estimated to increase by a factor of 40 upon P_{II} binding (Maheswaran et al. 2004). Using an optimized enzyme assay, an increase of catalytic efficiency of eightfold was recently determined (Beez et al. 2009) due to the fact that the previously used assay underestimated the activity of free NAGK. It is unequivocally clear from all assays that the sensitivity of NAGK to feedback inhibition by arginine is relieved upon binding of P_{II} by approximately 10-fold (Maheswaran et al. 2004; Beez et al. 2009).

Biochemical studies such as surface plasmon resonance (SPR) spectroscopy and enzyme assays demonstrated that complex formation between P_{II} and NAGK is highly sensitive to the ligand binding status of P_{II}: *S. elongatus* P_{II} binds readily to NAGK in the absence of effector molecules and in the presence of Mg-ATP. However, P_{II} bound with 2-oxoglutarate/ATP is unable to bind NAGK. SPR analysis also revealed that ADP is a highly potent inhibitor of P_{II}–NAGK interaction. For example, addition of 1 mM ADP to a P_{II}–NAGK complex causes an almost immediate dissociation of the complex (Maheswaran et al. 2004).

5.3.1.2 The Structure of the P_{II}–NAGK Complex

Recently, the crystal structure of the NAGK–P_{II} complex from the cyanobacterium *S. elongatus* has been resolved at 2.75 Å resolution (Llacer et al. 2007). The complex consists of two P_{II} trimers sandwiching a hexameric NAGK toroid. The NAGK toroid is a trimer of dimers, arranged in such a manner that the toroid has threefold symmetry and two identical faces on top and bottom. In the complex, the threefold axes of P_{II} and NAGK are aligned, such that each P_{II} subunit contacts one opposing NAGK subunit. The contacts map in the inner circuit of the NAGK toroid, in which two regions of the P_{II} protein protrude. One region is on the part of the P_{II} body which consists of the B-loop and nearby residues (the β1-α1connection). The second part is the distal region of the T-loop. The T-loop undergoes a large conformational change when the P_{II} protein binds to NAGK. In the complex it adopts a compact conformation, resembling a tightly flexed leg, whereas it is extended in the absence of NAGK. T-loop residues R45, E50 and S49 are of particular importance in complex formation by forming an ion pair and hydrogen-bonding network. Interestingly, these residues are signature residues for cyanobacterial P_{II} proteins and are also present in most P_{II} proteins from plants and green and red algae. The observed alteration of NAGK enzymatic properties can be explained by structural consequences upon complex formation: A widening of the arginine inhibition site on NAGK by P_{II} binding leads to decreased affinity towards this allosteric inhibitor, and the ion-pair network involving the T-loop of P_{II} tightens the catalytic centre of NAGK, leading to increased enzyme activity.

5.3.1.3 The Physiological Consequences of P_{II}–NAGK Interaction

The ultimate rationale for the sophisticated control of NAGK by P_{II} appears to be the need to tightly control arginine synthesis in response to the central metabolic state of the cells, since the synthesis of arginine demands high energy and combined nitrogen supply. In addition to its role as an essential amino acid for protein synthesis and as a precursor for polyamine synthesis, arginine plays an important role as a combined nitrogen buffer and for the synthesis of nitrogen storage compounds (Llacer and Rubio 2009). Figure 5.4 depicts schematically the control of NAGK by P_{II}. Association between P_{II} and NAGK is favoured by high ATP concentrations and excess nitrogen (low 2-oxogutarate) supply. Under these conditions, the P_{II} protein is in the non-phosphorylated state, which is a prerequisite for complex formation. Furthermore, the cellular level of the antagonistic metabolites 2-oxoglutarate and ADP is low under these conditions. Complexed with P_{II}, NAGK is in a state of high activity and decreased sensitivity towards arginine. Consequently, the flux into the arginine pathway will increase.

This hypothesis was verified experimentally (Maheswaran et al. 2006). When nitrogen-starved *Synechocystis* cells were suddenly exposed to excess ammonia, cells of the wild-type strain rapidly increased levels of arginine, which was used for the synthesis of the nitrogen reserve polymer cyanophycin (multi-L-arginyl-poly-L-aspartic acid). In contrast, a P_{II}-deficient mutant was unable to accumulate

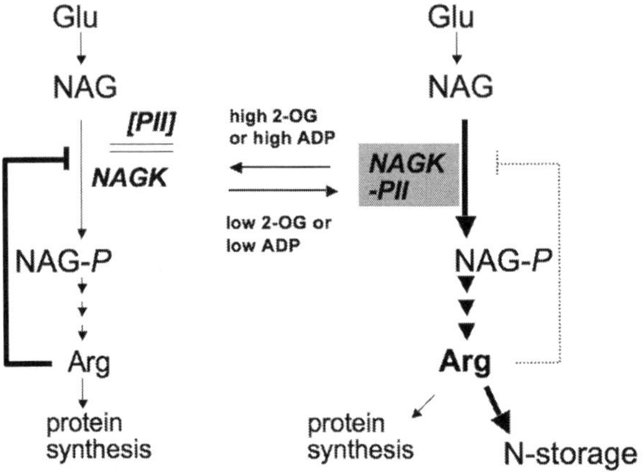

Fig. 5.4 Schematic representation of the control of arginine synthesis by P_{II}–NAGK complex formation. *Left side*: under conditions of low adenylylate charge (high ADP/ATP ratio) or high cellular 2-oxoglutarate levels, no complex between P_{II} and NAGK is formed. NAGK is in a state of low activity and is highly feedback inhibited by arginine. *Right side*: under ATP-replete and low 2-oxoglutarate conditions, P_{II} complexes NAGK, resulting in kinetic activation and alleviation from arginine inhibition. The increased levels of arginine enable the synthesis of nitrogen storage compounds (for details see text)

cyanophycin and to increase arginine levels due to a permanently low activity of NAGK, since it lacked P_{II} activation. When the nitrogen source becomes exhausted, the cellular level of 2-oxoglutarate increases (Muro-Pastor et al. 2001), promoting P_{II}–NAGK dissociation. NAGK now returns to its highly sensitive state of arginine feedback inhibition, which limits arginine synthesis to that needed for protein biosynthesis. When such conditions of elevated 2-oxoglutarate levels are sustained for prolonged periods, the P_{II} protein will become phosphorylated (see above), which robustly impairs any complex formation with NAGK. In agreement, we found that a PphA-deficient *Synechocystis* mutant is unable to accumulate arginine and cyanophycin, because it retains P_{II} in the phosphorylated state (Maheswaran et al. 2006). In the wild type, NAGK can be activated again after prolonged N-limitation only under conditions of sustained nitrogen excess where P_{II} becomes dephosphorylated.

In this regard, P_{II} phosphorylation seems to be a mechanism to sustain the signal of nitrogen-limiting conditions, protecting NAGK from transient activation in response to short-term fluctuations of cellular metabolite levels. A second possible in vivo scenario that should be considered is a rapid decrease in cellular ATP levels in nitrogen-sufficient cells, e.g., in response to sudden light–dark transitions. Under such conditions, the ADP levels will suddenly increase, leading to immediate dissociation of NAGK–P_{II} complexes irrespective of the 2-oxoglutarate levels. The activity of NAGK and thus arginine biosynthesis will immediately respond to the low-energy signal, a reasonable response considering the high-energy demand for arginine synthesis.

5.3.1.4 NAGK–P_{II} Interaction Is Conserved in Cyanobacteria and Plants

The initial indication of a conserved interaction between P_{II} and NAGK in chloroplasts of higher plants came from yeast two-hybrid interaction assays, using the corresponding genes from *Arabidopsis thalina* (Burillo et al. 2004). Subsequently, the specificity of this interaction and its localization in the chloroplasts were reported (Sugiyama et al. 2004). Biochemical and structural analyses revealed a high degree of similarity to the NAGK–P_{II} interaction in the cyanobacterium *S. elongatus* (Chen et al. 2006; Mizuno et al. 2007; Beez et al. 2009). In particular, NAGK sensitivity towards arginine inhibition is consistently alleviated upon complex formation. Initial experiments suggested that the *A. thaliana* complex was not antagonized by 2-oxoglutarate. However, the use of an optimized enzyme assay demonstrated that the P_{II}-dependent protection of NAGK activity from arginine inhibition is indeed antagonized by 2-oxoglutarate (Beez et al. 2009).

A striking difference between cyanobacterial and plant P_{II}–NAGK is the degree of arginine sensitivity, being by an order of magnitude higher in *S. elongatus*. Deletion of the non-conserved C-terminal 11 amino acids of *S. elongatus* NAGK led to a marked reduction in arginine sensitivity, with the C-terminal-truncated mutant having similar properties to the *A. thalina* enzyme. Strikingly, the P_{II} protein from *A. thaliana* was fully able to regulate the cyanobacterial NAGK enzyme, at least in vitro, whereas the *S. elongatus* P_{II} protein could only partially replace the higher plant P_{II} protein in NAGK activation, indicating that P_{II} in *Arabidopsis* has conserved all the requirements for the regulation of the cyanobacterial NAGK enzyme and furthermore has acquired additional features not represented in *Synechococcus* P_{II}, which contribute to the fully functional P_{II}–NAGK complex. This high degree of functional conservation, matched by conservation of key residues in P_{II} and NAGK, indicates that P_{II} is pivotal for the regulation of arginine synthesis in plants and in cyanobacteria.

5.3.2 *P_{II}–PipX Interaction*

5.3.2.1 In Vitro Properties of the P_{II}–PipX Complex

A second target of P_{II} was also identified by yeast two-hybrid screening using *S. elongatus glnB* as bait (Burillo et al. 2004). The protein, a small basic peptide of 89 amino acids, with a predicted pI of 8.97, displayed no homology to hitherto characterized proteins and was thus termed PipX (P_{II} interaction protein X) (Espinosa et al. 2006). However, genes encoding PipX homologues are found in all cyanobacteria, but not in other bacteria or in plants, indicating a conserved role in cyanobacteria (Espinosa et al. 2009). Further yeast two-hybrid screening using PipX as bait retrieved the transcription factor NtcA as a second interaction partner of PipX. Subsequent biochemical analysis showed that the interaction of PipX with P_{II} and NtcA was indeed specific and responded to the metabolite 2-oxoglutarate. In the absence of the effector molecule 2-oxoglutarate, PipX associates with P_{II}, but not with NtcA. The presence of 2-oxoglutarate allows PipX binding to NtcA, with 2-oxoglutarate concentrations in the millimolar range being required for optimal interaction. This requirement is probably caused by an effect of 2-oxoglutarate on NtcA, since it has already been shown by in vitro transcription and gel-shift analysis that 2-oxoglutarate alters the properties of NtcA with respect to DNA binding and transcriptional activation (Tanigawa et al. 2002; Paz-Yepez et al. 2003). With respect to P_{II} interactions, 2-oxoglutarate in concert with ATP leads to a decreased affinity of PipX–P_{II} interaction (Espinosa et al. 2006). Thus PipX provides a mechanistic link between transcriptional control by NtcA and P_{II} signal transduction, which had been previously suggested based on physiological studies (Aldehni et al. 2003; Paz-Yepes et al. 2003).

5.3.2.2 PipX Is a Co-activator of NtcA-Dependent Gene Expression

The physiological role of PipX was revealed by *pipX* mutant analysis, which showed a clear defect in the ability of PipX deletion mutants to induce NtcA-activated promoters. However, P_{II}-controlled functions such as ammonium inhibition of nitrate uptake or NAGK activation were not affected in the PipX mutant (Espinosa et al. 2007). Conversely, overexpression of *pipX* in a *S. elongatus* wild-type background caused a latent enhancement of NtcA-dependent gene expression. These results indicate that PipX is required for activation of NtcA-dependent gene expression, but not for modulation of other P_{II} functions. PipX–NtcA complexes bind to DNA, but PipX does not modify the promoter specificity of NtcA, nor does it change the affinity of NtcA towards its binding sites (Espinosa et al. 2008; Rasch, Espinosa and Forchhammer, unpublished results). Thus, we suggest that PipX somehow affects the interaction of NtcA with the initiation complex of RNA polymerase, leading to an enhancement of transcriptional initiation.

The postulated roles of P_{II} and PipX in modulating NtcA-dependent gene expression are schematically depicted in Fig. 5.5. Under conditions of intermediate 2-oxoglutarate levels, especially when the ATP levels decrease, P_{II} and NtcA compete for binding of PipX. Given the role of PipX as a co-activator of NtcA, it follows that P_{II} should decrease the activation of NtcA-dependent gene expression by sequestering PipX. The interaction of P_{II} with PipX would be required to release PipX from NtcA and tune down gene expression under conditions of dropping 2-oxoglutarate or ATP levels (corresponding to nitrogen excess or energy limitation). Indeed, this conclusion agrees with a reported phenotype of a P_{II} deletion mutant of *S. elongatus*: Using *luxAB* reporter fusion analysis on various *glnA* and *glnN* promoter constructs, it could be shown that the P_{II}-deficient strain overexpressed these constructs under non-inducing conditions (Aldehni et al. 2003; Aldehni and Forchhammer 2006).

However, this analysis and previous studies also showed that P_{II}-deficient *S. elongatus* strains were unable to fully turn on the expression of NtcA-dependent genes under inducing conditions (nitrogen deprivation). This phenotype was in apparent conflict with the idea of P_{II} acting as an antagonist of NtcA co-activation. This contradiction could recently be resolved by re-investigation of the genetic background

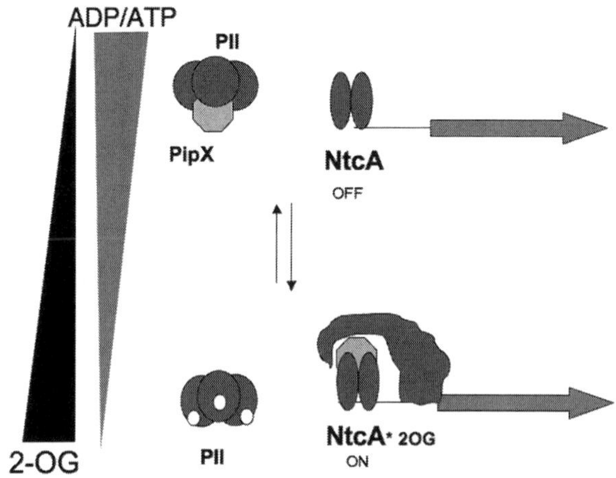

Fig. 5.5 Suggested model of the P_{II}–PipX–NtcA interaction network leading to NtcA activation. Under low 2-oxoglutarate or low ATP to ADP conditions, P_{II} is predicted to bind PipX, whereas NtcA is in an inactive state (*top*). Increasing 2-oxoglutarate levels under ATP-replete conditions would lead to partner swapping of PipX from P_{II} to NtcA. PipX enables NtcA to fully activate the transcription of NtcA-dependent genes, perhaps by promoting a productive interaction with RNA polymerase (*bottom*)

of P_{II}-deficient strains. Espinosa et al. (2009) demonstrated that the reported P_{II}-null mutants had secondary mutations in *pipX* alleles, leading either to reduced expression or to a complete loss of PipX function. From this and similar findings, such as that P_{II} can be easily mutated in a *pipX*-deficient background but not in the wild type, the authors postulated that a P_{II}-null mutation is toxic in the presence of wild-type levels of PipX. Therefore, P_{II} may be necessary to sequester PipX under normal growth conditions. If not sequestered by P_{II} under nitrogen-rich conditions, PipX may be lethal because of inappropriate activation of NtcA or by causing some other critical interactions which would normally not occur in the P_{II}-bound state.

5.3.3 PamA, an Enigmatic P_{II} Target

Yeast two-hybrid screening with a library of *Synechocystis* PCC 6803 genomic DNA as prey identified a potential membrane protein as a further P_{II} receptor. The protein, encoded by the sll0985 open reading frame, was termed PamA (*P_{II}-a*ssociated *m*embrane protein *A*) and is predicted to contain seven membrane-spanning helices (Osanai et al. 2005a). The three C-terminal membrane-spanning regions of PamA are homologous to transmembrane channel proteins of the MscS family. In vitro experiments confirmed the interaction of P_{II} with PamA and showed that P_{II} binding to PamA is antagonized by 2-oxoglutarate/ATP, whereas the phosphorylation state of P_{II} did not seem to affect the interaction. It was suggested that binding of P_{II} to PamA may regulate the transport activity of this potential membrane channel. However, the physiological substrate transported by PamA has not so far been elucidated. A *pamA* defect mutant showed reduced expression levels of a subset of nitrogen-controlled genes, such as genes of the *nrt* operon, *nblA* and *sigE* (Osanai et al. 2005b). SigE is an alternative sigma factor whose expression is normally increased under nitrogen-poor conditions. It positively regulates the expression of sugar catabolic genes (Osanai and Tanaka 2007). Consistent with this, a *pamA* mutant exhibits reduced expression of sugar catabolic genes and is unable to grow in glucose-containing medium. From these results it can be concluded that PamA indirectly regulates sugar catabolic genes through modulating *sigE* expression levels. However, how the binding of P_{II} contributes to this function of PamA remains elusive to date.

5.3.4 Membrane Binding of Non-phosphorylated P_{II} Protein

Previous work in the author's lab showed that the P_{II} protein in *Synechocystis* binds to the membrane fraction in a phosphorylation-dependent manner, as shown in Fig. 5.2 (Kloft and Forchhammer, unpublished data). When the cells were grown in presence of nitrate as combined nitrogen source, approximately 50% of the detectable P_{II} protein, the majority of the non-phosphorylated form of P_{II}, sedimented with the membrane fraction, whereas the phosphorylated forms remained soluble. Following treatment with 5 mM ammonium chloride, P_{II} became completely dephosphorylated, as expected. Half of the dephosphorylated P_{II} protein remained soluble, the remainder was again associated with the membrane. By contrast, in the PphA-deficient mutant MPphA, the P_{II} protein was highly phosphorylated and was never associated with the membrane fraction, apparently due to lack of non-phosphorylated P_{II}. This suggests that there might be a specific receptor protein in the membrane for non-phosphorylated P_{II} protein. This receptor is unlikely to be PamA, since PamA association was reported to be independent of P_{II} phosphorylation (see earlier). Given the phenotype of the PphA-deficient and P_{II}-deficient mutants, both of which excrete nitrite when utilizing nitrate and both of which lack the non-phopshorylated P_{II} protein, a component of the nitrate utilization system may be a candidate for such an interaction. However, attempts to reveal any binding of P_{II} to components of the NRT system have so far failed.

5.4 Concluding Remarks

We are just beginning to unveil a highly complicated network of P_{II} interactions, with P_{II} placed in the middle of a complex signalling node (Fig. 5.6). On the one hand, P_{II} integrates energy and C/N signals by forming a complex array of different signal-input states, based on composite occupation of interdependent ligand binding sites. On the other hand, P_{II} binds to various receptors with affinities which depend on these signal-input states, and it can be anticipated that the various receptors will compete for P_{II} binding. Further investigations are required to reveal additional P_{II}

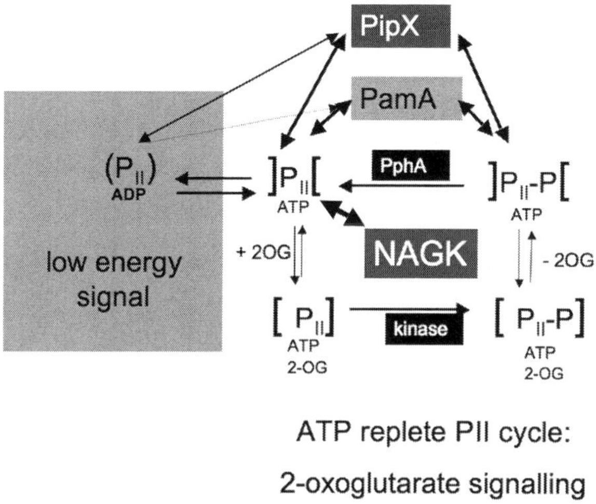

Fig. 5.6 Network of the known interactions of the P_{II} signalling protein. Under ATP-replete conditions, P_{II} cycles between a phosphorylated and a non-phosphorylated state, which is driven by fluctuations in cellular 2-oxoglutarate (2OG) levels and the activity of P_{II} kinase and phosphatase. Depending on the phosphorylation state and 2-oxoglutarate binding, P_{II} interacts with NAGK, PipX or PamA. Under low ATP/ADP conditions, P_{II} is non-phosphorylated and does not bind 2-oxoglutarate. However, PipX is able to bind to this form of P_{II}. Whether PamA interacts with this state of P_{II} remains to be elucidated

receptors and to understand the dynamics of P_{II} signalling in a living cell, with respect to the mutual interplay between these receptors, their competition for P_{II} binding and the subcellular targeting and sequestering of P_{II}.

References

Aldehni MF, Sauer J, Spielhaupter C, Schmid R, Forchhammer K (2003) The Signal transduction protein P_{II} is required for NtcA-regulated gene expression during nitrogen deprivation in the cyanobacterium *Synechococcus elongatus* Strain PCC 7942. J Bacteriol 185:2582–2591

Aldehni FA, Forchhammer K (2006) Analysis of a non-canonical NtcA-dependent promoter in *Synechococcus elongatus* and its regulation by NtcA and P_{II}. Arch Microbiol 184:378–386

Arcondéguy T, Jack R, Merrick M (2001) P_{II} signal transduction proteins, pivotal players in microbial nitrogen control. Microbiol Mol Biol Rev 65:80–105

Barford D, Das AK, Egloff MP (1998) The structure and mechanism of protein phosphatases: insights into catalysis and regulation. Annu Rev Biophys Biomol Struct 27:133–164

Beez S, Fokina O, Herrmann C et al (2009) N-Acetyl-L-Glutamate kinase (NAGK) from oxygenic phototrophs: P_{II} signal transduction across domains of life reveals novel insights into NAGK control. J Mol Biol 389:748–758

Benelli EM, Buck, Polikarpov I et al (2002) *Herbaspirillum seropedicae* signal transducer protein P_{II} is structurally similar to the enteric GlnK. Eur J Biochem 269:3296–3303

Burillo S, Luque I, Fuentes et al (2004) Interactions between the nitrogen signal transduction protein P_{II} and N-acetyl glutamate kinase in organisms that perform oxygenic photosynthesis. J Bacteriol 186:3346–3354

Chen YN, Ferrar T, Lohmeir-Vogel E et al (2006) The P_{II} signal transduction protein of *Arabidopsis thaliana* forms a metabolite regulated complex with plastid N-acetyl glutamate kinase. J Biol Chem 281:5726–5733

Conroy MJ, Durand A, Lupo D et al (2007) The crystal structure of the *Escherichia coli* AmtB-GlnK complex reveals how GlnK regulates the ammonia channel. Proc Natl Acad Sci USA 104:1213–1218

Das AK, Helps NR, Cohen PTW et al (1996) Crystal structure of the protein serine/threonine phosphatase 2C at 2.0 Å resolution. EMBO J. 15:6798–6809

Espinosa J, Forchhammer K, Burillo S et al (2006) Interaction network in cyanobacterial nitrogen regulation: PipX, a protein that interacts in a 2-oxoglutarate dependent manner with P_{II} and NtcA. Mol Microbiol 61:457–466

Espinosa J, Forchhammer K, Contreras, A (2007) Role of the *Synechococcus* PCC 7942 nitrogen regulator protein PipX on NtcA controlled processes. Microbiology 153:711–718

Espinosa J, Castells MA, Forchhammer K et al (2008) Regulatory interactions involved in nitrogen signalling in cyanobacteria: partner swapping and the role of Pipx-P_{II} complexes in Synechococcus sp. PCC 7942. In S. Netherlands (ed.), Biological nitrogen fixation: towards poverty alleviation through sustainable agriculture Vol 42. Springer, New York, pp 361–362

Espinosa J, Castells MA, Laichoubi KB, Contreras A (2009) Mutations at *pipX* suppress lethality of P_{II}-deficient mutants of *Synechococcus elongatus* PCC 7942. J Bacteriol 191:4863–4869

Forchhammer K (1999) The P_{II} protein in *Synechococcus* sp. PCC 7942 senses and signals 2-oxoglutarate under ATP replete conditions. In: Peschek G, Löffelhardt W, Schmetterer G (eds) The Phototrophic Prokaryotes. Kluwer Academic/Plenum, New York, pp 549–553

Forchhammer K (2004) Global carbon/nitrogen control by P_{II} signal transduction in cyanobacteria: from signals to targets. FEMS Microbiol Rev 28:319–333

Forchhammer K (2008) P_{II} signal transducers: Novel functional and structural insights. Trends Microbiol 16:65–72

Forchhammer K, Tandeau de Marsac N (1994) The P_{II} protein in the cyanobacterium *Synechococcus* sp. strain PCC 7942 is modified by serine phosphorylation and signals the cellular N-status. J Bacteriol 176:84–91

Forchhammer K, Tandeau de Marsac N (1995a) Functional analysis of the phosphoprotein P_{II} (*glnB* gene product) in the cyanobacterium *Synechococcus* sp. strain PCC 7942. J Bacteriol 177:2033–2040

Forchhammer K, Tandeau de Marsac N (1995b) Phosphorylation of the P_{II} protein (*glnB* gene product) in the cyanobacterium *Synechococcus* sp. strain PCC 7942: Analysis of in vitro kinase activity J Bacteriol 177:5812–5817

Forchhammer K, Hedler A (1997) Phosphoprotein P_{II} from Cyanobacteria: Functional homology to the P_{II} signal transduction protein from *Escherichia coli*. Eur J Biochem 244:869–875

Forchhammer K, Irmler A, Kloft N et al (2004) P_{II} signalling in unicellular cyanobacteria: analysis of redox-signals and energy charge. Physiol Plantarum 120:51–54

Gruswith F, O'Connell J, Stroud RM (2007) Inhibitory complex of the transmembrane ammonia channel, AmtB, and the cytosolic regulatory protein, GlnK, at 1.96 Å. Proc Natl Acad Sci USA 104:42–47

Heinrich, A, Maheswaran M, Ruppert U et al (2004) The *Synechococcus elongatus* P_{II} signal transduction protein controls arginine synthesis by complex formation with N-acetyl-L-glutamate kinase. Mol Micorbiol 52:1303–1314

Irmler A, Sanner S, Dierks H et al (1997) Dephosphorylation of the phosphoprotein P_{II} in *Synechococcus* PCC 7942: identification of an ATP and 2-oxoglutarate-regulated phosphatase activity. Mol Microbiol 26:81–90

Irmler A, Forchhammer K (2001) A PP2C-type phosphatase dephosphorylates the P_{II} signaling protein in the cyanobacterium Synechocystis PCC 6803. Proc Natl Acad Sci 98: 12978–12983

Kloft N, Forchhammer K (2005) Signal Transduction Protein P_{II} Phosphatase PphA Is Required for Light-Dependent Control of Nitrate Utilization in *Synechocystis* sp. strain PCC 6803. J Bacteriol 187:6683–6690

Kloft N, Rasch G, Forchhammer K (2005) Protein phosphatase PphA from *Synechocystis* sp. PCC 6803: the physiological framework of P_{II}-P dephosphorylation. Microbiology 151:1275–1283

Lee HM, Flores E, Forchhammer K et al (2000) Phosphorylation of the signal transducer P_{II} protein and an additional effector are required for the P_{II}-mediated regulation of nitrate and nitrite uptake in the cyanobacterium *Synechococcus* sp. PCC 7942. Eur J Biochem 267:591–600

Leigh JA, Dodsworth JA (2007) Nitrogen regulation in bacteria and archa. Annu Rev Microbiol 61:349–377

Llacer JL, Contreras A, Forchhammer K et al (2007) The crystal structure of the complex of P_{II} and acetylglutamate kinase reveals how P_{II} controls the storage of nitrogen as arginine. Proc Natl Acad Sci USA 104:17644–17649

Llacer JL, Rubio V (2009) Arginine and nitrogen storage. Curr Opin Struct Biol 18:673–681

Maheswaran M, Urbanke C, Forchhammer K (2004) Complex formation and catalytic activation by the P_{II} signaling protein of N-acetyl-L-glutamate kinase from *Synechococcus elongatus* strain PCC 7942. J Biol Chem 279:55202–55210

Maheswaran M, Ziegler K, Lockau W et al (2006) P_{II}-Regulated Arginine Synthesis Controls the Accumulation of Cyanophycin in *Synechocystis* sp. Strain PCC 6803. J Bacteriol 188: 2730–2734

Mizuno Y, Moorhead GBG, Ng, KKS (2007) Structural basis for the regulation of N-acetylglutamate kinase by P_{II} in *Arabidopsis thaliana*. J Biol Chem 282:35733–35740

Moorhead GBG, Ferrar TS, Chen YM et al (2007) The higher plant P_{II} signal transduction protein: structure, function and properties. Canad J Botany 85:533–537

Muro-Pastor MI, Reyes JC, Florencio FJ (2001) Cyanobacteria perceive nitrogen status by sensing intracellular 2-oxoglutarate levels. J Biol Chem 276:38320–38328

Ninfa AJ, Jiang P (2005) P_{II} signal transduction proteins: sensors of [alpha]-ketoglutarate that regulate nitrogen metabolism. Curr Opin Microbiol8:168–173

Osanai T, Sato S, Tabata S et al (2005a) Identification of PamA as a P_{II}-binding membrane protein important in nitrogen-related and sugar-catabolic gene expression in *Synechocystis* sp. PCC 6803. J Biol Chem 280:34684–34690

Osanai T, Kanesaki Y, Nakano T et al (2005b) Positive regulation of sugar catabolic pathways in the cyanobacterium *Synechocystis* sp. PCC 6803 by the group 2 factor SigE. J Biol Chem 280:30653–30659

Osanai T, Tanaka K (2007) Keeping in touch with P_{II}: P_{II}-interacting proteins in unicellular cyanobacteria. Plant Cell Physiol 48:908– 914

Palinska KA, Laloui W, Bedu S et al (2002) The signal transducer P_{II} and bicarbonate acquisition in *Prochlorococcus marinus* PCC 9511, a marine cyanobacterium naturally deficient in nitrate and nitrite assimilation. Microbiology 148:2505–2412

Paz-Yepes J, Flores E, Herrero A (2003) Transcriptional effects of the signal transduction protein P_{II} (*glnB* gene product) on NtcA-dependent genes in *Synechococcus* sp. PCC 7942. FEBS Lett 543:42–46

Ruppert U, Irmler A, Kloft N, Forchhammer K (2002) The novel protein phosphatase PphA from *Synechocystis* PCC 6803 controls dephosphorylation of the signaling protein P_{II}. Mol Microbiol 44:855–864

Sant' Anna FH, Trentini DB, de Souto Weber S et al (2009) The P_{II} Superfamily Revised: A Novel Group and Evolutionary Insights. J Mol Evol 68:322–336

Scanlan D, Ostrowski M, Mazard S et al (2009) Ecological genomics of marine picocyanobacteria. Microbiol Mol Biol Rev 73:249–299

Schlicker C, Fokina O, Kloft N et al (2008). Structural analysis of the PP2C Phosphatase tPphA from *Thermosynechococcus elongatus:* a flexible flap subdomain controls access to the catalytic site. J Mol Biol 376:570–581

Stadtman ER (1990) Discovery of glutamine synthetase cascade. Methods Enzymol 182:793–809

Sugiyama K, Hayakawa T, Kudo T et al (2004) Interaction of *N*-acetylglutamate kinase with a P_{II}-Like protein in rice. Plant Cell Physiol 45:1768–1778

Tanigawa R, Shirokane M, Maeda S-I et al (2002) Transcriptional activation of NtcA-dependent promoters of *Synechococcus* sp. PCC 7942 by 2-oxoglutarate in vitro. Proc Natl Acad Sci USA 99:4251–4255

Tsinoremas NF, Castets AM, Harrison MA et al (1991) Photosynthetic electron transport controls nitrogen assimilation in cyanobacteria by means of post-translational modification of the *glnB* gene product. Proc Natl Acad Sci USA 88:4565–4569

Xu Y, Chea E, Carr PD et al (1998) GlnK, a P_{II}-homologue: Structure reveals ATP binding site and indicates how the T-loop may be involved in molecular recognition. J Mol Biol 282:149–165

Xu Y, Carr PD, Huber T et al (2001) The structure of the P_{II}-ATP complex. Eur J Biochem 268:2028–2037

Xu Y, Carr PD, Clancy P et al (2003) The structures of the P_{II} proteins from the cyanobacteria *Synechococcus* sp. PCC 7942 and *Synechocystis* sp. PCC 6803. Acta Crystallogr D 59: 2183–2190

Zhang W, Shi L (2004) Evolution of the PPM-family protein phosphatases in *Streptomyces*: duplication of catalytic domain and lateral recruitment of additional sensory domains. Microbiology 150:4189–4197

Zhang Y, Pu H, Wang Q et al (2007) P_{II} is important in regulation of nitrogen metabolism but not required for heterocyst formation in the cyanobacterium *Anabaena* sp. PCC 7120. J Biol Chem 282:33641–33648

Chapter 6
Pathway and Importance of Photorespiratory 2-Phosphoglycolate Metabolism in Cyanobacteria

Martin Hagemann, Marion Eisenhut, Claudia Hackenberg, and Hermann Bauwe

Abstract Cyanobacteria invented oxygenic photosynthesis about 3.5 billion years ago. The by-product molecular oxygen initiated the oxygenase reaction of RubisCO, the main carboxylating enzyme in photosynthetic organisms. During oxygenase reaction, the toxic side product 2-phosphoglycolate (2-PG) is produced and must be quickly metabolized. Photorespiratory 2-PG metabolism is used for this purpose by higher plants. The existence of an active 2-PG metabolism in cyanobacteria has been the subject of controversy since these organisms have evolved an efficient carbon-concentrating mechanism (CCM), which should considerably reduce the oxygenase activity of RubisCO. Based on emerging cyanobacterial genomic information, we have found clear indications for the existence of many genes possibly involved in the photorespiratory 2-PG metabolism. Using a genetic approach with the model *Synechocystis* sp. strain PCC 6803, we generated and characterized defined mutants in these genes to verify their function. Our results show that cyanobacteria perform an active photorespiratory 2-PG metabolism, which employs three routes in *Synechocystis*: a plant-like cycle, a bacterial-like glycerate pathway, and a complete decarboxylation branch. In addition to the detoxification of 2-PG, this essential metabolism helps cyanobacterial cells acclimate to high light conditions.

6.1 Introduction

6.1.1 Cyanobacterial Photosynthesis and the Carbon-Concentrating Mechanism

Cyanobacteria invented oxygenic photosynthesis about 3.5 billion years ago. This capability was later, about 1.2 billion years ago, transferred via endosymbiosis into eukaryotes leading to the emergence of eukaryotic algae and plants

M. Hagemann (✉)
University of Rostock, Institute of Biological Sciences and Plant Physiology, Albert-Einstein-Strasse 3, D-18051, Rostock, Germany
e-mail: martin.hagemann@biologie.uni-rostock.de

P.C. Hallenbeck (ed.), *Recent Advances in Phototrophic Prokaryotes*, Advances in Experimental Medicine and Biology 675, DOI 10.1007/978-1-4419-1528-3_6,

(Mereschkowsky 1905; Margulis 1970; Deusch et al. 2008). Over the long term, the activity of these photosynthetic organisms generated the present-day oxygen-containing atmosphere. In parallel, the amount of CO_2 was considerably diminished, which resulted in a situation in which today CO_2 is a limiting factor for photosynthetic organisms. Cyanobacteria, like algae and plants, fix most of the inorganic carbon (C_i) via the Calvin–Benson cycle with ribulose 1,5-bisphosphate carboxylase/oxygenase (RubisCO) as the carboxylating enzyme. However, cyanobacterial RubisCO has a rather low affinity for CO_2 (e.g., the K_m for CO_2 is about 300 μM for the RubisCO of *Anabaena variabilis*, Badger 1980), which is theoretically not sufficient to efficiently fix CO_2 at the low concentrations prevailing in aquatic environments.

The challenge of fixing CO_2 at low environmental concentrations was solved by the evolution of an efficient inorganic carbon-concentrating mechanism (CCM) among cyanobacteria (for reviews, see Kaplan and Reinhold 1999; Badger et al. 2006). CCM activity allows them to cope with low C_i concentrations especially in the presence of oxygen. Compared to the CCM existing in eukaryotic algae, the molecular biology and functionality of the cyanobacterial CCM is much better understood (Giordano et al. 2005). It comprises two principal mechanisms: active uptake systems for C_i and the compartmentalization of RubisCO in the prokaryotic carboxysome organelle.

In the model strain *Synechocystis* sp. PCC 6803 (henceforth *Synechocystis*) and the related strain *Synechococcus* sp. PCC 7002, three transport systems for bicarbonate, the predominate C_i source in alkaline waters such as oceans, have been identified. First, an ATP-dependent ABC-type transport system for bicarbonate (called Cmp) was identified, which has a very high affinity for bicarbonate and becomes induced when cyanobacteria are shifted to low C_i. Later, two symport systems (called SbtA and BicA) for bicarbonate and H^+ or Na^+, respectively, were discovered. SbtA system is a high-affinity transporter that is induced at low C_i, while BicA is a constitutively expressed low-affinity system (reviewed in Badger et al. 2006). Moreover, CO_2 is actively taken up by cyanobacteria. Initially the gas diffuses into the cells and is subsequently converted to bicarbonate at specialized NDH1 complexes (Zhang et al. 2004), which keeps the CO_2 concentration gradient steep enough to favor further import. There is one constitutive system of rather low affinity and a second inducible NDH1 subcomplex showing much higher CO_2 affinity. The joint action of all transporters allows an about 1,000-fold accumulation of C_i in the form of bicarbonate inside the cyanobacterial cell. Only the inactivation of all five transporters results in a high C_i-requiring (HCR) phenotype of the corresponding mutants, while defects of single transporters are compensated for (Xu et al. 2008).

The accumulated bicarbonate diffuses into the carboxysome, a prokaryotic organelle in which RubisCO and carbonic anhydrase are enclosed by a protein shell (Kerfeld et al. 2005; Yeates et al. 2008). Inside the carboxysome the bicarbonate is transformed into CO_2 by the action of carbonic anhydrase leading to high CO_2 concentrations in the vicinity of RubisCO. Recently, it has been shown that RubisCO, carbonic anhydrase, and shell proteins form a defined complex inside the carboxysome to facilitate CO_2 release near the active center of RubisCO and to avoid leakage of CO_2 into the cytoplasm (Cot et al. 2008). As a consequence of the important function of the carboxysome, mutations affecting its structural integrity also result in a HCR phenotype (Kaplan and Reinhold 1999). It is believed that carboxysomes evolved among cyanobacteria about 700 million years ago (after the primary endosymbiotic event leading to chloroplasts), probably by two independent lateral gene transfer events (Badger et al. 2006).

In addition to having a low affinity for CO_2, in the presence of atmospheric oxygen RubisCO also exhibits oxygenase activity, a side reaction in which molecular oxygen instead of CO_2 is bound to the acceptor ribulose 1,5-bisphosphate. Oxygenase activity further diminishes the carboxylating activity of RubisCO and results in the appearance of the toxic by-product 2-phosphoglycolate (2-PG), which inhibits Calvin–Benson cycle enzyme activities. In higher plants, 2-PG is rapidly metabolized by the photorespiratory 2-PG metabolism. For the conversion of 2-PG and other toxic compounds into the Calvin–Benson cycle intermediate glycerate 3-phosphate, plants require at least 10 different enzymes (Ogren 1984; Tolbert 1997; Bauwe and Kolukisaoglu 2003). In higher plants, this pathway is an essential partner of oxygenic photosynthesis, since photorespiratory mutants have a HCR phenotype (Bauwe and Kolukisaoglu 2003). In contrast, because they have CCM, it was believed that oxygenase function of RubisCO is rather low or absent inside cyanobacterial cells. Therefore, the existence of an active 2-PG metabolism in these organisms was neglected.

6.1.2 Previous Attempts to Identify Photorespiratory 2-PG Metabolism in Cyanobacteria

Historically, the oxygenase function of RubisCO and photorespiratory 2-PG metabolism in plants were described earlier than CCM in cyanobacteria. Because of the close evolutionary relation between photosynthetic mechanisms in cyanobacteria and plants, many efforts were initially undertaken to provide evidence for a functional 2-PG metabolism in cyanobacteria. It was shown that the cyanobacterial RubisCO has oxygenase activity and glycolate can be detected among the early carbon fixation products (Fig. 6.1). In the pre-genomic era basically two types of experiments were used: on the one hand detection of enzyme activities involved in the 2-PG metabolism known from plants and on the other hand glycolate release from cyanobacteria grown under conditions known to promote 2-PG production. In order to show that a 2-PG metabolic pathway similar to that described for plants exists, the activities of some enzymes characteristic of this pathway were measured in crude extracts from cyanobacterial cells. These experiments indicated the presence of 2-PG phosphatase, glycolate oxidase (or rather glycolate dehydrogenase), glutamate-glyoxylate aminotransferase, and hydroxypyruvate reductase (Norman and Colman 1988; Renström and Bergman 1989). In contrast, serine hydroxymethyltransferase activity was not observed. These results suggested

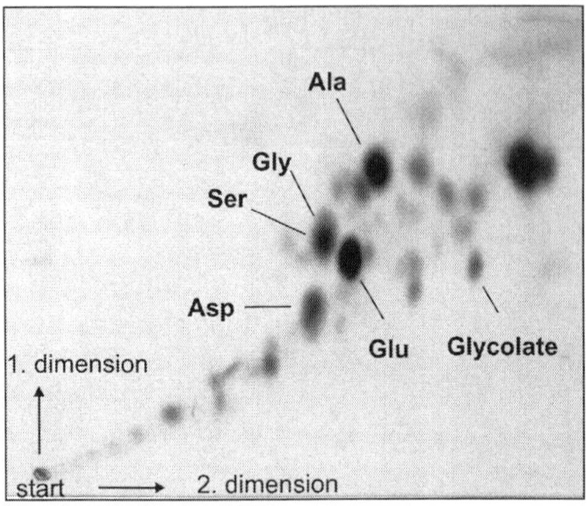

Fig. 6.1 ^{14}C-labeling pattern of *Synechocystis* sp. strain PCC 6803. The cells were incubated with $NaH^{14}CO_3$ in the light for 5 min. Low molecular mass substances were isolated with hot 80% ethanol and separated by two-dimensional thin layer chromatography. The labeled spots were identified by co-chromatography with non-labeled reference substances

that the glycolate pathway is incomplete and that an alternative pathway must exist to metabolize the glycolate formed (Norman and Colman 1988). A likely candidate was the bacterial glycerate pathway, which is known from *E. coli* and other heterotrophic bacteria to convert glycolate into glycerate as C-source. It was indeed shown that extracts of *Anabaena cylindrica* catalyzed the decarboxylation of glyoxylate to tartronic semialdehyde, by glyoxylate carboligase followed by the subsequent reduction of tartronic semialdehyde to glycerate by tartronic semialdehyde reductase (Codd and Stewart 1973).

Moreover, glycolate excretion could be detected with some cyanobacterial strains, e.g., *Plectonema boryanum*, *A. cylindrica*, *A. variabilis* or *Nostoc* sp. 73102, but not with other strains. It should be mentioned that the glycolate amounts were usually rather low, but could be significantly enhanced when cells were either transferred from high CO_2 to air, exposed to elevated oxygen concentrations up to 100%, or treated with inhibitors for enzymes involved in 2-PG metabolism (Renström and Bergman 1989). At the same time, glycolate excretion was also found among green algae such as *Chlamydomonas* (Kaplan and Berry 1981; Moroney et al. 1986). These findings gave rise to the still popular hypothesis that both algae and cyanobacteria do not perform a 2-PG metabolism since the toxic photorespiratory intermediate is simply excreted into the liquid medium.

In summary, it was not possible with the experimental attempts of the pre-genomic era to elucidate the fate of intermediates of the 2-PG pathway in cyanobacteria. The discovery of the efficient CCM during the same time led to the widely accepted view that cyanobacteria do not carry out a complete photorespiratory 2-PG metabolism despite some strain-specific differences (Colman 1989). Recently, we have reanalyzed the mode of 2-PG metabolism and its possible diverse function with the cyanobacterial model strain *Synechocystis* based on emerging genomic information (Kaneko et al. 1996) and the well-developed genetic tools for this cyanobacterial strain. In the following sections we review and summarize the results of this work.

6.2 New Results Verify the Essential Function of 2-PG Metabolism for Cyanobacteria

6.2.1 Searching of Cyanobacterial Genomes for Enzymes of 2-PG Metabolism

During the last 13 years, about 40 genomes of different cyanobacterial strains have been completely sequenced and deposited in public databases such as CyanoBase. This data set represents a blueprint of the capacity of cyanobacteria to perform specific metabolic pathways including photorespiratory 2-PG metabolism. In our bioinformatic analyses, we used the available information from the model plant *Arabidopsis* to search for corresponding genes in the genome of *Synechocystis* using the Blast algorithm (Altschul et al. 1997). For this purpose, protein sequences from functionally characterized enzymes of the photorespiratory cycle from *Arabidopsis thaliana* were extracted from the TAIR database (http://www.arabidopsis.org/) and compared with predicted proteins from the *Synechocystis* sp. PCC 6803 genome at CyanoBase (http://genome.kazusa.or.jp/cyanobase/) as described by Hagemann et al. (2005). With the exception of three enzymes, 2-PG phosphatase, glycolate oxidase, and glycerate kinase, it was possible to identify candidate genes in the *Synechocystis* genome for each protein involved in the 2-PG metabolism in *Arabidopsis*, which showed rather high degrees of similarities on the protein sequence level including a homolog to serine hydroxymethyltransferases (Hagemann et al. 2005). In some cases, e.g., for hydroxypyruvate reductase, not only one but several candidate proteins seem to exist in *Synechocystis* (Table 6.1).

The existence of the majority of enzymes for a 2-PG cycle similar to that in *Arabidopsis* made it very likely that the gaps could be closed by enzymes of corresponding activities but not homologous to those from *Arabidopsis*. Indeed, using proteins from *E. coli* and other heterotrophic bacteria in Blast searches, candidate genes were identified for the 2-PG phosphatase and the glycerate kinase, while, instead of glycolate oxidase, a glycolate dehydrogenase seemed to exist in *Synechocystis* (Eisenhut et al. 2006). Moreover, when we included proteins known to be involved in additional routes for glycolate metabolism from heterotrophic bacteria, it became obvious that the *Synechocystis* genome also harbors putative glyoxylate carboligase and tartronic semialdehyde reductase genes for the direct conversion of glyoxylate to glycerate (Eisenhut et al. 2006). These genome analyses clearly indicated that *Synechocystis* most probably has the capacity for active 2-PG metabolism, which possibly employs two different routes: the plant-like glycolate cycle and the bacterial-like glycerate pathway (Fig. 6.2).

The extension of such searches to other cyanobacterial genomes revealed strain-specific differences. For example, in the larger genome of the filamentous *Anabaena* (*Nostoc*) sp. PCC 7120, candidate proteins were detected for a plant-like glycerate kinase (Boldt et al. 2005) and a plant-like glycolate oxidase (Eisenhut et al. 2008a). Therefore, this and many other cyanobacterial strains seem to harbor a plant-like

Table 6.1 Overview on the enzymes proven to be involved in photorespiratory 2-PG metabolism in the cyanobacterium *Synechocystis* sp. strain PCC 6803 or in the plant *A. thaliana*

Enzyme	ORF in 6803	Mutant of 6803	Recombinant protein	Enzymatic active	Reference	Present in A. thal.	Cyanobacterial origin?
2-Phosphoglycolate	slr0458	Segregated	Non-tested	Non-tested	Eisenhut et al. (2006)	–	
Phosphatase	sll1349	Segregated	Non-tested	Non-tested	Eisenhut et al. (2006)	–	
Glycolate DH							
GlcD1	sll0404	Segregated	Yes	Yes	Eisenhut et al. (2006)	+	At5g06580
GlcD2	slr0806	Segregated	Non-tested	Non-tested	Eisenhut et al. (2008a)	–	
Glycolate oxidase	–	–	–	–	–	+	**At4g18360**
Serine/glyoxylate Aminotransferase	sll1559	Non-segregated	Yes	Yes	Eisenhut et al. (2006)	+	
Glycine decarboxylase							
P-protein, GcvP	slr0293	Segregated	Yes	Yes	Hagemann et al. (2005); Hasse et al. (2007)	+	**At4g33010**
T-protein, GcvT	sll0171	Segregated	Yes	Non-tested	Hagemann et al. (2005)	+	At3g16950
H-protein, GcvH	slr0879	Segregated	Yes	Yes	Hagemann et al. (2005); Hasse et al. (2007)	+	
L-protein	slr1096	Non-segregated	Yes	Non-tested	Hagemann et al. (2005); Engels and Pistorius (1997)	+	
Serine hydroxymethyl transferase	sll1931	Non-segregated	Yes	Yes	Hagemann et al. (2005); Eisenhut et al. (2006)	+	At4g32520
Glutamate/glyoxylate aminotransferase	sll0006	Non-tested	Non-tested	Non-tested		–	
Hydroxypyruvate	sll1908	Non-tested	Non-tested	Non-tested		+	At1g17745
Reductase	sll0891	Segregated	Non-tested	Non-tested	Unpubl.	+	

Table 6.1 (continued)

Enzyme	ORF in 6803	Mutant of 6803	Recombinant protein	Enzymatic active	Reference	Present in A. thal.	Cyanobacterial origin?
Glycerate kinase							
Bacterial-type	slr1840	Segregated	Yes	Yes	Bartsch et al. (2008)	–	
Plant-type	–					+	**At1g80380**
Glyoxylate	sll1981	Segregated	Non-tested	Non-tested	Eisenhut et al. (2006)	–	
Carboligase	slr2088	Non-segregated	Non-tested	Non-tested	Eisenhut et al. (2006)		
Tartronic semi-aldehyde reductase	slr0229	Segregated	Yes	Yes	Eisenhut et al. (2006), unpubl.	+	
Hydroxyacid reductase	sll1556	Segregated	Yes	Non-tested	Unpubl.	+	
Oxalate decarboxylase	sll1358	Segregated	Yes	Yes	Eisenhut et al. (2008a); Tottey et al. (2008)	–	
Formate dehydrogenase	sll1359	Non-tested	Non-tested	Non-tested		–	

Gene IDs of the corresponding genes from the two genomes of these organisms are shown. Moreover, attempts to mutate the *Synechocystis* genes or over-express them in *E. coli* in order to verify the proposed function are included. The last column summarizes putative genes in *A. thaliana* which were probably conveyed by the primary endosymbiotic event from cyanobacteria into present-day higher plants such as *Arabidopsis* (Unpubl. – unpublished own results)

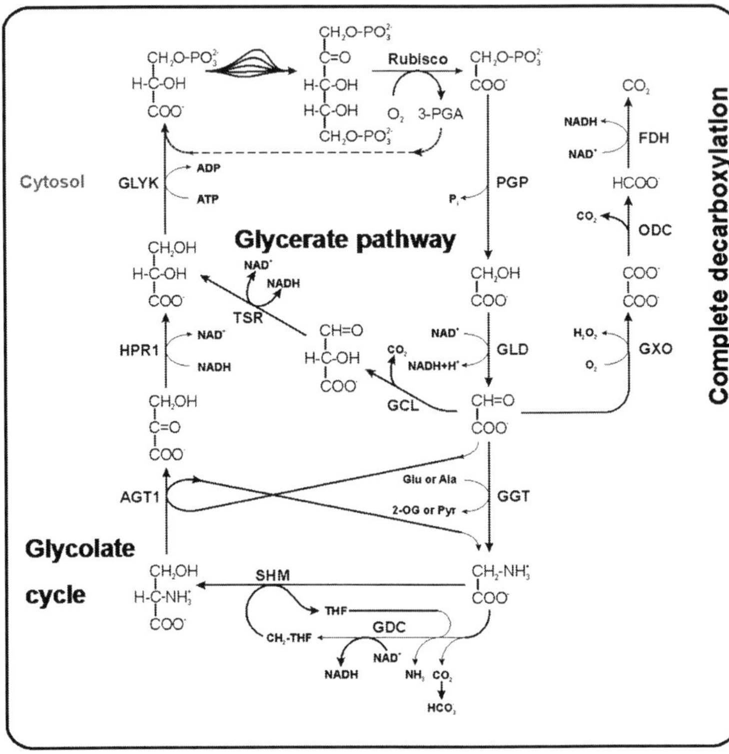

Fig. 6.2 Schematic drawing of the complete photorespiratory 2-PG metabolism in cells of *Synechocystis* sp. strain PCC 6803. 2-PG metabolism is branched into three routes: plant-like glycolate cycle, bacterial-like glycerate pathway, and complete decarboxylation branch (PGP – 2-PG phosphatase, GLD – glycolate dehydrogenase, GGT – glycine/glutamate aminotransferase, GDC – glycine decarboxylase, SHM – serine hydroxymethyltransferase, AGT1 – alanine/glyoxylate aminotransferase, HPR1 – hydroxypyruvate reductase, GLYK – glycerate kinase, GCL – glyoxylate carboligase, TSR – tartronic semi-aldehyde reductase, GXO – glyoxylate oxidase, ODC – oxalate decarboxylase, FDH – formate dehydrogenase)

2-PG cycle that is more similar to *Arabidopsis* than the predicted pathway in our model strain *Synechocystis*. However, none of the cyanobacterial genomes harbored a plant-like 2-PG phosphatase. Among the cyanobacterial genomes, those from the picoplanktonic *Prochlorococcus* and *Synechococcus* strains are of special interest, since their genomes are much smaller than those from model freshwater cyanobacteria. Their genome reduction is believed to represent an adaptation to the nutrient-poor and constant oceanic environment, in which non-essential genes were successively deleted. Systematic searches in 16 genomes from picoplanktonic *Prochlorococcus* and *Synechococcus* strains indicated that their reduced genomes also retained the capacity to possess a plant-like 2-PG cycle and a bacterial-like glycerate pathway (Scanlan et al. 2009). These genome searches made it very likely that all present-day cyanobacteria should be able to perform an active 2-PG metabolism. Similar searches have been carried out for the recently published genomes of two diatom species, which indicate that a 2-PG metabolism exists in these eukaryotic algae but shows considerable differences compared to that of higher plants (Kroth et al. 2008).

6.2.2 Systematic Mutation of Candidate Genes for 2-PG Metabolism

In order to test whether or not these candidate genes really code for proteins active in 2-PG metabolism, basically two strategies were used. First, for almost all of the candidate genes, specific mutants were generated, and second, selected genes were over-expressed in *E. coli* to prove the enzymatic activity of purified recombinant proteins (Table 6.1). Single mutants defective in almost all candidate genes for 2-PG metabolism of *Synechocystis* were obtained by interposon mutagenesis. Briefly, the coding sequences of selected genes were amplified by PCR, and drug-resistant cartridges were introduced at selected restriction sites. These constructs were transferred into *Synechocystis* for homologous recombination of the wild-type (WT) gene by the inactivated gene. Drug-resistant clones were isolated and analyzed by PCR in order to determine whether the WT gene copy was completely absent (completely segregated mutants) or if WT as well as mutated gene copies coexisted (non-segregated mutants). WT and mutated cells of *Synechocystis* were cultivated in BG11 medium (Rippka et al. 1979) buffered to pH 8 in a CO_2-enriched atmosphere (5% CO_2 in air defined as high CO_2 conditions – HC) or at ambient air (about 0.035% CO_2 in air defined as low CO_2 conditions – LC). In most cases, the mutants contained a completely segregated genome, i.e., the WT gene copy could be completely inactivated. This was the first indication that these genes are not essential for the viability of *Synechocystis* under standard growth conditions at LC. There were only four exceptions: glyoxylate/serine aminotransferase, L-protein subunit of glycine decarboxylase complex, serine hydroxymethyltransferase, and one of the two possible glyoxylate carboligases, where the mutant genotype was found not to be segregated, i.e., the WT copy was retained in addition to the mutated gene copy (Table 6.1). Incomplete segregation of a *Synechocystis* mutant can be taken as an indication that the corresponding protein is essential for the viability of the cell. Since it was possible to knock out the majority of putative 2-PG metabolism genes completely, these four essential proteins are probably also involved in other vital processes.

For the L-protein subunit of the glycine decarboxylase complex, the essential nature of the gene has been shown before and is due to the involvement of L-protein in another enzyme complex, the pyruvate decarboxylase complex (Engels and Pistorius 1997), which is essential in contrast to the glycine decarboxylase (Hagemann et al. 2005). The serine hydroxymethyltransferase is known to produce activated C1 subunits for many biosynthetic processes during serine/glycine conversion, which explains its essential character (Bauwe and Kolukisaoglu 2003). The putative glyoxylate carboligase Slr2088 has been shown to be involved in de novo amino acid biosynthesis. A corresponding mutant could be only obtained when the cells were grown on media supplemented with amino acids (Kouhen and Joset 2002). Probably, the essential aminotransferase Sll1559 is also involved in the basic amino acid metabolism, thus preventing its complete knockout.

The successful generation and complete segregation of many mutants under ambient conditions indicated that these genes are not required for the normal growth of *Synechocystis*. This assumption was mostly supported by growth measurements, in which all analyzed mutants, with the exception of Δ*glcD1* mutant, grew at HC like WT (Hagemann et al. 2005; Eisenhut et al. 2006). The GlcD1 represents a glycolate dehydrogenase, which catalyzes the step just before the splitting into the plant-like 2-PG cycle and the bacterial-like glycerate pathway. During analyses of cellular metabolites, we observed an accumulation of low levels of glycolate in the mutant Δ*glcD1*, which might be the reason for its slower growth. This finding also supports the assumption that the gene *glcD1* (*sll*0404) indeed codes for a glycolate dehydrogenase (Table 6.1). Finally and more importantly, glycolate accumulation in this mutant indicates that oxygenase reaction of RubisCO also occurs in *Synechocystis* under HC despite the existence of an intact CCM (Eisenhut et al. 2006).

The absence of marked growth differences between most single mutants and WT cells of *Synechocystis* could be explained by the existence of two different routes for 2-PG metabolism and the existence of more than one candidate protein for certain steps, compensating for the knockout mutations. To examine this possibility, a series of double mutants were generated. The enzyme 2-PG phosphatase catalyzes the initial reaction of 2-PG metabolism. Its mutation led to a clear HCR phenotype in *Arabidopsis* (Schwarte and Bauwe 2007). In *Synechocystis*, at least two genes code for possible 2-PG phosphatases. A double mutant initially showed a HCR phenotype, but revertants quickly appeared making a clear conclusion difficult (Eisenhut et al. 2006). Another double mutant, Δ*gcvT*/Δ*tsr* (defective in T-protein subunit of glycine decarboxylase and tartronic semialdehyde reductase, see Fig. 6.2), was generated in order to simultaneously block both the plant-like glycolate cycle and the bacterial-like glycerate pathway. This completely segregated double mutants and selected single mutants were compared with regard to their growth rates under LC, which should favor the oxygenase reaction of RubisCO. These experiments revealed that the Δ*glcD1* and Δ*gcvT* mutants showed slower growth rates than WT at LC, while the Δ*tsr* mutant grew like WT. Interestingly, the double mutant Δ*gcvT*/Δ*tsr*, defective in the two possible routes for 2-PG metabolism, showed the largest growth reduction of all investigated mutants at LC (Eisenhut et al. 2006). The changed phenotype of mutants defective in enzymes of the 2-PG metabolism at LC indicates that (1) this metabolic process occurs in *Synechocystis*, (2) the plant-like 2-PG cycle seems to be of greater importance for 2-PG recycling than the bacterial-like glycerate pathway, and (3) the two routes cooperate in 2-PG metabolism.

6.2.3 DNA Microarray Analysis to Search for New Routes in 2-PG Metabolism

The above-mentioned results demonstrate that 2-PG metabolism is active in *Synechocystis*; however, its inactivation did not lead to the HCR phenotype that is characteristic of photorespiratory plant mutants. This result could indicate that either CCM is sufficient to compensate for the absence of the 2-PG metabolism or additional routes exist for 2-PG detoxification. It should be mentioned here that we made repeated attempts to show glycolate release or uptake from *Synechocystis* WT or the mutant cells, because glycolate export could easily explain the non-essential character of 2-PG metabolism. However, in contrast to previous reports dealing with other cyanobacterial strains (Renström and Bergman 1989), we were unable to detect any glycolate release into the medium or any significant glycolate uptake activity. Therefore, we examined the global gene expression pattern of *Synechocystis* WT and mutant cells shifted from HC to LC using DNA microarray technology. These experiments gave a complete picture of the transcriptional regulation of HC/LC acclimation and also helped to identify possible additional pathways for 2-PG metabolism (Eisenhut et al. 2007). We found that about 200 genes were up-regulated during HC/LC shifts, as previously reported (Wang et al. 2004). Among them, genes for the bicarbonate transporters Cmp and SbtA showed the highest induction ratios, while genes for carboxysomes were only slightly affected. Those genes that were identified as necessary for 2-PG metabolism via the plant-like cycle and the bacterial-like glycerate pathway did not show any significant alterations in their transcript levels. However, among the many LC-induced genes encoding for hypothetical proteins, we found promising candidates for enzymes possibly forming a third route for 2-PG breakdown: the complete decarboxylation of glycolate via oxalate and formate to CO_2 (Eisenhut et al. 2007). The existence of such an additional decarboxylating branch (Fig. 6.2, Table 6.1) could explain the missing HCR phenotype of our previous double mutant, Δ*gcvT*/Δ*tsr*.

Using the same strategy as before, the function of the complete decarboxylation branch was proven by an additional round of mutagenesis. By inactivating oxalate decarboxylase, the single mutant Δ*odc* was generated, which is blocked only in this branch. This single mutant did not show any differences in growth compared to WT cells at HC as well as LC. The double mutant, Δ*odc*/Δ*gcvT*, grew like WT at HC, but slower at LC. Finally, we generated a triple mutant Δ*odc*/Δ*gcvT*/Δ*tsr*, which could be obtained under HC and was found to be completely segregated in all three mutated genes. This triple mutant was able to grow at HC, but bleached after transfer to LC. Therefore, a complete block in all three postulated routes resulted in the HCR phenotype known from photorespiratory plant mutants (Eisenhut et al. 2008a). It should be mentioned, however, that the triple mutant was unstable, but the HCR phenotype was nevertheless reproducibly observed with freshly generated triple mutant clones.

However, the HCR phenotype of the triple mutant did not correspond with the finding that the mutant Δ*glcD1* could grow at LC, albeit with lower rates

(Eisenhut et al. 2006). This indicated the presence of an additional bypass, because, in the absence of such a bypass, the mutation in Δ*glcD1* should block all three branching routes for 2-PG metabolism and should result in the phenotype of the triple mutant. Searches for another candidate protein able to convert glycolate into glyoxylate indeed pointed to the presence of a second glycolate dehydrogenase in *Synechocystis*, GlcD2 (Table 6.1). The completely segregated double mutant Δ*glcD1*/Δ*glcD2*, now fully defective in oxidation of glycolate to glyoxylate, showed the expected HCR phenotype (Fig. 6.3), which further supported the conclusion that an active 2-PG metabolism is essential for *Synechocystis* and possibly other cyanobacteria as well (Eisenhut et al. 2008a). The HCR phenotype was previously only found with cyanobacterial mutants defective in CCM. In order to rule out that secondary mutations in our mutants affected CCM, we compared the C_i affinity of the double mutant Δ*glcD1*/Δ*glcD2* with WT cells. These experiments verified that the mutant showed almost the same C_i affinity as WT cells (Eisenhut et al. 2008a), revealing that the absence of an active photorespiratory 2-PG metabolism resulted in a HCR phenotype despite an intact CCM in *Synechocystis*.

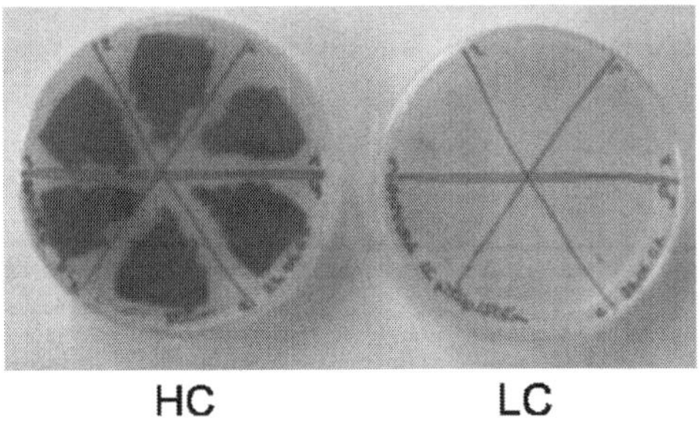

Fig. 6.3 Growth of several clones of the double mutant Δ*glcD1*/Δ*glcD2* of *Synechocystis* sp. strain PCC 6803 on plates incubated under CO_2-enriched air (5% CO_2, high carbon dioxide – HC) or ambient air (0.035% CO_2, low carbon dioxide – LC)

6.2.4 Function of Photorespiratory 2-PG Metabolism in Cyanobacteria

The function of photorespiratory 2-PG metabolism for higher plants is still a matter of discussion. Its main function is related to scavenging of organic carbon. Here, two molecules of 2-PG are converted to one molecule of glycerate 3-phosphate, which is necessary to refill the Calvin–Benson cycle. Moreover, the cycle detoxifies critical intermediates such as 2-PG and synthesizes valuable metabolites for biosynthetic processes, e.g., activated C1 units. Last but not least, it has been hypothesized to play an important role during stress acclimation, since high amounts of energy and reducing equivalents are used during the re-fixation of released CO_2 and NH_3. This helps to regenerate acceptors of the photosynthetic electron chain (Kozaki and Takeba 1996; Wingler et al. 2000; Bauwe and Kolukisaoglu 2003), thus preventing acceptor limitation and subsequent photoinhibition. Our studies on cyanobacterial 2-PG metabolism clearly suggest an essential function for this pathway in cyanobacteria. Most probably, the cell lysis observed with double- and triple mutants under LC is based on the accumulation of intermediates, which become toxic above a certain threshold concentration. This has been shown for the accumulation of glycine, e.g., in the mutant Δ*gcvT*. Glycine toxicity was caused by a depletion of Mg^{2+} inside *Synechocystis* cells due to the complexation of Mg^{2+} by glycine. Correspondingly, the toxic glycine effects could be compensated for by supplementation of the corresponding mutants with Mg^{2+} (Eisenhut et al. 2007). The regeneration of Calvin–Benson intermediates seems to be less important in the cyanobacterium *Synechocystis*, since the double mutant Δ*gcvT*/Δ*tsr* was still able to grow at LC, albeit with slower rates (Eisenhut et al. 2006). In this mutant, only the decarboxylation branch (see Fig. 6.2) remained active leading to the complete conversion of glyoxylate to CO_2 without regeneration of Calvin–Benson cycle intermediates.

In higher plants, photorespiratory metabolism also contributes to high light acclimation and diminishes photoinhibition (Kozaki and Takeba 1996; Takahashi et al. 2007). Cyanobacteria have evolved many strategies to cope with high light conditions. Over-reduction of the electron chain is initially avoided by the dissipation of excess light energy absorbed by chlorophyll, mainly via carotenoids, and other non-photochemical quenching mechanisms (Havaux et al. 2005; Kirilovsky 2007). In addition, a substantial part of electrons can be transferred from photosystem I to molecular oxygen, which results in photoreduction of O_2 via superoxide anion to H_2O_2 in plant chloroplasts, i.e., the Mehler reaction (Mehler 1951; Asada 1999). For the model cyanobacterium *Synechocystis*, it was shown that O_2 is reduced directly to water in a single reaction mediated by flavodiiron proteins (Vicente et al. 2002). Thereby, the photoreduction of O_2 acts as electron sink under stress conditions, e.g., high light or low C_i, which helps to prevent photodamage to photosystem II and is regarded as an important protection system in all photosynthetic organisms (Asada 1999; Badger et al. 2000; Helman et al. 2003). In order to evaluate the importance of photorespiratory 2-PG metabolism for high light acclimation of *Synechocystis*, we generated combined mutations in *gcvT* (T-protein subunit of glycine decarboxylase blocking the main branch) and genes for the flavodiiron proteins Flv1 (Sll1521) and Flv3 (Sll0550), which are apparently involved in light-dependent O_2 reduction activity (Helman et al. 2003). The analyses of the mutant genotypes revealed that a combination of mutations in *gcvT* and *flv3* resulted in non-segregated mutants, while the single mutants and a double mutant Δ*gcvT*/Δ*flv*1 were completely segregated (Hackenberg et al. 2009). This finding by itself indicates that photorespiratory 2-PG metabolism and photoreduction of O_2 performed by Flv3 somehow cooperate in an essential function. Subsequently, single- and double mutants were characterized regarding their photosynthetic parameters and high light acclimation. These experiments revealed changed performance of photosynthesis and growth at high and changing light conditions for the single mutant Δ*flv3* and to a significantly greater extent for the non-segregated double mutant Δ*gcvT*/Δ*flv3* (Hackenberg et al. 2009). In summary, these results clearly indicate that, similar to plants, photorespiratory

2-PG metabolism supports the resistance of cyanobacteria to over-reducing conditions such as high light or low C_i despite the presence of additional protection mechanisms.

Another possible function of 2-PG metabolism may be related to signaling of LC. The transfer of cyanobacterial cells from HC into LC induces a complex acclimation process, which includes mainly up-regulation of CCM activity. DNA-microarray experiments reveal that several hundred genes are differentially regulated to achieve this goal. Some transcriptional factors, such as CmpR or NdhR, are known to bind to promoters, e.g., that of the *cmp* gene operon. Their inactivation consequently abolished the up-regulation of the corresponding genes for C_i transporters (Wang et al. 2004). Recently, another transcriptional factor, named AbrB (Sll0822), was identified, which seems to represent a master repressor for induction of CCM at LC. A completely segregated Δ*abrB* mutant of *Synechocystis* expresses LC-inducible genes at high levels even under HC and, correspondingly, Δ*abrB* cells display an enhanced C_i affinity at HC, which is normally characteristic of LC-grown WT cells (Lieman-Hurwitz et al. 2009). It is still unknown by which mechanism these transcriptional factors are regulated to either repress or activate C_i-regulated genes. One attractive hypothesis includes intermediates of the photorespiratory 2-PG metabolism as potential metabolic signals to sense the internal C_i status of a cyanobacterial cell, since shifts to LC should be accompanied by at least a transient accumulation of corresponding intermediates (Kaplan and Reinhold 1999).

This hypothesis was recently supported by the finding that the DNA-binding capacity of the transcriptional factor CmpR is influenced by the 2-PG concentration (Nishimura et al. 2008). Our attempts to identify the pathway of 2-PG metabolism in *Synechocystis* also provide some indirect support for this hypothesis. Cells of the Δ*glcD1* mutant, which accumulate glycolate even under HC (Eisenhut et al. 2006), contained similarly high amounts of carboxysome at HC as found in WT cells only after a shift to LC (Eisenhut et al. 2007). Moreover, our recent metabolomic analyses of *Synechocystis* WT and mutant cells revealed marked changes in the metabolite pattern, when we compared HC- and LC-grown cells. Again, in cells of the Δ*glcD1* mutant some of the metabolic changes characteristic for LC-treated WT cells were already observed under HC (Eisenhut et al. 2008b). Therefore, the existence of an active photorespiratory 2-PG metabolism among cyanobacteria not only is essential for viability at LC but could be also involved directly or indirectly in their C_i sensing mechanism.

6.2.5 Evolution of Photorespiratory 2-PG Metabolism

A complete block of 2-PG metabolism results in a HCR phenotype not only in C3 plants but also in cyanobacteria. Recently, it was shown that 2-PG metabolism is also necessary for maize, a C4 plant (Zelitch et al. 2009). From all these results, it appears that 2-PG metabolism is essential for all organisms performing oxygenic photosynthesis in an oxygen-containing atmosphere. Hence, it is likely that an ancient 2-PG metabolism evolved in parallel with the invention of oxygenic photosynthesis. Even in the ancient, oxygen-free atmosphere, increasing amounts of molecular oxygen might have accumulated inside the cyanobacterial cell with consequent stimulation of the oxygenase reaction of RubisCO, which, at that time, was not yet sequestered inside the carboxysome. Moreover, the cyanobacterial mode of 2-PG metabolism includes one route (see Fig. 6.2) that is very similar to that occurring in higher plants of the C3 type. This suggests that ancient cyanobacteria possessed an early photorespiratory 2-PG metabolism which was conveyed to plants by endosymbiosis (Eisenhut et al. 2008a). This hypothesis is supported by phylogenetic comparisons of enzymes involved in cyanobacterial and plant 2-PG metabolism. In many cases, these comparisons reveal a close clustering of cyanobacterial and plant proteins, e.g., for glycerate kinase (Fig. 6.4), which implies a close phylogenetic relation. Such systematic comparisons predict that the plant enzymes glycolate oxidase, P- and T-proteins of glycine decarboxylase, glycerate kinase, and possibly hydroxypyruvate reductase as well as serine hydroxymethyltransferase evolved from cyanobacterial proteins acquired during endosymbiosis (Table 6.1, Eisenhut et al. 2008a).

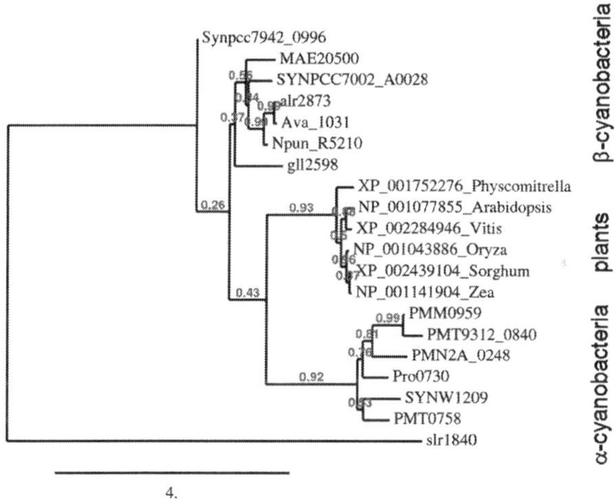

Fig. 6.4 Phylogenetic un-rooted maximum likelihood (ML) tree of glycerate kinases from cyanobacteria and plants. The plant glycerate kinase cluster is flanked with glycerate kinases of α- and β-cyanobacteria. The bacterial-type glycerate kinase from *Synechocystis* sp. PCC 6803 (Slr1840) served as out-group. For the analysis, amino acid sequences obtained from CyanoBase or NCBI were aligned by ClustalX. ML phylogenetic trees were inferred using phyml (Guindon and Gascuel 2003), with a Whelan and Goldman (WAG) evolutionary model. Numbers at nodes indicate bootstrap values for 1,000 replicates of the original data set

Interestingly, the two additional pathways for 2-PG breakdown, the bacterial-type glycerate pathway and the decarboxylation branch, seem to be missing in plants. The glycerate pathway is encoded in all cyanobacterial genomes and hence probably evolved early in cyanobacterial history. It can be speculated that this branch was also present in the primary endosymbiont, but was lost during the evolution of higher plants, or is now used for other purposes. It would be interesting to investigate whether or not the re-introduction of the cyanobacterial glycerate pathway into higher plants would have an effect on their photorespiratory 2-PG metabolism. In contrast, the decarboxylation branch is also missing in most of the more recent cyanobacterial genomes; probably it was acquired later during the evolution by only certain cyanobacterial strains such as *Synechocystis*.

6.3 Conclusions

We have shown that cyanobacteria perform an active photorespiratory metabolism despite the activity of the CCM. 2-PG metabolism uses three different routes in *Synechocystis*, while in the majority of cyanobacteria only the plant-like cycle and the glycerate pathway seem to exist. A complete block of 2-PG metabolism resulted in an HCR phenotype, which shows that this metabolism is essential for all organisms performing oxygenic photosynthesis in the present-day atmosphere. The main function of 2-PG metabolism in cyanobacteria is directed to the detoxification of critical intermediates. However, it is also involved in acclimation to high light conditions. These findings gave rise to the hypothesis that an early photorespiratory 2-PG metabolism was already present in ancient cyanobacteria and was conveyed endosymbiotically to plants.

Acknowledgments The authors would like to thank Dr. Aaron Kaplan, Hebrew University, Jerusalem, Israel, and Dr. Hans C. P. Matthijs, University of Amsterdam, Amsterdam, The Netherlands, for fruitful cooperation during the work on cyanobacterial photorespiration. This work was supported by grants from the DFG (Deutsche Forschungsgemeinschaft).

References

Altschul SF, Madden TL, Schäffer AA et al (1997) Gapped BLAST and PSI-BLAST: a new generation of protein database search programs. Nucleic Acids Res 25:3389–3402

Asada K (1999) The water-water cycle in chloroplasts: Scavenging of active oxygen and dissipation of excess photons. Annu Rev Plant Physiol Plant Mol Biol 50:601–639

Badger MR (1980) Kinetic properties of ribulose 1,5-bisphosphate carboxylase/oxygenase from *Anabaena variabilis*. Arch Biochem Biophys 201:247–255

Badger MR, von Caemmerer S, Ruuska S et al (2000) Electron flow in higher plants and algae: rates and control of direct photoreduction (Mehler reaction) and RubisCO oxygenase. Phil Trans R Soc Lond 355:1433–1446

Badger MR, Price GD, Long BM et al (2006) The environmental plasticity and ecological genomics of the cyanobacterial CO_2 concentrating mechanism. J Exp Bot 57:249–265

Bartsch O, Hagemann M, Bauwe H (2008) Only plant-type (GLYK) glycerate kinases produce D-glycerate 3-phosphate. FEBS Lett 582:3025–3028

Bauwe H, Kolukisaoglu U (2003) Genetic manipulation of glycine decarboxylation.J Exp Bot 54:1523–1535

Boldt R, Edner C, Kolukisaoglu U et al (2005) D-GLYCERATE 3-KINASE, the last unknown enzyme in the photorespiratory cycle in Arabidopsis, belongs to a novel kinase family. Plant Cell 17:2413–2420

Codd GA, Stewart WDP (1973) Pathways of glycollate metabolism in the blue-green alga *Anabaena cylindrica*. Arch Mikrobiol 94:11–28

Colman B (1989) Photosynthetic carbon assimilation and the suppression of photorespiration in the cyanobacteria. Aquatic Bot 34:211–231

Cot SS, So AK, Espie GS (2008) A multiprotein bicarbonate dehydration complex essential to carboxysome function in cyanobacteria. J Bacteriol 190:936–945
Deusch O, Landan G, Roettger M et al (2008) Genes of cyanobacterial origin in plant nuclear genomes point to a heterocyst-forming plastid ancestor. Mol Biol Evol 25:748–761
Eisenhut M, Kahlon S, Hasse D et al (2006) The plant-like C2 glycolate cycle and the bacterial-like glycerate pathway cooperate in phosphoglycolate metabolism in cyanobacteria. Plant Physiol 142:333–342
Eisenhut M, von Wobeser EA, Jonas L et al (2007) Long-term response towards inorganic carbon limitation in wild type and glycolate turnover mutants of the cyanobacterium *Synechocystis* sp. strain PCC 6803. Plant Physiol 144:1946–1959
Eisenhut M, Bauwe H, Hagemann M (2007) Glycine accumulation is toxic for the cyanobacterium *Synechocystis* sp. strain PCC 6803, but can be compensated by supplementation with magnesium ions. FEMS Microbiol Lett 277:232–237
Eisenhut M, Ruth W, Haimovich M et al (2008a) The photorespiratory glycolate metabolism is essential for cyanobacteria and might have been conveyed endosymbiontically to plants. Proc Natl Acad Sci USA 105:17199–17204
Eisenhut M, Huege J, Schwarz D et al (2008b) Metabolome phenotyping of inorganic carbon limitation in cells of the wild type and photorespiratory mutants of the cyanobacterium *Synechocystis* sp. strain PCC 6803. Plant Physiol 148:2109–2120
Engels A, Pistorius EK (1997) Characterization of a gene encoding dihydrolipoamide dehydrogenase of the cyanobacterium *Synechocystis* sp. PCC 6803. Microbiol 143:3543–3553
Giordano M, Beardall J, Raven JA (2005) CO_2 concentrating mechanisms in algae: mechanisms, environmental modulation, and evolution. Annu Rev Plant Biol 56:99–131
Guindon S, Gascuel O (2003) A simple, fast, and accurate algorithm to estimate large phylogenies by maximum likelihood. Syst Biol 52:696–704
Hackenberg C, Engelhardt A, Matthijs HCP et al (2009) Photorespiratory 2-phosphoglycolate metabolism and photoreduction of O_2 cooperate in high-light acclimation of *Synechocystis* sp. strain PCC 6803. Planta, epub ahead of press, doi 10.1007/s00425-009-0972-9
Hagemann M, Vinnemeier J, Oberpichler I et al (2005) The glycine decarboxylase complex is not essential for the cyanobacterium *Synechocystis* sp. strain PCC 6803. Plant Biol 7:15–22
Hasse D, Mikkat S, Thrun HA et al (2007) Properties of recombinant glycine decarboxylase P- and H-protein subunits from the cyanobacterium *Synechocystis* sp. strain PCC 6803. FEBS Lett 581:1297–1301
Havaux M, Guedeney G, Hagemann M et al (2005) The chlorophyll-binding protein IsiA is inducible by high light and protects the cyanobacterium *Synechocystis* PCC 6803 from photooxidative stress. FEBS Lett 579:2289–2293
Helman Y, Tchernov D, Reinhold L et al (2003) Genes encoding A-type flavoproteins are essential for photoreduction of O_2 in cyanobacteria. Curr Biol 13:230–235
Kaneko T, Sato S, Kotani H et al (1996) Sequence analysis of the genome of the unicellular cyanobacterium *Synechocystis* sp. strain PCC6803. II. Sequence determination of the entire genome and assignment of potential protein-coding regions. DNA Res 3:109–136
Kaplan A, Berry JA (1981) Glycolate excretion and the oxygen to carbon dioxide net exchange ratio during photosynthesis in *Chlamydomonas reinhardtii*. Plant Physiol 67:229–232
Kaplan A, Reinhold L (1999) CO_2 concentrating mechanisms in photosynthetic microorganisms. Annu Rev Plant Physiol Plant Mol Biol 50:539–570
Kerfeld CA, Sawaya MR, Tanaka S et al (2005) Protein structures forming the shell of primitive bacterial organelles. Science 309:936–938
Kirilovsky D (2007) Photoprotection in cyanobacteria: The orange carotenoid protein (OCP)-related non-photochemical-quenching mechanism. Photosynth Res 93:7–16
Kouhen OM, Joset F (2002) Biosynthesis of the branched-chain amino acids in the cyanobacterium *Synechocystis* PCC6803: existence of compensatory pathways. Curr Microbiol 45:94–98
Kozaki A, Takeba G (1996) Photorespiration protects C3 plants from photooxidation. Nature 384:557–560

Kroth PG, Chiovitti A, Gruber A et al (2008) A model for carbohydrate metabolism in the diatom *Phaeodactylum tricornutum* deduced from comparative whole genome analysis. PLoS One 3:e1426
Lieman-Hurwitz J, Haimovich M, Shalev-Malul G et al (2009) A cyanobacterial AbrB-like protein affects the apparent photosynthetic affinity for CO_2 by modulating low-CO_2-induced gene expression. Environ Microbiol 11:927–936
Margulis L (1970) Origin of eukaryotic cells. Yale University Press, New Haven.
Mehler AH (1951) Studies on reactions of illuminated chloroplasts. I. Mechanisms of the reduction of oxygen and other Hill reagents. Arch Biochem Biophys 33:65–77
Mereschkowsky C (1905) Über Natur und Ursprung der Chromatophoren im Pflanzenreiche. Biol Centralbl 25:593–604
Moroney JV, Wilson BJ, Tolbert NE (1986) Glycolate metabolism and excretion by *Chlamydomonas reinhardtii*. Plant Physiol 82:821–826
Nishimura T, Takahashi Y, Yamaguchi O et al (2008) Mechanism of low CO_2-induced activation of the *cmp* bicarbonate transporter operon by a LysR family protein in the cyanobacterium *Synechococcus elongatus* strain PCC 7942. Mol Microbiol 68:98–109
Norman EG, Colman B (1988) Evidence for an incomplete glycolate pathway in cyanobacteria. J Plant Physiol 132:766–768
Ogren WL (1984) Photorespiration – Pathways, regulation, and modification. Annu Rev Plant Physiol 35:415–442
Renström E, Bergman B (1989) Glycolate metabolism in cyanobacteria. I. Glycolate excretion and phosphoglycolate phosphatase activity. Physiol Plant 75:137–143
Rippka R, Deruelles J, Waterbury JB et al (1979) Generic assignments, strain histories and properties of pure cultures of cyanobacteria. J Gen Microbiol 111:1–61
Scanlan DJ, Ostrowski M, Mazard S et al (2009) Ecological genomics of marine picocyanobacteria. Microbiol Mol Biol Rev 73:249–299
Schwarte S, Bauwe H (2007) Identification of the photorespiratory 2-phosphoglycolate phosphatase, PGLP1, in Arabidopsis. Plant Physiol 144:1580–1586
Takahashi S, Bauwe H, Badger M (2007) Impairment of the photorespiratory pathway accelerates photoinhibition of photosystem II by suppression of repair but not acceleration of damage processes in *Arabidopsis*. Plant Physiol 144:487–494
Tolbert NE (1997) The C-2 oxidative photosynthetic carbon cycle. Annu Rev Plant Physiol Plant Mol Biol 48:1–25
Tottey S, Waldron KJ, Firbank SJ et al (2008) Protein-folding location can regulate manganese-binding versus copper- or zinc-binding. Nature 455:1138–1142
Vicente JB, Gomes CM, Wasserfallen A et al (2002) Module fusion in an A-type flavoprotein from the cyanobacterium *Synechocystis* condenses a multiple-component pathway in a single polypeptide chain. Biochem Biophys Res Comm 294:82–87
Wang HL, Postier BL, Burnap RL (2004) Alterations in global patterns of gene expression in *Synechocystis* sp. PCC 6803 in response to inorganic carbon limitation and the inactivation of *ndhR*, a LysR family regulator. J Biol Chem 279:5739–5751
Wingler A, Lea PJ, Quick WP et al (2000) Photorespiration: metabolic pathways and their role in stress protection. Philos Trans R Soc Lond B Biol Sci 355:1517–1529
Xu M, Bernát G, Singh A et al (2008) Properties of mutants of *Synechocystis* sp. strain PCC 6803 lacking inorganic carbon sequestration systems. Plant Cell Physiol 49:1672–1677
Yeates TO, Kerfeld CA, Heinhorst S et al (2008) Protein-based organelles in bacteria: carboxysomes and related microcompartments.Nat Rev Microbiol 6:681–691
Zelitch I, Schultes NP, Peterson RB et al (2009) High glycolate oxidase activity is required for survival of maize in normal air. Plant Physiol 149:195–204
Zhang P, Battchikova N, Jansen T et al (2004) Expression and functional roles of the two distinct NDH-1 complexes and the carbon acquisition complex NdhD3/NdhF3/CupA/Sll1735 in *Synechocystis* sp PCC 6803. Plant Cell 16:3326–3340

Chapter 7
Beyond the Genome: Functional Studies of Phototrophic Sulfur Oxidation

Thomas E. Hanson, Rachael M. Morgan-Kiss, Leong-Keat Chan, and Jennifer Hiras

Abstract The increasing availability of complete genomic sequences for cultured phototrophic bacteria and assembled metagenomes from environments dominated by phototrophs has reinforced the need for a "post-genomic" analytical effort to test models of cellular structure and function proposed from genomic data. Comparative genomics has produced a testable model for pathways of sulfur compound oxidation in the phototrophic bacteria. In the case of sulfide, two enzymes are predicted to oxidize sulfide: sulfide:quinone oxidoreductase and flavocytochrome *c* sulfide dehydrogenase. However, these models do not predict which enzyme is important under what conditions. In *Chlorobaculum tepidum*, a model green sulfur bacterium, a combination of genetics and physiological analysis of mutant strains has led to the realization that this organism contains at least two active sulfide:quinone oxidoreductases and that there is significant interaction between sulfide oxidation and light harvesting. In the case of elemental sulfur, an organothiol intermediate of unknown structure has been proposed to activate elemental sulfur for transport into the cytoplasm where it can be oxidized or assimilated, and recent approaches using classical metabolite analysis have begun to shed light on this issue both in *C. tepidum* and the purple sulfur bacterium *Allochromatium vinosum*.

7.1 Introduction

The cycling of inorganic sulfur compounds in the environment is driven to a significant extent by microbial action (Overmann et al. 1996; Canfield and Raiswell, 1999; Turchyn and Schrag 2006). Biologically mediated oxidation of sulfide and intermediate oxidation state sulfur compounds ($S_2O_3^{2-}$ and S^0) provides energy and reducing power for chemo- and photolithotrophic microorganisms (Friedrich

T.E. Hanson (✉)
College of Earth, Ocean and Environment and Delaware Biotechnology Institute and Department of Biological Sciences, University of Delaware, Newark, DE 19711, USA
e-mail: tehanson@udel.edu

P.C. Hallenbeck (ed.), *Recent Advances in Phototrophic Prokaryotes*, Advances in Experimental Medicine and Biology 675, DOI 10.1007/978-1-4419-1528-3_7,

et al. 2005). In some ecosystems, sulfide oxidation supports the bulk of primary production forming the base of food webs in sulfidic caves (Macalady et al. 2008), hot springs (Elshahed et al. 2007), marine hydrothermal vents (Goffredi et al. 2004), and marine cold seeps (Orphan et al. 2004; Omoregie et al. 2008). Sulfide oxidation also forms the basis for symbiotic associations, most notably the association between the Vestimentiferan tubeworm *Riftia pachyptila* with its sulfide-oxidizing gammaproteobacterial community found in the trophosome, a specialized organ (Arndt et al. 2001). Many other thiotrophic symbioses between eukaryotes, including metazoans, and prokaryotes are known (Bernhard et al. 2000, 2003, 2006; Arndt et al. 2001; Goffredi et al. 2004) as well as strictly microbial symbioses between sulfide-oxidizing non-motile phototrophic bacteria and motile heterotrophic bacteria (Overmann, 2006; Pfannes et al. 2007). In these ecosystems and symbioses, sulfide oxidation provides the reductant and energy for carbon dioxide fixation in the sulfide oxidizer that is then either shared with a symbiotic partner or consumed in higher trophic levels of a food web. Microbial sulfate reduction produces sulfide, and tightly coupled closed sulfur cycles where sulfur reduction and oxidation co-occur have been observed in marine sediments (Madrid et al. 2006), lakes and anoxic ocean basins (Overmann et al. 1996; Holmer and Storkholm, 2001), microbial mats (Baumgartner et al. 2006), and biofilms (Celis-Garcia et al. 2008).

This chapter will detail recent advances in functional (i.e., genetic, biochemical, and metabolite profiling) studies of sulfur oxidation in phototrophic bacteria that are refining metabolic models based exclusively on genome data.

7.2 Sulfide and Microbial Sulfur Cycling

Sulfide ($H_2S_{(g)}$, $HS^{-}_{(aq)}$) is the most reduced form of the element sulfur and is a strong reductant whose oxidation proceeds in the environment by both chemical and biological pathways (Preisler et al. 2007). Sulfide metabolism is thought to be an ancient process that has been maintained in a wide range of lineages throughout evolution (Martin et al. 2003; Theissen et al. 2003). Sulfide has recently been recognized as a gaseous signaling molecule in humans (Lefer 2007), acting as a vasorelaxant that can prevent ischemic reperfusion injury to myocardial cells (Elrod et al. 2007; Yang et al. 2008), and has been characterized as the "first inorganic substrate" for human cells (Goubern et al. 2007). Mitochondrial sulfide:quinone oxidoreductase (SQR) activity is a major pathway for the elimination of sulfide as a signal/substrate and also helps to prevent toxicity from sulfide generated by the metabolism of the human gut microbiome (Goubern et al. 2007). Microbial sulfide oxidation in part controls sulfide fluxes across anaerobic to aerobic transitions in diverse ecosystems including freshwater (Kamp et al. 2006), groundwater (Macalady et al. 2008), estuaries (Weston et al. 2006), coastal ecosystems and sediments (Bruchert et al. 2003), deep ocean sediments (Wang et al. 2008), sewers (Yang et al. 2005), and engineered systems that remove sulfide from industrial supply feeds and effluents (Arnirfakhri et al. 2006; Janssen et al. 2009).

7.2.1 Enzymatic Routes of Sulfide Oxidation

In sulfide-oxidizing organisms including the phototrophic sulfur bacteria, one of the two enzymes catalyzes the oxidation: sulfide:quinone oxidoreductase (SQR, E.C. 1.8.5.) (Shahak et al. 1999; Theissen et al. 2003) or flavocytochrome *c* (FCC, also known as flavocytochrome *c* sulfide dehydrogenase) (Chen et al. 1994; Reinartz et al. 1998). Many phototrophic bacteria contain genes that encode both enzymes, and the most recent models of sulfur oxidation in both the green sulfur and purple sulfur bacteria indicate that these enzymes are alternate routes that result in the production of either polysulfide (green sulfur) or protein-encapsulated elemental sulfur globules (purple sulfur) in the periplasm (Dahl, 2008; Frigaard and Bryant, 2008a, 2008b).

Electrons liberated from sulfide are passed into the electron transport chain at different levels depending on whether SQR or FCC catalyzes the oxidation. SQR donates electrons from sulfide to the electron transport chain at the level of the quinone pool, upstream of the cytochrome b/c_1 complex (menaquinol:cytochrome *c* oxidoreductase), while FCC donates electrons at the level of cytochrome *c* downstream of the b/c_1 complex (Oh-oka and Blankenship, 2004). Theoretically, the energy yield should be greater for organisms utilizing SQR compared to those utilizing FCC, as proton motive force is generated when electrons are passed through the b/c_1 complex en route either to the reaction center in phototrophic bacteria (Oh-oka and Blankenship, 2004) or to terminal electron-accepting complexes that reduce oxygen (Celis-Garcia et al. 2008; Shahak, 2008), nitrate (Zopfi et al. 2008), or arsenate (Hollibaugh et al. 2006) in chemotrophic sulfide oxidizers. Furthermore, while reverse electron transport to NAD^+ will be required regardless of whether SQR or FCC is utilized (Klamt et al. 2008), the difference in redox potential between quinol (+40 mV relative to the standard hydrogen electrode) and NAD^+ (–320 mV) is significantly smaller than between reduced cytochrome *c* (+240 mV) and NAD^+. Thus, about 19 kJ less energy per mole of electrons should be required for NADH generation via reverse electron transport in an organism utilizing SQR relative to an organism dependent on FCC.

Beyond thermodynamics, additional evidence suggests that SQR is more important than FCC for chemo- and phototrophic sulfide oxidation in bacteria. First, sequence analyses indicate that homologs of FCC seem to be confined to autotrophs that can utilize thiosulfate in addition to sulfide as an energy source (Theissen et al. 2003; Frigaard and Bryant, 2008b). The oxidation of sulfur/polysulfides produced by sulfide oxidation in purple sulfur phototrophic bacteria (PSB) requires the action of gene products encoded by the dissimilatory sulfite oxidoreductase (Dsr) gene cluster (Dahl et al. 1999, 2005; Chan et al. 2008). In the Chlorobiaceae, the Dsr system is also likely involved in elemental sulfur oxidation, but this has not been experimentally demonstrated as yet (Frigaard and Bryant, 2008a, 2008b).

In contrast, homologs of SQR are widely distributed in bacteria and similar proteins have been identified in the archaea (Theissen et al. 2003; Frigaard and Bryant, 2008b) and eukaryotes (Shahak, 2008). Second, mutation of the *fccAB* genes in the PSB *A. vinosum* did not inhibit its ability to oxidize sulfide and grow photolithoautotrophically, although the specific growth rates and biomass yields were not reported (Reinartz et al. 1998). *A. vinosum* also contains SQR and the Dsr and Sox systems. Third, in the chemolithotrophic sulfur oxidizer *Acidithiobacillus ferrooxidans* NASF-1, *sqr* transcripts were 3-fold more abundant in sulfide- than iron-grown cells (Wakai et al. 2004). Finally, sulfide oxidation activities directly linked to energy production or detoxification have been demonstrated with the purified SQR proteins from the proteobacterium *Rhodobacter capsulatus* (Schutz et al. 1997) and the cyanobacteria *Oscillatoria limnetica* and *Aphanothece halophytica* (Arieli et al. 1994; Bronstein et al. 2000). *R. capsulatus* contains neither Dsr nor Sox sulfur oxidation systems and while genomic information is not available for *O. limnetica* and *A. halophytica*, other cyanobacterial genome sequences (Kaneko and Tabata, 1997; Meeks et al. 2001; Nakamura et al. 2002, 2003) do not contain the Dsr or Sox systems.

7.2.2 SQR in the Chlorobiaceae (Green Sulfur Bacteria)

While the function of SQR in sulfide oxidation has been demonstrated in the facultative anoxygenic cyanobacterium *O. limnetica* (Arieli et al. 1994; Bronstein et al. 2000) and the purple nonsulfur bacterium *R. capsulatus* (Schutz et al. 1997), in the Chlorobiaceae SQR function has only been examined to a limited extent in *Chlorobium limicola* (Shahak et al. 1992). The Chlorobiaceae are phototrophs that are commonly found in freshwaters and estuaries where reduced sulfur compounds and light coexist. With the exception of *Chlorobium ferrooxidans*, all characterized GSB, including *C. tepidum*, can utilize sulfide as an electron donor to support phototrophic growth. Biochemically, it was shown that the membrane of *C. limicola* forma *thiosulfatophilum* catalyzes electron transfer from sulfide to plastoquinone in the dark (Shahak et al. 1992). All sequenced GSB strains contain orthologs for FCC, SQR, or both; some strains like *C. tepidum* can have more than one gene encoding distinct forms of SQR and FCC in their genomes, complicating functional prediction from sequence alone. SQR-encoding genes can even be found in organisms like *C. ferrooxidans* that cannot grow on sulfide, where it may confer resistance to sulfidic environments (Heising et al. 1999). The genome of *C. ferroxidans* lacks many other proposed sulfur oxidation genes (such as those encoding the Dsr system), which correlates with its ability to oxidize Fe^{2+} and H_2 but not sulfur.

Three SQR homologs, CT0117, CT0876, and CT1087, have been identified in the genome of *C. tepidum* (Fig. 7.2) (Chan et al. 2009). As with many other homologs within and between bacterial strains, amino acid sequences of *C. tepidum*'s three SQRs are highly divergent, with pairwise sequence identity averaging in the 20% range. However, the N termini of proteins CT0117, CT0876 and CT1087 are highly homologous to other SQRs found in eubacteria. On the basis of sequence comparisons, CT0117 has been proposed to be the bona fide SQR in *C. tepidum* while CT0876 and CT1087 have been labeled SQR-like proteins (SQRLPs) (Frigaard and Bryant, 2008b). The role of these three homologs has now been examined in more detail by our group.

Both mutation of the CT1087 gene in *C. tepidum* and its expression in *E. coli* suggest that its protein product is indeed an active SQR in vitro and in vivo, while the case of CT0876 is not yet resolved (Chan et al. 2009). While strains lacking CT0117, CT1087, or both exhibit clear defects in sulfide-dependent growth at high sulfide concentrations, sulfide still appears to be oxidized by these strains, including in the double mutant (data not shown). However, it is currently unclear whether or not CT0876 is required for this oxidation activity, or whether this can be attributed to FCC. Furthermore, the growth properties of the existing strains have only been characterized under a limited set of permissive conditions. To fully understand the functions of the multiple SQR homologs, mutant growth and sulfide oxidation properties must be examined under a full range of light intensities and sulfide concentrations.

SQR assays conducted in the single mutant *C. tepidum* strains lacking either CT0117 or CT1087 suggest that there may be kinetic differences between these two SQR homologs (Table 7.1). Strain CT1087::TnOGm, which contains CT0117 and CT0876, displays an extremely low apparent K_m and low V_{max}. Strain CT0117::TnOGm, which contains CT1087 and CT0876, displays a dramatic increase in K_m and V_{max}. Results obtained with strain CT0876::TnOGm are somewhat inconsistent. CT0876 is not active judged from the lack of detectable SQR activity in strain CT0117::TnOGm, CT1087::TnEm and the fact that SQR specific activity is statistically indistinguishable in strain CT0876::TnOGm from the wild type at the two sulfide concentrations shown. However, the kinetic parameters are quite different from the wild type. This reflects the preliminary nature of the kinetic data and the fact that they were determined in membrane fractions rather than using purified proteins and therefore were not controlled to contain equal concentrations of SQR-active sites in the assays. Current efforts are focused on illuminating the kinetic differences between SQRs hinted at here and to understand the mechanism of sulfide oxidation in *C. tepidum* at various levels of organization from purified enzymes to intact cells by a combination of microbial genetics, biochemistry, and analytical electrochemistry.

Table 7.1 SQR activity in *C. tepidum* wild-type and mutant strains

	Specific activity (nmol dUQ_{red} mg protein^{-1} min^{-1})			
Strain	1.0 mM Na_2S	0.5 mM Na_2S	$^{**}K_m$(mM)	$^{**}V_{max}$
Wild type	87.0 ± 6.9	63.0 ± 4.5	0.2	102
CT0117::TnOGm	31.5 ± 4.5*	7.5 ± 2.6*	5.0	200
CT0876::TnOGm	82.5 ± 2.6	60.0 ± 6.9	1.4	238
CT1087::TnOGm	52.5 ± 5.2*	45.0 ± 9.0*	0.1	75
CT0117::TnOGm	N.D.[a]	N.D.[a]	N.D.[a]	N.D.[a]
CT1087::TnEm	N.D.[a]	N.D.[a]	N.D.[a]	N.D.[a]

$^{*}P < 0.05$ ($n \geq 3$) compared to the wild type assayed under the same conditions **Apparent K_m and V_{max} were estimated from the SQR activity obtained in the membrane fraction of each strain, not from purified protein [a]No SQR activity could be detected

7.2.3 Interactions Between Sulfide Oxidation and Light Harvesting

C. tepidum was originally characterized as having an upper growth limit of 4 mM for sulfide. However, with improved culturing practices, *C. tepidum* growth yield in batch cultures increases linearly at a rate of 9 g protein/mol sulfide in the range of 2–8 mM sulfide, but *C. tepidum* does not grow well with 10 mM sulfide as the sole electron donor (Chan et al. 2009). This wide range of sulfide tolerance coupled with the already observed regulation of the CT1087 SQR suggested that sulfide may be an important regulator of cellular processes in *C. tepidum* and this was investigated by characterizing the properties of cultures grown at various sulfide concentrations. This led to the observation of at least two global responses to sulfide concentration.

First, the color of the cultures is notably altered when they are grown at different sulfide concentrations, and this corresponds to a shift in the Bchl *c* Qy peak observed in intact cells (Fig. 7.1a), which is determined by the alkylation state of Bchl *c* in the chlorosome. Works by the PI's group (Morgan-Kiss et al. 2009) and others (Borrego and Garcia-Gil, 1994; Saga et al. 2005; Chew et al. 2007) have demonstrated that these Qy shifts are accompanied by changes in whole cell fluorescence and chlorosome volume fraction that correspond to altered excitation energy flow between the chlorosome and reaction center, apparently in response to sulfide. Increasing sulfide concentration imparts a more negative redox potential to the medium and suggests the hypothesis that *C. tepidum* tunes its light-harvesting capabilities to redox potential. It is currently unknown how rapidly *C. tepidum* can adjust the alkylation state of Bchl *c* within chlorosomes. This is not a trivial consideration as Bchl *c* exists primarily in a non-protein bound, paracrystalline state inside the chlorosome isolated from most biosynthetic enzymes by a unilamellar membrane (Hohmann-Marriott and Blankenship, 2007; Sorensen et al. 2008). Details of Bchl transport into the chlorosome are still lacking, but this adaptive response may provide additional insights into this process.

Second, a suite of at least six polypeptide bands < 30 kDa displayed altered staining intensity in varied sulfide concentrations or when compared to a control extract from cells growing on thiosulfate by SDS-PAGE (Fig. 7.1b). These polypeptides may arise from a number of processes including degradation of damaged proteins, increased expression of proteins required for resistance to increased sulfide, or proteins involved in adjusting chlorosome structure and function. Current efforts are directed at discriminating between these distinct possibilities.

7.3 Low Molecular Weight Thiols and Phototrophic Sulfur Oxidation

All cellular organisms are dependent on thiols, compounds containing reduced sulfhydryl (R–SH) groups, as catalysts and carbon carriers in central metabolism (i.e., lipoic acid, coenzyme A) and to maintain redox balance and regulation. Glutathione (L-γ-glutamyl-L-cysteinyl-glycine, GSH) is perhaps the best-known

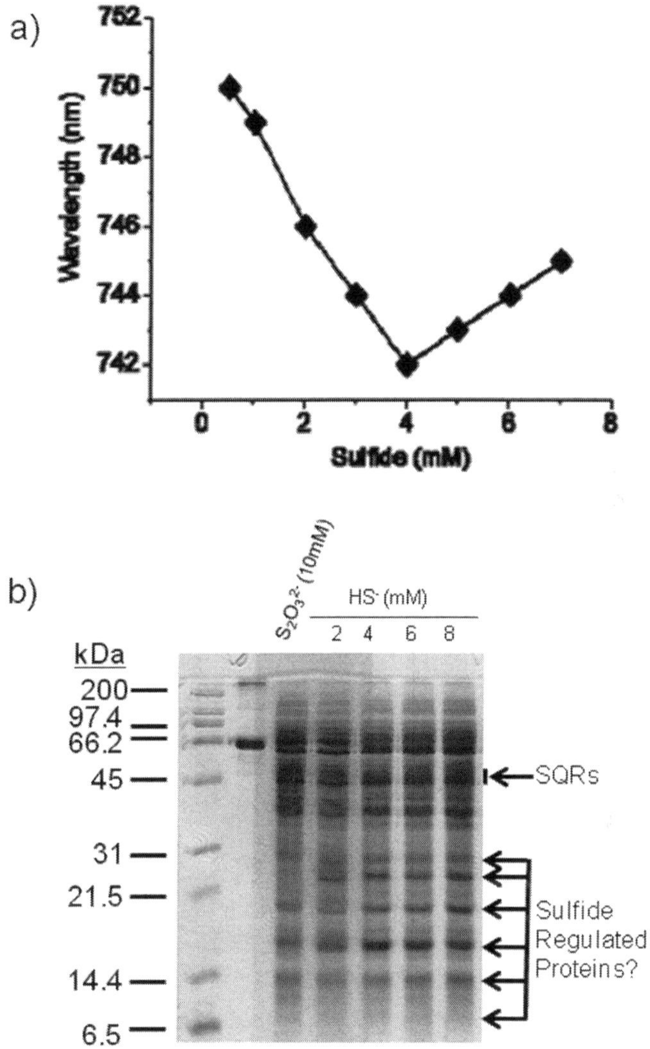

Fig. 7.1 Properties of *C. tepidum* grown with varying concentrations of sulfide. (**a**) Wavelength maximum of the Bchl *c* absorption peak in vivo. (**b**) One-dimensional SDS-PAGE protein gels of whole cell extracts of cultures from selected sulfide concentrations. Polypeptide bands enhanced in high sulfide cultures are noted with *arrows* as are the expected position of SQR polypeptides

low molecular weight (LMW) thiol found in a variety of bacteria and higher eukaryotes, including humans. GSH protects cells against oxidative stresses by mediating cellular redox balance via cycling between reduced thiol (GSH) and oxidized disulfide (GSSG) forms (Fahey, 2001). LMW thiols are kept in their reduced form by specific disulfide reductases with GSSG reductase (GSR) as the relevant enzyme

Table 7.2 Distribution of redox-balancing LMW thiols

LMW thiol	Phylogenetic distribution
Glutathione	Many eukaryotes, proteobacteria, and all cyanobacteria
Glutathione amide	*C. gracile, A. vinosum?*
Mycothiol	Actinomycetes
Ergothioneine	Fungi and actinomycetes
Trypanothione	Pathogenic protozoa: *Trypanosoma* and *Leishmania*
Coenzyme A	*Staphylococcus aureus, Pyrococcus horikoshii*

in glutathione-containing organisms. Organisms that employ alternative thiols produce an enzyme equivalent to GSR whose preferred substrate is the disulfide of the predominant thiol in the cell. Redox-balancing thiols and their cognate disulfide oxidoreductases vary among organisms (Table 7.2). Bartsch and co-workers proposed that a form of GSH metabolism may have been present in photosynthetic anaerobes prior to the evolution of oxygenic photosynthesis (Bartsch et al. 1996), and it seems clear that eukaryotes obtained GSH metabolism via endosymbiosis as both proteobacteria (pre-mitochondria) and cyanobacteria (pre-chloroplasts) contain GSH as the major LMW thiol. LMW thiols have been proposed as an active player in the oxidation of elemental sulfur in phototrophic bacteria, whether it is stored in the periplasm (Chromatiaceae) or extracellularly (Chlorobiaceae, Ectothiorhodospiraceae).

In purple sulfur bacteria, specifically the Chromatiaceae, glutathione amide (GASH) is produced at millimolar levels by the purple photosynthetic anaerobe *A. vinosum* and is maintained in the reduced state by a disulfide reductase utilizing reduced nicotinamide adenine dinucleotide (NADH), previously thought to be GSH reductase (Fahey 2001). The *Chromatium gracile* glutathione amide reductase enzyme exhibited a 70-fold lower K_m for the disulfide of GASH than for GSSG and a 150-fold preference for NADH over NADPH. GASH has also been found only in anaerobic *Chromatium* species (Vergauwen et al. 2001) and was observed in both the thiol and the perthiol (GASSH) forms when *C. gracile* was grown on sulfide (Bartsch et al. 1996). This observation suggested that GASSH could be involved in sulfur globule formation and/or degradation (Bartsch et al. 1996). However, very recent work failed to detect GASH in *A. vinosum* cells during elemental sulfur oxidation (Franz et al. 2009). Instead, the cultures appeared to contain significant pools of polysulfide terminated with sulfone groups both internally and in the surrounding medium. These contradictory results clearly reveal a need for more functional studies to resolve the importance of GASH in the purple sulfur bacteria. In particular, the sulfur oxidation phenotype of GASH-less *A. vinosum* mutants needs to be determined.

In the green sulfur bacteria, prior work has demonstrated that the Chlorobiaceae contain LMW thiols whose structure is currently unknown. The existence of a novel thiol, named U11, was found during a study of microbial LMW thiol diversity that included *C. limicola* (Fahey et al. 1987). *C. limicola* also lacked detectable amounts of glutathione and all other common LMW thiols. Consistent with these data, when Chlorobiaceae genomes have been analyzed, no biosynthetic pathways for redox-balancing LMW thiols (Table 7.1), no cognate disulfide oxidoreductases, and no glutathione-dependent enzymes have been found. Given that the genomic sequences have proven to be uninformative, our group has applied classical bimane labeling methods for LMW thiol detection to *C. tepdium* and a single LMW thiol has been detected (Fig. 7.2). This thiol exhibits a unique retention time relative to a selection of known thiols and the previously described U11 from *C. limicola*. During growth with both polysulfide and thiosulfate in batch culture, the peak area of the *C. tepidum* LMW thiol increases (Fig. 7.2). Indeed, after normalization for biomass, it appears that the pool size of this thiol metabolite increases as the cells grow into stationary phase (J. Hiras and T. E. Hanson, in preparation). Our current efforts are focused on the structural characterization of this metabolite and its distribution throughout the Chlorobiaceae. We believe that this is the previously reported U11 thiol as shorter retention times for most thiols are observed in our hands relative to those previously reported (Fahey et al. 1987).

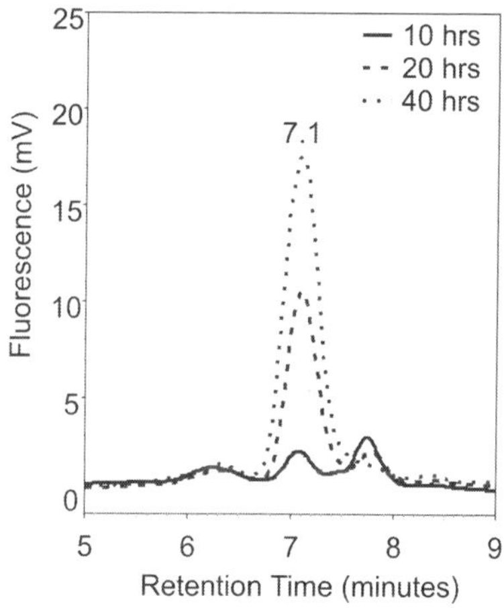

Fig. 7.2 HPLC detection of a bimane derivatized LMW thiol extracted from *C. tepidum*. A portion of a full chromatogram is shown focusing on the 7.1 min peak as extracted from cultures growing with both thiosulfate and sulfide as electron donors at the indicated times after inoculation

7.4 Concluding Remarks

As metabolic models arise from increasing amounts of genomic sequence data, particularly for pathways like sulfur oxidation that display great diversity in the microbial realm, concerted efforts that rely on classical biochemistry, physiology, genetics, and metabolite analysis need to be increasingly applied to test and validate these models. This will require significant efforts to develop genetic methods for previously genetically inaccessible organisms for which genome sequences are available and their cultivation under rigorously defined conditions for biochemical and metabolic analyses. In a sense, this means the field needs to come full circle and what was once old has to become new again to tease apart and truly appreciate the biological meaning lurking in the current morass of genomic data.

References

Arieli B, Shahak Y, Taglicht D, Hauska G, Padan E (1994) Purification and characterization of sulfide-quinone reductase, a novel enzyme driving anoxygenic photosynthesis in *Oscillatoria limnetica*. J Biol Chem 269:5705–5711

Arndt C, Gaill F, Felbeck H (2001) Anaerobic sulfur metabolism in thiotrophic symbioses. J Exp Biol 204:741–750

Arnirfakhri J, Vossoughi M, Soltanieh M (2006) Assessment of desulfurization of natural gas by chemoautotrophic bacteria in an anaerobic baffled reactor (ABR). Chem Eng Process 45:232–237

Bartsch RG, Newton GL, Sherrill C, Fahey RC (1996) Glutathione amide and its perthiol in anaerobic sulfur bacteria. J Bacteriol 178:4742–4746

Baumgartner LK, Reid RP, Dupraz C, Decho AW, Buckley DH, Spear JR, Przekop KM, Visscher PT (2006) Sulfate reducing bacteria in microbial mats: changing paradigms, new discoveries. Sediment Geol 185:131–145

Bernhard JM (2003) Potential symbionts in bathyal foraminifera. Science 299: 861

Bernhard JM, Buck KR, Farmer MA, Bowser SS (2000) The Santa Barbara Basin is a symbiosis oasis. Nature 403: 77–80

Bernhard JM, Habura A, Bowser SS (2006) An endobiont-bearing *Allogromiid* from the Santa Barbara Basin: implications for the early diversification of foraminifera. J Geophys Res-Biogeosci 111:G03002 doi: 10.1029/2005JG000158

Borrego CM, Garcia-Gil LJ (1994) Rearrangement of light harvesting bacteriochlorophyll homologues as a response of green sulfur bacteria to low light intensities. Photosynth Res 45:21–30

Bronstein M, Schutz M, Hauska G, Padan E, Shahak Y (2000) Cyanobacterial sulfide-quinone reductase: cloning and heterologous expression. J Bacteriol 182:3336–3344

Bruchert V, Jorgensen BB, Neumann K, Riechmann D, Schlosser M, Schulz H (2003) Regulation of bacterial sulfate reduction and hydrogen sulfide fluxes in the central Namibian coastal upwelling zone. Geochim Cosmochim Acta 67:4505–4518

Canfield DE, Raiswell R (1999) The evolution of the sulfur cycle. Am J Sci 299:697–723

Celis-Garcia LB, Gonzalez-Blanco G, Meraz M (2008) Removal of sulfur inorganic compounds by a biofilm of sulfate reducing and sulfide oxidizing bacteria in a down-flow fluidized bed reactor. J Chem Technol Biotechnol 83:260–268

Chan L-K, Morgan-Kiss R, Hanson TE (2008) Genetic and proteomic studies of sulfur oxidation in *Chlorobium tepidum* (syn. *Chlorobaculum tepidum*). In: Hell R, Dahl C, Knaff D and Leustek T (eds) Sulfur metabolism in phototrophic organisms pp 363–379. Springer, New York

Chan LK, Morgan-Kiss RM, Hanson TE (2009) Functional analysis of three sulfide:quinone oxidoreductase homologs in *Chlorobaculum tepidum*. J Bacteriol 191:1026–1034

Chen ZW, Koh M, Van Driessche G, Van Beeumen JJ, Bartsch RG, Meyer TE, Cusanovich MA, Mathews FS (1994) The structure of flavocytochrome c sulfide dehydrogenase from a purple phototrophic bacterium. Science 266:430–432

Chew AGM, Frigaard NU, Bryant DA (2007) Bacteriochlorophyllide *c* c-8^2 and c-12^1 methyltransferases are essential for adaptation to low light in *Chlorobaculum tepidum*. J Bacteriol 189:6176–6184

Dahl C (2008) Inorganic sulfur compounds as electron donors in purple sulfur bacteria. In: Hell R, Dahl C, Knaff D, Leustek T (eds) Sulfur metabolism in phototrophic organisms pp 289–317. Springer, Dordrecht, The Netherlands

Dahl C, Engels S, Pott-Sperling AS, Schulte A, Sander J, Lubbe Y, Deuster O, Brune DC (2005) Novel genes of the *dsr* gene cluster and evidence for close interaction of Dsr proteins during sulfur oxidation in the phototrophic sulfur bacterium *Allochromatium vinosum*. J Bacteriol 187:1392–1404

Dahl C, Rakhely G, Pott-Sperling AS, Fodor B, Takacs M, Toth A, Kraeling M, Gy"orfi K, Kovacs A, Tusz J, Kovacs KL (1999) Genes involved in hydrogen and sulfur metabolism in phototrophic sulfur bacteria. FEMS Microbiol Lett 180:317–324

Elrod JW, Calvert JW, Morrison J, Doeller JE, Kraus DW, Tao L, Jiao X, Scalia R, Kiss L, Szabo C, Kimura H, Chow CW, Lefer DJ (2007) Hydrogen sulfide attenuates myocardial ischemia-reperfusion injury by preservation of mitochondrial function. Proc Natl Acad Sci U S A 104:15560–15565

Elshahed MS, Youssef NH, Luo QW, Najar FZ, Roe BA, Sisk TM, Buhring SI, Hinrichs KU, Krumholz LR (2007) Phylogenetic and metabolic diversity of planctomycetes from anaerobic, sulfide- and sulfur-rich Zodletone Spring, Oklahoma. Appl Environ Microbiol 73: 4707–4716

Fahey RC (2001) Novel thiols of prokaryotes. Annu Rev Microbiol 55:333–356

Fahey RC, Buschbacher RM, Newton GL (1987) The evolution of glutathione metabolism in phototrophic microorganisms. J Mol Evol 25:81–88

Franz B, Gehrke T, Lichtenberg H, Hormes J, Dahl C, Prange A (2009) Unexpected extracellular and intracellular sulfur species during growth of *Allochromatium vinosum* with reduced sulfur compounds. Microbiol 155: 2766–2774

Friedrich CG, Bardischewsky F, Rother D, Quentmeier A, Fischer J (2005) Prokaryotic sulfur oxidation. Curr Opin Microbiol 8:253–259

Frigaard N-U, Bryant DA (2008a) Genomic and evolutionary perspectives on sulfur metabolism in green sulfur bacteria. In: Dahl C, Friedrich CG (eds) Microbial sulfur metabolism pp 60–76. Springer, New York

Frigaard N, Bryant D (2008b) Sulfur oxidation in green sulfur bacteria. In: Hell R, Dahl C, Knaff D, Leustek T (eds) Sulfur metabolism in phototrophic organisms pp 363–379. Springer, New York

Goffredi SK, Waren A, Orphan VJ, Van Dover CL, Vrijenhoek RC (2004) Novel forms of structural integration between microbes and a hydrothermal vent gastropod from the Indian Ocean. Appl Environ Microbiol 70:3082–3090

Goubern M, Andriamihaja M, Nubel T, Blachier F, Bouillaud F (2007) Sulfide, the first inorganic substrate for human cells. FASEB J 21:1699–1706

Heising S, Richter L, Ludwig W, Schink B (1999) *Chlorobium ferrooxidans* sp. nov., a phototrophic green sulfur bacterium that oxidizes ferrous iron in coculture with a "*Geospirillum*" sp. strain. Arch Microbiol 172:116–124

Hohmann-Marriott MF, Blankenship RE (2007) Hypothesis on chlorosome biogenesis in green photosynthetic bacteria. FEBS Lett 581:800–803

Hollibaugh JT, Budinoff C, Hollibaugh RA, Ransom B, Bano N (2006) Sulfide oxidation coupled to arsenate reduction by a diverse microbial community in a soda lake. Appl Environ Microbiol 72:2043–2049

Holmer M, Storkholm P (2001) Sulphate reduction and sulphur cycling in lake sediments: A review. Freshw Biol 46:431–451

Janssen AJ, Lens PN, Stams AJ, Plugge CM, Sorokin DY, Muyzer G, Dijkman H, Van Zessen E, Luimes P, Buisman CJ (2009) Application of bacteria involved in the biological sulfur cycle for paper mill effluent purification. Sci Total Environ 407:1333–1343

Kamp A, Stief P, Schulz-Vogt HN (2006) Anaerobic sulfide oxidation with nitrate by a freshwater beggiatoa enrichment culture. Appl Environ Microbiol 72: 4755–4760

Kaneko T, Tabata S (1997) Complete genome structure of the unicellular cyanobacterium *Synechocystis* sp. PCC 6803. Plant Cell Physiol 38: 1171–1176

Klamt S, Grammel H, Straube R, Ghosh R, Gilles ED (2008) Modeling the electron transport chain of purple non-sulfur bacteria. Mol Syst Biol 4:156 doi:10.1038/msb4100191

Lefer DJ (2007) A new gaseous signaling molecule emerges: Cardioprotective role of hydrogen sulfide. Proc Natl Acad Sci U S A 104:17907–17908

Macalady JL, Dattagupta S, Schaperdoth I, Jones DS, Druschel GK, Eastman D (2008) Niche differentiation among sulfur-oxidizing bacterial populations in cave waters. ISME Journal 2:590–601

Madrid VM, Aller RC, Aller JY, Chistoserdov AY (2006) Evidence of the activity of dissimilatory sulfate-reducing prokaryotes in nonsulfidogenic tropical mobile muds. FEMS Microbiol Ecol 57:169–181

Martin W, Rotte C, Hoffmeister M, Theissen U, Gelius-Dietrich G, Ahr S, Henze K (2003) Early cell evolution, eukaryotes, anoxia, sulfide, oxygen, fungi first (?), and a tree of genomes revisited. IUBMB Life 55:193–204

Meeks JC, Elhai J, Thiel T, Potts M, Larimer F, Lamerdin J, Predki P, Atlas R (2001) An overview of the genome of *Nostoc punctiforme*, a multicellular, symbiotic cyanobacterium. Photosynth Res 70:85–106

Morgan-Kiss RM, Chan LK, Modla S, Weber TS, Warner M, Czymmek KJ, Hanson TE (2009) *Chlorobaculum tepidum* regulates chlorosome structure and function in response to temperature and electron donor availability. Photosynth Res 99:11–21

Nakamura Y, Kaneko T, Sato S, Ikeuchi M, Katoh H, Sasamoto S, Watanabe A, Iriguchi M, Kawashima K, Kimura T, Kishida Y, Kiyokawa C, Kohara M, Matsumoto M, Matsuno A, Nakazaki N, Shimpo S, Sugimoto M, Takeuchi C, Yamada M, Tabata S (2002) Complete genome structure of the thermophilic cyanobacterium *Thermosynechococcus elongatus* BP-1. DNA Res 9:123–130

Nakamura Y, Kaneko T, Sato S, Mimuro M, Miyashita H, Tsuchiya T, Sasamoto S, Watanabe A, Kawashima K, Kishida Y, Kiyokawa C, Kohara M, Matsumoto M, Matsuno A, Nakazaki N, Shimpo S, Takeuchi C, Yamada M, Tabata S (2003) Complete genome structure of *Gloeobacter violaceus* PCC 7421, a cyanobacterium that lacks thylakoids. DNA Res 10:137–145

Oh-oka H, Blankenship RE (2004) Green bacteria: Secondary electron donor (cytochromes). In: Lennarz WJ, Lane MD (eds) Encyclopedia of biochemistry pp 321–324. Elsevier, Boston

Omoregie EO, Mastalerz V, de Lange G, Straub KL, Kappler A, Roy H, Stadnitskaia A, Foucher JP, Boetius A (2008) Biogeochemistry and community composition of iron- and sulfur-precipitating microbial mats at the chefren mud volcano (Nile deep sea fan, eastern Mediterranean). Appl Environ Microbiol 74:3198–3215

Orphan VJ, Ussler W, Naehr TH, House CH, Hinrichs KU, Paull CK (2004) Geological, geochemical, and microbiological heterogeneity of the seafloor around methane vents in the Eel River basin, offshore California. Chem Geol 205:265–289

Overmann J (2006) Symbiosis between non-related bacteria in phototrophic consortia. Prog Mol Subcell Biol 41:21–37

Overmann J, Beatty JT, Krouse HR, Hall KJ (1996) The sulfur cycle in the chemocline of a meromictic salt lake. Limnol Oceanogr 41:147–156

Pfannes KR, Vogl K, Overmann J (2007) Heterotrophic symbionts of phototrophic consortia: Members of a novel diverse cluster of betaproteobacteria characterized by a tandem *rrn* operon structure. Environ Microbiol 9:2782–2794

Preisler A, de Beer D, Lichtschlag A, Lavik G, Boetius A, Jorgensen BB (2007) Biological and chemical sulfide oxidation in a *Beggiatoa* inhabited marine sediment. ISME Journal 1:341–353

Reinartz M, Tschape J, Bruser T, Truper HG, Dahl C (1998) Sulfide oxidation in the phototrophic sulfur bacterium *Chromatium vinosum*. Arch Microbiol 170:59–68

Saga Y, Osumi S, Higuchi H, Tamiaki H (2005) Bacteriochlorophyll-c homolog composition in green sulfur photosynthetic bacterium *Chlorobium vibrioforme* dependent on the concentration of sodium sulfide in liquid cultures. Photosynth Res 86:123–130

Schutz M, Shahak Y, Padan E, Hauska G (1997) Sulfide-quinone reductase from *Rhodobacter capsulatus*. Purification, cloning, and expression. J Biol Chem 272:9890–9894

Shahak Y (2008) Sulfide oxidation from cyanobacteria to humans: sulfide-quinone oxidoreductase (SQR). In: Hell R, Dahl C, Knaff D, Leustek T (eds) Sulfur metabolism in phototrophic organisms pp 319–335. Springer, Dordrecht, The Netherlands

Shahak Y, Arieli B, Padan E, Hauska G (1992) Sulfide quinone reductase (SQR) activity in *Chlorobium*. FEBS Lett 299: 127–130

Shahak Y, Schutz M, Bronstein M, Griesbeck C, Hauska G, Padan E (1999) Sulfide-dependent anoxygenic photosynthesis in prokaryotes- sulfide-quinone reductase (SQR), the initial step. In: Peschek GA, Loffelhardt WL, Schmetterer G (eds) The phototrophic prokaryotes pp 217–228. Kluwer Academic/ Plenum, New York

Sorensen PG, Cox RP, Miller M (2008) Chlorosome lipids from *Chlorobium tepidum*: characterization and quantification of polar lipids and wax esters. Photosynth Res 95:191–196

Theissen U, Hoffmeister M, Grieshaber M, Martin W (2003) Single eubacterial origin of eukaryotic sulfide : quinone oxidoreductase, a mitochondrial enzyme conserved from the early evolution of eukaryotes during anoxic and sulfidic times. Mol Biol Evol 20:1564–1574

Turchyn AV, Schrag DP (2006) Cenozoic evolution of the sulfur cycle: insight from oxygen isotopes in marine sulfate. Earth Planet Sci Lett 241:763–779

Vergauwen B, Pauwels F, Jacquemotte F, Meyer TE, Cusanovich MK, Bartsch RG, Van Beeumen JJ (2001) Characterization of glutathione amide reductase from *Chromatium gracile* – identification of a novel thiol peroxidase (Prx/Grx) fueled by glutathione amide redox cycling. J Biol Chem 276:20890–20897

Wakai S, Kikumoto M, Kanao T, Kamimura K (2004) Involvement of sulfide:quinone oxidoreductase in sulfur oxidation of an acidophilic iron-oxidizing bacterium, *Acidithiobacillus ferrooxidans* NASF-1. Biosci Biotechnol Biochem 68:2519–2528

Wang GZ, Spivack AJ, Rutherford S, Manor U, D′Hondt S (2008) Quantification of co-occurring reaction rates in deep subseafloor sediments. Geochim Cosmochim Acta 72:3479–3488

Weston NB, Porubsky WP, Samarkin VA, Erickson M, Macavoy SE, Joye SB (2006) Porewater stoichiometry of terminal metabolic products, sulfate, and dissolved organic carbon and nitrogen in estuarine intertidal creek-bank sediments. Biogeochem 77: 375–408

Yang G, Wu L, Jiang B, Yang W, Qi J, Cao K, Meng Q, Mustafa AK, Mu W, Zhang S, Snyder SH, Wang R (2008) H_2S as a physiologic vasorelaxant: hypertension in mice with deletion of cystathionine gamma-lyase. Science 322:587–590

Yang W, Vollertsen J, Hvitved-Jacobsen T (2005) Anoxic sulfide oxidation in wastewater of sewer networks. Water Sci Technol 52:191–199

Zopfi J, Bttcher ME, Jorgensen BB (2008) Biogeochemistry of sulfur and iron in *Thioploca*-colonized surface sediments in the upwelling area off central Chile. Geochim Cosmochim Acta 72:827–843

Chapter 8
Multicellularity in a Heterocyst-Forming Cyanobacterium: Pathways for Intercellular Communication

Vicente Mariscal and Enrique Flores

Abstract The filamentous, heterocyst-forming cyanobacteria are among the simplest multicellular prokaryotes, and *Anabaena* sp. strain PCC 7120 is being used as a model for studying multicellularity in these organisms. In the absence of combined nitrogen two interdependent cell types are present in an *Anabaena* filament: vegetative cells and heterocysts. Vegetative cells perform oxygenic photosynthesis and supply carbon compounds to the heterocysts, which are specialized in the assimilation of atmospheric N_2 and supply nitrogenous compounds to the vegetative cells. In this chapter, we discuss two possible pathways for the exchange of metabolites and regulatory signals between vegetative cells and heterocysts: the continuous periplasm that surrounds the cells in the filament and some septal proteinaceous complexes that could allow the direct intercellular transfer of small molecules.

8.1 Introduction

Bacteria are generally viewed as unicellular organisms. However, some unicellular bacteria communicate through signals that make them behave as a multicellular organism (Hooshangi and Bentley 2008), and other bacteria are, as discussed later, truly multicellular. Evolution to multicellularity implies several features, including the fact that cells must stick together, which may involve specific adhesion molecules, and the acquisition of a barrier that isolates the organism as a whole from the external medium. In contrast to colonial organisms, true multicellular organisms exhibit a division of labor: different cells within the organism specialize in different functions. This requires cell signaling, in which signal molecules and regulatory metabolites are exchanged between cells, and signal transduction mechanisms, which translate the signals into different patterns of gene activity.

E. Flores (✉)
Instituto de Bioquímica Vegetal y Fotosíntesis, C.S.I.C. and Universidad de Sevilla, Américo Vespucio 49, E-41092 Seville, Spain
e-mail: flores@cica.es

P.C. Hallenbeck (ed.), *Recent Advances in Phototrophic Prokaryotes*, Advances in Experimental Medicine and Biology 675, DOI 10.1007/978-1-4419-1528-3_8,

The filamentous, heterocyst-forming cyanobacteria are true multicellular organisms. They have been described as reproducing by random trichome breakage, implying that their unit of growth is the filament (Rippka et al. 1978). *Anabaena* sp. strain PCC 7120, hereafter referred to as *Anabaena*, is a filamentous heterocyst-forming cyanobacterium that is used as a model for studying multicellularity. It consists of strings of cells joined together and surrounded by a continuous Gram-negative type of outer membrane (Rippka et al. 1978; Flores et al. 2006). Under nitrogen deprivation some vegetative cells in the filament differentiate into heterocysts, which are specialized for N_2 fixation and exchange nutrients with the vegetative cells of the filament.

8.2 Morphology of Heterocyst-Forming Cyanobacteria

Historically, the cyanobacteria were thought to be green algae because of their chlorophyll *a*-dependent photosynthesis. However, electron microscopy studies have shown that they are prokaryotes carrying walls that bear close structural resemblance to the walls of Gram-negative bacteria (Wolk 1973; Stanier and Cohen-Bazire 1977). Filamentous cyanobacteria consist of filaments that can be hundreds of cells long. The general organization of the cells is very similar to that of unicellular cyanobacteria: the nuclear material, ribosomes, glycogen granules, and carboxysomes are centrally located, and the thylakoid membranes are generally located in the periphery of the cell. All these materials are enclosed by the cytoplasmic membrane, which surrounds each cell in the filament. External to the cytoplasmic membrane is a medium electron-dense layer of peptidoglycan (murein), which can vary in thickness among species and which enters into the septum between cells. The filament is surrounded by an outer membrane, which exhibits special features when compared with some other well-characterized Gram-negative bacteria such as the presence of carotenoids and specific porins (Ris and Singh 1961; Moslavac et al. 2005; Flores et al. 2006). Some cyanobacterial strains also possess an S-layer (a surface layer made of proteins attached to the outermost portion of the cell wall), similar to many other bacteria (Hoiczyk and Hansel 2000).

Cell division in filamentous cyanobacteria resembles the septal division found in most unicellular bacteria, which divide by forming a septum derived from the innermost wall layer (cytoplasmic membrane and peptidoglycan). However, whereas in unicellular Gram-negative bacteria the completion of this septum is followed by the invagination of the outer membrane and the separation of the daughter cells, in filamentous cyanobacteria the outer membrane does not invaginate (at least between most of the cells in the filament) and the daughter cells remain enclosed by the continuous outer membrane (Ris and Singh 1961; Hoiczyk and Baumeister 1995). Thus, the septum that separates the cells in the filamentous cyanobacteria contains the medium electron-dense layer of peptidoglycan lying between the cytoplasmic membranes of the daughter cells.

In the septum between cells, some structures, termed microplasmodesmata, are present in the form of thin strings perpendicular to the cytoplasmic membranes

(Pankratz and Bowen 1963). In freeze-fracture electron microscopy, they have been identified as pits and protrusions seen in the different leaflets of the cytoplasmic membranes of adjacent cells (Giddings and Staeheling 1978). However, the term "microplasmodesmata" may not be appropriate for describing such structures, since plasmodesmata refer to "cytoplasmic bridges delimited by cytoplasmic membranes," whereas in this case, no cytoplasmic membrane continuity between adjacent cells appears to exist (Flores et al. 2006). At the connection of a heterocyst with a vegetative cell, the septum is also traversed by the interconnecting structures found between vegetative cells, although the number of such structures is less (Giddings and Staeheling 1978). In spite of the narrowing at the heterocyst cell poles, the outer membrane is still continuous, delimiting a continuous periplasmic space (Lang and Fay 1971; Sherman et al. 2000; Flores et al. 2006).

8.3 Heterocyst Differentiation in *Anabaena*

When combined nitrogen is withdrawn from the external medium, 8–10% of the vegetative cells in a filament of *Anabaena* differentiate into heterocysts, which are distributed in a semiregular pattern along the filaments (Golden and Yoon 2003). Heterocysts are terminally differentiated cells which contain the oxygen-sensitive enzyme complex nitrogenase that carries out nitrogen fixation. The metabolic and morphological changes associated with heterocyst differentiation underline the need for a microoxic environment for nitrogenase expression and function. Metabolic changes include (i) the loss of photosystem II activity, thus avoiding photosynthetic oxygen production, (ii) an increased respiration rate for eliminating free oxygen, (iii) the loss of photosynthetic CO_2 fixation, and (iv) the expression of the N_2-fixing system (Golden and Yoon 2003; Herrero et al. 2004). Heterocysts contain glutamine synthetase, but lack glutamate synthase, implying that, for production of glutamine with the ammonium resulting from N_2 fixation, glutamate has to be transferred from the vegetative cells to the heterocysts (Thomas et al. 1977; Martín-Figueroa et al. 2000).

Several characteristic morphological changes can be observed in heterocysts, such as (i) the loss of cellular inclusions (carboxysomes and glycogen granules), (ii) a marked rearrangement of thylakoids to the cell poles, where they develop a heavily contorted form (Lang and Fay 1971), and (iii) the deposition of supplemental glycolipid and polysaccharide layers in the cell envelope to hamper the influx of gases. The heterocyst special envelope, formed externally to the outer membrane of the preexisting cell wall, is composed of two distinct layers: an outermost homogeneous polysaccharide layer and an innermost laminated glycolipid layer. The outermost layer is loosely organized and of irregular thickness, and it is expanded toward the heterocyst poles to resemble a bottleneck. The innermost laminated layer is thinner and is attached to the outer membrane (Lang and Fay 1971). At the point of connection with the vegetative cell, the heterocyst acquires a shallow cup-like structure. Inside these constrictions an electron opaque material called cyanophycin can be found. Cyanophycin is a polymer of arginine and aspartate that represents a nitrogen reserve material in many cyanobacteria (Lang and Fay 1971; Picossi et al. 2004).

The differentiation of heterocysts in the *Anabaena* filament results from a tightly controlled developmental program that involves positive and negative regulatory signals, in which NtcA and HetR act as master positive regulators (Buikema and Haselkorn 1991; Frías et al. 1994; Wei et al. 1994). NtcA is a transcriptional regulator that operates global nitrogen control (Herrero et al. 2004), and HetR has been reported to have protease and DNA-binding activity (Zhou et al. 1998; Huang et al. 2004). The *ntcA* gene is induced several fold during the early steps of heterocyst differentiation in a HetR-dependent and autoregulated manner, and *hetR* induction during heterocyst differentiation is NtcA dependent (Muro-Pastor et al. 2002; Olmedo-Verd et al. 2008).

Some genes that negatively affect heterocyst differentiation have also been identified. The *patS* gene governs de novo pattern formation when filaments are placed in the absence of combined nitrogen (Yoon and Golden 1998). Expression of *patS* increases after nitrogen step-down, in a patterned way, in cells that will become heterocysts, but after differentiation is complete, expression of *patS* returns to pre-induction levels. Thus, *patS* is mainly expressed in proheterocysts. The *patS* gene encodes a short peptide that inhibits heterocyst differentiation in an unknown manner (Yoon and Golden 1998). A *patS* null mutant exhibits a phenotype of multiple contiguous heterocysts (Mch) and shortened vegetative cell intervals. It has been proposed that the PatS peptide diffuses away from differentiating proheterocysts to create a gradient of an inhibitory signal along the filament (Yoon and Golden 1998; Wu et al. 2004).

PatS and fixed nitrogen produced by the heterocysts have been proposed as the major diffusible signals regulating the frequency and spacing of heterocysts (Yoon and Golden 2001). Heterocysts rely on vegetative cells as sources of fixed carbon compounds; in return, they supply the surrounding vegetative cells with fixed nitrogen compounds (Wolk 1968; Wolk et al. 1974, 1976; Thomas et al. 1977; Curatti et al. 2002). Glutamine is an N-containing metabolite that can be exported out of the heterocysts and made available to the vegetative cells, but the possibility that other amino acids are also transferred should be considered (Wolk et al. 1976; Thomas et al. 1977; Picossi et al. 2005; Pernil et al. 2008). This cellular interdependence implies the existence of communication mechanisms through which metabolites and regulatory signals move along the filament (Flores et al. 2006). In the remainder of this chapter we discuss possible pathways and mechanisms of intercellular molecular transfer in *Anabaena*.

8.4 Periplasmic Molecular Transfer

Although the mechanism of intercellular transfer of compounds in the *Anabaena* filament is unknown, two different conduits have been suggested: a continuous periplasm (Flores et al. 2006; Mariscal et al. 2007) and some cell-to-cell joining structures (Mullineaux et al. 2008).

8.4.1 GFP Diffusion Through the Periplasm

To provide a conduit for the movement of molecules between different types of cells in the N_2-fixing filament of *Anabaena*, the periplasm should be not only physically continuous but also functionally continuous between vegetative cells and between vegetative cells and heterocysts. We have experimentally addressed this question taking advantage of the cell specificity of the expression of *patS* (Yoon and Golden 1998, 2001) and using the *patS* promoter to drive the expression of engineered versions of the *gfp* gene in proheterocysts of *Anabaena* (Mariscal et al. 2007).

An *Anabaena* strain (CSVM18) that expresses a green fluorescent protein (GFP) with a signal peptide for translocation through the twin-arginine translocation (TAT) system and cleavage by signal peptidase I has been constructed (Mariscal et al. 2007). The TAT system is known to translocate folded proteins across the cytoplasmic membrane (Robinson and Bolhuis 2004). When this strain is subjected to nitrogen deprivation, GFP is produced in proheterocysts, targeted by the TAT system, and released into the periplasm, where it diffuses away from the producing cell (Mariscal et al. 2007). Figure 8.1 (a) shows the GFP fluorescence of a colony of strain CSVM18 grown for several days on solid medium in the absence of combined nitrogen. The GFP fluorescence is observed in the periphery of the cells along the entire filament.

Substitution of the TAT signal peptide containing a cleavage site for signal peptidase I (as in strain CSVM18) by a TAT signal peptide containing a cleavage site for signal peptidase II (as in strain CSVM19) results in the anchoring of the GFP to the cytoplasmic membrane (Mariscal et al. 2007). For this, we have used the signal peptide from *Anabaena* Alr0608, a periplasmic nitrate-binding protein (NrtA, also known as the 48-kDa protein) that in *Synechococcus* sp. PCC 7942 (Maeda and Omata 1997) and *Synechocystis* sp. PCC 6803 (Huang et al. 2006) is bound to the cytoplasmic membrane. In strain CSVM19, GFP accumulates in the periphery of the proheterocysts, presumably at the cytoplasmic membrane, and does not pass to the adjacent vegetative cells (Mariscal et al. 2007). Figure 8.1(b) shows a colony of strain CSVM19 grown on solid media in the absence of combined nitrogen. The GFP fluorescence is observed only in the periphery of the producing cells. Comparison of the results obtained with *Anabaena* strains CSVM18 (periplasmic soluble GFP) and CSVM19 (cytoplasmic membrane-anchored GFP) indicates that the soluble GFP moves in the periplasm from proheterocysts to vegetative cells in the diazotrophic cyanobacterial filament.

Because the anchored GFP can rapidly diffuse within the cytoplasmic membrane (Mariscal et al. 2007), lack of passage to the neighboring vegetative cells suggests that there is no cytoplasmic membrane continuity between vegetative cells and heterocysts. This is consistent with the notion that plant-like plasmodesmata are not present in filamentous cyanobacteria (Flores et al. 2006).

Recently Zhang et al. (2008) have used a similar approach to conclude that the GFP does not traverse the septa between cells because of the existence of periplasmic barriers that impede GFP diffusion. They observed GFP fluorescence in the periphery of the producing cells, but were unable to detect a GFP signal spreading

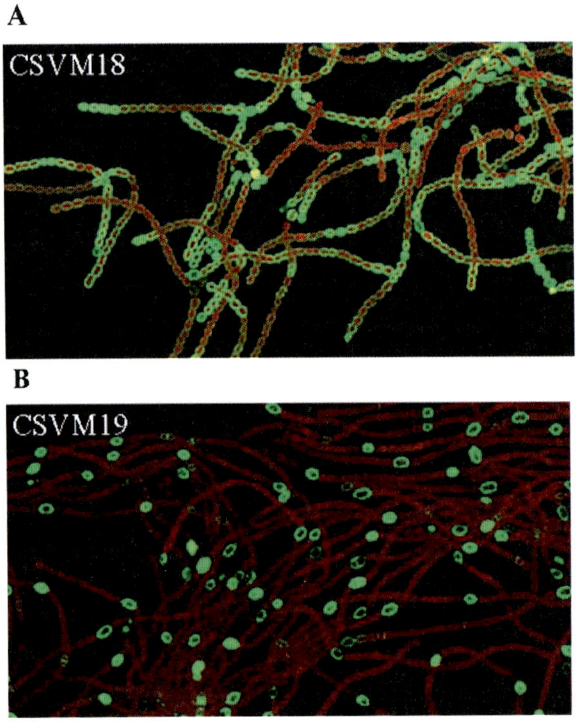

Fig. 8.1 Movement of periplasmic GFP. In the *Anabaena* strains shown, the *gfp* gene is expressed from the promoter of the *patS* gene, which is mostly active in proheterocysts (for details, see Mariscal et al. 2007). (**a**) In strain CSVM18, the GFP is produced as a soluble periplasmic protein. (**b**) In strain CSVM19, the GFP is located in the periplasmic face, anchored to the cytoplasmic membrane. Small blocks of agar ($BG11_0$ medium containing no source of combined nitrogen) carrying the filaments were cut and placed in a sample holder with a glass cover slip on top. Cells were imaged using a Leica HCX PLAN-APO 63X 1.4 NA oil immersion objective attached to a Leica TCS SP2 confocal laser-scanning microscope. GFP was excited at 488 nm using an argon ion laser. Fluorescent emission was monitored by collection across windows of 500–530 nm (GFP imaging) and 650–700 nm (cyanobacterial autofluorescence). The images are overlays of the green GFP fluorescence and the red cyanobacterial autofluorescence. Heterocysts have a reduced or no autofluorescence

to the neighboring vegetative cells. These results are similar to our results with strain CSVM19, in which the GFP is anchored to the plasma membrane. Zhang et al. fused the GFP to the signal peptide of the *Escherichia coli* TorA protein (Zhang et al. 2008). The utilization of a heterologous TAT signal peptide, which might not be efficiently processed in *Anabaena*, and the utilization of a promoter weaker than the one used in our work could be reasons for the reported discrepancy. The presence of periplasmic barriers that slow diffusion of the GFP along the filament has, however, also been considered by us (Mariscal et al. 2007).

We conclude that the periplasm of *Anabaena* is not only common to all the cells in the filament but also functionally continuous, thus representing a putative pathway for intercellular molecular transfer.

8.4.2 Amino Acid Transporters in Anabaena *and Their Role in Diazotrophy*

The model for the transfer of metabolites through the periplasm in the diazotrophic cyanobacterial filament calls for the existence of permeases that mediate the differential export and uptake of transferred metabolites in the different cell types of the filament (Flores et al. 2006). For amino acids exported from the heterocyst, specific amino acid uptake permeases are expected to be present in vegetative cells. Initial attempts to identify amino acid permeases in *Anabaena* relied on the isolation of mutants resistant to toxic amino acid analogs and the measurement of the activity of uptake of ^{14}C-labeled amino acids. The amino acid transport activities identified were interpreted to correspond to at least five transporters: active transporters N-I and N-II with overlapping, but not identical, specificities for neutral amino acids; one high-affinity, active transporter for basic amino acids; one low-affinity, passive system for basic amino acids; and a transporter for acidic amino acids (Herrero and Flores 1990; Montesinos et al. 1995).

More recently, the identification of genes encoding amino acid permeases was achieved through the inactivation of ORFs selected from the *Anabaena* genomic sequence by homology with heterologous permeases (Picossi et al. 2005; Pernil et al. 2008). The three amino acid transporters that have thus been identified on a molecular level in *Anabaena* belong to the ABC superfamily. ABC transporters constitute one of the largest and most ancient superfamilies of transporters and utilize the energy of ATP hydrolysis to fuel transport of substrates across cellular membranes. Five proteins or domains generally compose ABC uptake transporters: a periplasmic-binding protein, two transmembrane polypeptides, and two cytoplasmic ATPase subunits (Rees et al. 2009).

The N-I transporter is the product of the *natA*, *natB*, *natC*, *natD,* and *natE* genes encoding ATPase subunits (NatA and NatE), transmembrane polypeptides (NatC and NatD), and a periplasmic substrate-binding protein (NatB), respectively (Picossi et al. 2005). It mainly transports Pro and hydrophobic amino acids, but can also transport some neutral polar amino acids including Gln. The genes encoding this transporter are expressed in vegetative cells but not in heterocysts (Picossi et al. 2005).

The N-II transporter is the product of the *natF*, *natG*, *natH*, and *bgtA* genes encoding an ATPase subunit (BgtA), transmembrane polypeptides (NatG and NatH), and a periplasmic substrate-binding protein (NatF), respectively (Pernil et al. 2008). It mainly transports Asp and Glu, but can also transport some neutral polar amino acids including Gln. The N-II system is present in both vegetative cells and heterocysts; in the latter, it is the only system identified so far responsible for the uptake of Glu (Pernil et al. 2008).

The basic amino acid transport system Bgt is the product of the *bgtB* and *bgtA* genes. The BgtB protein consists of a periplasmic domain fused to a transmembrane domain, and BgtA is an ATPase subunit that is shared with the N-II system. Bgt mainly transports Arg and Lys, but can also transport His and Gln. This transporter is present in both vegetative cells and heterocysts (Pernil et al. 2008).

Inactivation of each of the three amino acid transporters has a different effect on the diazotrophic growth of *Anabaena*. Mutants of N-I, and to lesser extent of N-II, are specifically impaired in diazotrophic growth (Picossi et al. 2005; Pernil et al. 2008). However, inactivation of Bgt does not result in any noticeable effect on growth under standard laboratory conditions (Pernil et al. 2008). Therefore, amino acid transporters N-I and N-II could have a role in the vegetative cell uptake of some amino acids produced by the N_2-fixing heterocysts (Pernil et al. 2008). Gln is a recognized candidate vehicle for the transfer of N from heterocysts to vegetative cells (Wolk et al. 1976; Thomas et al. 1977), and Ala has also been proposed to have a role in N transfer (Picossi et al. 2005).

8.5 Direct Communication Between Cells

Intercellular molecular exchange could also involve the intercellular joining structures that seem to connect the cells in the *Anabaena* filament. These structures might allow molecules to be exchanged from cytoplasm to cytoplasm without passing through the periplasm (Lang and Fay 1971; Giddings and Staeheling 1978).

8.5.1 Rapid Diffusion of Calcein Between Cells

The fluorescein derivative calcein is available as a non-fluorescent acetoxymethyl ester (AM) that is sufficiently hydrophobic to traverse cell membranes, so that it can be loaded into the cells. In the cytoplasm, the ester groups are hydrolyzed by esterases to produce a fluorescent, hydrophilic, and negatively charged product of 623 Da, which can no longer traverse the cytoplasmic membrane (Mullineaux et al. 2008).

Intercellular exchange of calcein has been observed and quantified in *Anabaena* filaments by fluorescence recovery after photobleaching (FRAP) experiments, which permit exchange coefficients to be derived. Mechanistically, calcein appears to be exchanged by free diffusion, and although diffusion can be detected between vegetative cells and heterocysts as well as between vegetative cells, it is slower in the former case. Moreover, the transfer is nitrogen regulated so that the exchange coefficient increases about 10-fold in filaments grown under diazotrophic conditions compared to those grown with nitrate (Mullineaux et al. 2008). Calcein transfer seems to be restricted to the filamentous, heterocyst-forming cyanobacteria, since no exchange has been detected in *Oscillatoria*, which is a filamentous, non-heterocyst-forming cyanobacterium. Transfer of calcein may also give an indication of the specificity, in terms of charge and size discrimination, of the channels that are putatively involved in the process. Thus, whereas the negatively charged calcein is exchanged, transfer of the larger GFP (27 kDa) between the cytoplasms of adjacent cells has not been observed (Yoon and Golden 1998; Mariscal et al. 2007).

8.5.2 SepJ and Its Role in Calcein Transfer

The SepJ protein (also known as FraG) is the product of open reading frame *alr2338* that is located upstream of the *hetR* gene in the genome of *Anabaena* (Flores et al. 2007; Nayar et al. 2007). SepJ is a 751-amino-acid protein that comprises a 200-residue N-terminal coiled-coil region; an internal 211-amino-acid sequence that is rich in Pro and Ser and shows similarity to plant extensins and human titin protein, which is an elastic protein also known as connectin (Means 1998); and a 340-residue C-terminal region that is homologous to proteins in the drug/metabolite exporter (DME) family of bacteria and Archaea (Jack et al. 2001). Protein topology and subcellular localization predictions suggest that SepJ bears a large extra-cytoplasmic domain which could be involved in bridging the gap between adjacent cells in the filament and that the C-terminal region is a cytoplasmic membrane domain.

GFP tagging of SepJ shows that it is located at the cell poles in the intercellular septa (Flores et al. 2007). We have named this gene *sepJ* according to the particular localization of its protein product (Sep for septal protein; Flores et al. 2007). Figure 8.2(a) shows the very focalized location of the SepJ–GFP fusion in the middle of the septum. Figure 8.2(b) shows SepJ–GFP immunolocalization in an *Anabaena* filament. The preparation of samples for immunolocalization involves the fixation of samples to a slide and a step of dehydration; thus, cells remain in the original place on the slide but because of shrinking, they become smaller. SepJ–GFP is clearly observed at the cell poles indicating the presence of SepJ in the cytoplasmic membranes of the two adjacent cells in a septum, but the SepJ–GFP proteins from the two cells are so close together that they are seen as a single fluorescence spot in the intercellular septa of the native filaments (Fig. 8.2a). Additionally, SepJ–GFP is seen to position at a ring, similar to a Z ring, in the middle of the cells where cell division starts (Fig. 8.2b; see also Flores et al. 2007).

Although substantial levels of SepJ–GFP are observed in cells grown with combined nitrogen, *sepJ* expression increases during acclimation to diazotrophic growth (Flores et al. 2007). A *sepJ* null mutant shows extensive fragmentation of the filament and is incapable of diazotrophic growth (Flores et al. 2007; Nayar et al. 2007). Filament fragmentation of the *sepJ* mutant increases when it is incubated in the absence of combined nitrogen, resulting in very short filaments with two to three cells per filament, and heterocyst differentiation in these filaments aborts at an early stage.

The *sepJ* mutant is also hampered in intercellular calcein transfer (Mullineaux et al. 2008). Because of its particular location in the septa and its implication in calcein transfer, SepJ has been proposed to be a component of a septum protein complex that permits intercellular transfer of calcein (Mullineaux et al. 2008). Other fragmentation mutants of *Anabaena* that form short filaments and that therefore

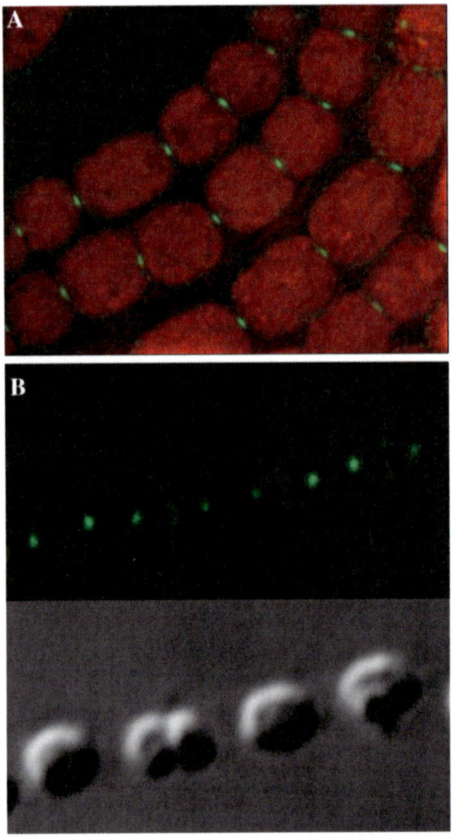

Fig. 8.2 Subcellular location of SepJ–GFP. (**a**) *Anabaena* strain CSAM137 (*sepJ–gfp*) grown on BG11 solid medium (with nitrate as nitrogen source). Small blocks of agar carrying the filaments were cut and placed in a sample holder with a glass cover slip on top. Cells were imaged using a Leica DM6000B fluorescence microscope and an ORCA-ER camera (Hamamatsu) and convolved with the Leica Application Suite Advanced Fluorescence software. The image is an overlay of the green GFP fluorescence and the red cyanobacterial autofluorescence. (**b**) Visualization of SepJ–GFP by immunofluorescence. BG11-grown filaments were collected and fixed to a poly-L-lysine-coated slide. Samples were immersed in methanol during 10 min at −20°C and dried at room temperature. Slides were then immersed in PBS buffer (pH 7.4) at room temperature for 10 min and incubated for 2 h with a 1:500 dilution of anti-GFP antibody, AlexaFluor 488-conjugate (Santa Cruz Biotechnology). Samples were washed three times for 10 min with PBS buffer and mounted with a cover slip. Cells were imaged using a Leica HCX PLAN-APO 63X 1.4 NA oil immersion objective attached to a Leica TCS SP2 confocal laser-scanning microscope. GFP was excited at 488 nm using an argon ion laser. The *panels* show the fluorescent emission monitored by collection across a window of 500–530 nm (*top*) and a bright field micrograph of the same field (*bottom*)

might lack additional proteins essential for the integrity of the filament have been identified (Bauer et al. 1995). It would be interesting to know whether the septal machinery that permits calcein transfer between *Anabaena* cells is a multiprotein complex of SepJ and some Fra proteins.

8.6 Concluding Remarks

Although it is well known that in *Anabaena* heterocysts and vegetative cells are mutually dependent and that its ability to survive and thrive under diazotrophic conditions strictly depends on an exchange of nutrients and regulatory compounds between these two cell types, the mechanisms of intercellular molecular transfer are not fully understood. The study of the fine structure of *Anabaena* led to the identification of two possible transfer pathways: the continuous periplasmic space and the septal tiny structures that connect adjacent cells. Recent studies have addressed the functionality of these structures in *Anabaena*. Engineered GFP proteins that are transported into the periplasm and whose expression is cell specific have demonstrated the existence of a functionally continuous periplasm. On the other hand, a direct and nitrogen-regulated transfer of small compounds which are dependent on SepJ has been shown using calcein as a probe. Due to the particular location of SepJ at the septum between cells, this protein might be a component of a multiprotein cell–cell joining complex at the septa. However, the nature of the compounds transferred by each pathway, the periplasm and the cell–cell joining complexes, and their relevance in the diazotrophic physiology of *Anabaena* remain unknown.

Acknowledgments We thank Antonia Herrero for a critical reading of the manuscript. Work on this project was made possible by grant no. BFU2008-03811 from Ministerio de Ciencia e Innovación and Proyecto de Excelencia CVI1896 from Junta de Andalucía (Spain).

References

Bauer CC, Buikema WJ, Black K et al (1995) A short-filament mutant of *Anabaena* sp. strain PCC 7120 that fragments in nitrogen-deficient medium. J Bacteriol 177:1520–1526

Buikema WJ, Haselkorn R (1991) Characterization of a gene controlling heterocyst differentiation in the cyanobacterium *Anabaena* 7120. Genes Dev 5:321–330

Curatti L, Flores E, Salerno G (2002) Sucrose is involved in the diazotrophic metabolism of the heterocyst-forming cyanobacterium *Anabaena* sp. FEBS Lett 513:175–178

Flores E, Herrero A, Wolk CP et al (2006) Is the periplasm continuous in filamentous multicellular cyanobacteria? Trends Microbiol 14:439–443

Flores E, Pernil R, Muro-Pastor AM et al (2007) Septum-localized protein required for filament integrity and diazotrophy in the heterocyst-forming cyanobacterium *Anabaena* sp. strain PCC 7120. J Bacteriol 189:3884–3890

Frías JE, Flores E, Herrero A (1994) Requirement of the regulatory protein NtcA for the expression of nitrogen assimilation and heterocyst development genes in the cyanobacterium *Anabaena* sp. PCC 7120. Mol Microbiol 14:823–832

Giddings TH, Staeheling LA (1978) Plasma membrane architecture of *Anabaena cylindrica*: occurrence of microplasmodesmata and changes associated with heterocyst development and the cell cycle. Cytobiologie 16:235–249

Golden JW, Yoon HS (2003) Heterocyst development in *Anabaena*. Curr Opin Microbiol 6: 557–563

Herrero A, Flores E (1990) Transport of basic amino acids by the dinitrogen-fixing cyanobacterium *Anabaena* PCC 7120. J Biol Chem 265:3931–3935

Herrero A, Muro-Pastor AM, Valladares A et al (2004) Cellular differentiation and the NtcA transcription factor in filamentous cyanobacteria. FEMS Microbiol Rev 28:469–487

Hoiczyk E, Baumeister W (1995) Envelope structure of four gliding filamentous cyanobacteria. J Bacteriol 177:2387–2395

Hoiczyk E, Hansel A (2000) Cyanobacterial cell walls: news from an unusual prokaryotic envelope. J Bacteriol 182:1191–1199

Hooshangi S, Bentley WE (2008) From unicellular properties to multicellular behavior: bacteria quorum sensing circuitry and applications. Curr Opin Biotechnol 19:550–555

Huang X, Dong Y, Zhao J (2004) HetR homodimer is a DNA-binding protein required for heterocyst differentiation, and the DNA-binding activity is inhibited by PatS. Proc Natl Acad Sci USA 101:4848–4853

Huang F, Fulda S, Hagemann M et al (2006) Proteomic screening of salt-stress-induced changes in plasma membranes of *Synechocystis* sp. strain PCC 6803. Proteomics 6:910–920

Jack DL, Yang NM, Saier MH Jr. (2001) The drug/metabolite transporter superfamily. Eur J Biochem 268:3620–3639

Lang NJ, Fay P (1971) The heterocysts of blue-green algae II. Details of ultrastructure. Proc R Soc Lond B 178:193–203

Maeda S, Omata T (1997) Substrate-binding lipoprotein of the cyanobacterium *Synechococcus* sp. strain PCC 7942 involved in the transport of nitrate and nitrite. J Biol Chem 272:3036–3041

Mariscal V, Herrero A, Flores E (2007) Continuous periplasm in a filamentous, heterocyst-forming cyanobacterium. Mol Microbiol 65:1139–1145

Martín-Figueroa E, Navarro F, Florencio FJ (2000) The GS-GOGAT pathway is not operative in the heterocysts. Cloning and expression of *glsF* gene from the cyanobacterium *Anabaena* sp. PCC 7120. FEBS Lett 476:282–286

Means AR (1998) The clash in titin. Nature 395:846–847

Montesinos ML, Herrero A, Flores E (1995) Amino acid transport systems required for diazotrophic growth in the cyanobacterium *Anabaena* sp. strain PCC 7120. J Bacteriol 177: 3150–3157

Moslavac S, Bredemeier R, Mirus O et al (2005) Proteomic analysis of the outer membrane of *Anabaena* sp. strain PCC 7120. J Proteome Res 4:1330–1338

Mullineaux CW, Mariscal V, Nenninger A et al (2008) Mechanism of intercellular molecular exchange in heterocyst-forming cyanobacteria. EMBO J 27:1299–1308

Muro-Pastor AM, Valladares A, Flores E et al (2002) Mutual dependence of the expression of the cell differentiation regulatory protein HetR and the global nitrogen regulator NtcA during heterocyst development. Mol Microbiol 44:1377–1385

Nayar AS, Yamaura H, Rajagopalan R et al (2007) FraG is necessary for filament integrity and heterocyst maturation in the cyanobacterium *Anabaena* sp. strain PCC 7120. Microbiology 153:601–607

Olmedo-Verd E, Valladares A, Flores E et al (2008) Role of two NtcA-binding sites in the complex ntcA gene promoter of the heterocyst-forming cyanobacterium *Anabaena* sp. strain PCC 7120. J Bacteriol 190:7584–7590

Pankratz HS, Bowen CC (1963) Cytology of blue-green algae. I. The cells of *Symploca muscorum*. Am J Bot 50:387–399

Pernil R, Picossi S, Mariscal V et al (2008) ABC-type amino acid uptake transporters Bgt and N-II of *Anabaena* sp. strain PCC 7120 share an ATPase subunit and are expressed in vegetative cells and heterocysts. Mol Microbiol 67:1067–1080

Picossi S, Valladares A, Flores E et al (2004) Nitrogen-regulated genes for the metabolism of cyanophycin, a bacterial nitrogen reserve polymer: expression and mutational analysis of two cyanophycin synthetase and cyanophycinase gene clusters in heterocyst-forming cyanobacterium *Anabaena* sp. PCC 7120. J Biol Chem 279:11582–11592

Picossi S, Montesinos ML, Pernil R et al (2005) ABC-type neutral amino acid permease N-I is required for optimal diazotrophic growth and is repressed in the heterocysts of *Anabaena* sp. strain PCC 7120. Mol Microbiol 57:1582–1592

Rees DC, Johnson E, Lewinson O (2009) ABC transporters: the power to change. Nat Rev Mol Cell Biol 10:218–227

Rippka R, Deruelles J, Waterbury JB et al (1978) Generic assignments, strain stories and properties of pure cultures of cyanobacteria. J Gen Microbiol 111:1–61
Ris H, Singh RN (1961) Electron microscope studies on blue-green algae. J Biophys Biochem Cytol 9:63–80
Robinson C, Bolhuis A (2004) Tat-dependent protein targeting in prokaryotes and chloroplasts. Biochim Bipphys Acta 1694:135–147
Sherman DM, Tucker D, Sherman LA (2000) Heterocyst development and localization of cyanophycin in N_2-fixing cultures of *Anabaena* sp. PCC 7120 (cyanobacteria). J Phycol 36:932–941
Stanier RY, Cohen-Bazire G (1977) Phototrophic prokaryotes: the cyanobacteria. Annu Rev Microbiol 31:225–274
Thomas J, Meeks JC, Wolk CP et al (1977) Formation of glutamine from [^{13}N]ammonia, [^{13}N]dinitrogen, and [^{14}C]glutamate by heterocysts isolated from *Anabaena cylindrica*. J Bacteriol 129:1545–1555
Wei TF, Ramasubramanian TS, Golden JW (1994) *Anabaena* sp. strain PCC 7120 *ntcA* gene required for growth on nitrate and heterocyst development. J Bacteriol 176:4473–4482
Wolk CP (1968) Movement of carbon from vegetative cells to heterocysts in *Anabaena cylindrica*. J Bacteriol 96:2138–2143
Wolk CP (1973) Physiology and cytological chemistry blue-green algae. Bacteriol Rev 37:32–101
Wolk CP, Austin SM, Bortins J et al (1974) Autoradiographic localization of ^{13}N after fixation of ^{13}N-labeled nitrogen gas by a heterocyst-forming blue-green alga. J Cell Biol 61:440–453
Wolk CP, Thomas J, Shaffer PW et al (1976) Pathway of nitrogen metabolism after fixation of ^{13}N-labeled nitrogen gas by the cyanobacterium, *Anabaena cylindrica*. J Biol Chem 251:5027–5034
Wu X, Liu D, Lee MH et al (2004) patS minigenes inhibit heterocyst development of *Anabaena* sp. strain PCC 7120. J Bacteriol 186:6422–6429
Yoon HS, Golden JW (1998) Heterocyst pattern formation controlled by a diffusible peptide. Science 282:935–938
Yoon HS, Golden JW (2001) PatS and products of nitrogen fixation control heterocyst pattern. J Bacteriol 183:2605–2613
Zhang LC, Chen YF, Chen WL et al (2008) Existence of periplasmic barriers preventing green fluorescent protein diffusion from cell to cell in the cyanobacterium *Anabaena* sp. strain PCC 7120. Mol Microbiol 70: 814–823
Zhou R, Wei X, Jiang N et al (1998) Evidence that HetR protein is an unusual serine-type protease. Proc Natl Acad Sci USA 95: 4959–4963

Part III
Bioenergetics, Proteins, and Genomics

Chapter 9
The Photoactive Orange Carotenoid Protein and Photoprotection in Cyanobacteria

Diana Kirilovsky

Abstract Photoprotective mechanisms have been evolved by photosynthetic organisms to cope with fluctuating high light conditions. One of these mechanisms downregulates photosynthesis by increasing thermal dissipation of the energy absorbed by the photosystem II antenna. While this process has been well studied in plants, the equivalent process in cyanobacteria was only recently discovered. In this chapter we describe the results leading to its discovery and the more recent advances in the elucidation of this mechanism. The light activation of a soluble carotenoid protein, the orange carotenoid protein (OCP), binding hydroxyechinenone, is the key inducer of this photoprotective mechanism. Light causes structural changes within both the carotenoid and the protein, leading to the conversion of an orange inactive form into a red active form. The activated red form induces an increase of energy dissipation leading to a decrease in the fluorescence of the phycobilisomes, the cyanobacterial antenna, and thus of the energy arriving to the reaction centers. The OCP, which senses light and triggers photoprotection, is a unique example of a photoactive protein containing a carotenoid as the photoresponsive chromophore.

9.1 Introduction

By harvesting solar energy and converting it into chemical energy, plants, algae, and cyanobacteria provide organic carbon molecules and oxygen that are essential for life on earth. However, excess light can be lethal because harmful reactive oxygen species are generated when the photochemical reaction centers cannot use the incoming energy fast enough. Cyanobacteria, like plants, have developed

---

D. Kirilovsky (✉)
Commissariat à l'Energie Atomique (CEA), Institut de Biologie et Technologies de Saclay (iBiTecS) and Centre National de la Recherche Scientifique (CNRS), 91191 Gif sur Yvette, France
e-mail: diana.kirilovsky@cea.fr

P.C. Hallenbeck (ed.), *Recent Advances in Phototrophic Prokaryotes*, Advances in Experimental Medicine and Biology 675, DOI 10.1007/978-1-4419-1528-3_9,

physiological mechanisms allowing acclimation and survival in a wide range of environmental conditions, especially under high light. High light intensities induce irreversible inactivation of photosystem II (PS II) due to damage and degradation of the D1 protein, an essential constituent of the PS II (for reviews, see Prasil et al. 1992; Aro et al. 1993).

It has long been assumed that cyanobacteria do not use a mechanism dissipating the excess energy as heat at the level of antennae to decrease the energy arriving to the reaction centers to protect themselves from high light intensities (Campbell et al. 1998). Several recent experiments refute this view. Mechanisms mediated by the phycobilisomes (PBs), the soluble antenna of cyanobacteria (El Bissati et al. 2000; Rakhimberdieva et al. 2004; Scott et al. 2006; Wilson et al. 2006), or by empty (not attached to PS I) complexes of the chlorophyll–protein IsiA (iron-starvation-inducible protein) (Yeremenko et al. 2004; Ihalainen et al. 2005), or by HLIPs (high-light-inducible proteins) (Funk and Vermaas 1999; He et al. 2001; Havaux et al. 2003) were recently described. Each of these mechanisms appears to be a photoprotective process.

This mini-review will focus on the characteristics of the phycobilisome-related light-induced photoprotective mechanism. This mechanism decreases the energy arriving at the photosystem II (PS II) reaction center, thus reducing the generation of reactive oxygen species and the probability of PS II damage (Horton et al. 1996; Niyogi 1999). In plants, an equivalent mechanism involving energy dissipation as heat in the antenna (resulting in a detectable quenching of the chlorophyll fluorescence) is induced by the low luminal pH generated during photosynthesis under high irradiance. In cyanobacteria, this process is not triggered by a lowering of the luminal pH or a change in the redox state of the plastoquinone pool (Rakhimberdieva et al. 2004; Scott et al. 2006; Wilson et al. 2006). Instead, the decrease of energy reaching the reaction centers and the quenching of the antenna fluorescence is induced by the light activation of the orange carotenoid protein (OCP) (Wilson et al. 2006), a soluble 35 kDa protein containing a single non-covalently bound carotenoid (Holt and Krogmann 1981; Wu and Krogmann 1997; Kerfeld 2004a, b). This chapter will focus on the function and characteristics of the OCP. Additional information and more details about the blue light-induced photoprotective mechanism are described in Karapetyan (2007), Kirilovsky (2007), Bailey and Grossman (2008), and Kerfeld et al. (2009).

9.2 The Light-Induced Phycobilisome-Related Protective Mechanism in Cyanobacteria: qE_{cya}

To harvest light, most cyanobacteria use a particular type of photosynthetic antenna, the phycobilisome (PB), a large membrane extrinsic complex composed of several types of chromophorylated phycobiliproteins and of linker peptides needed for the structural organization and functioning of the PBs (for reviews, see Glazer 1984; Grossman et al. 1993; MacColl 1998; Tandeau de Marsac 2003; Adir 2005). These complexes, which are attached to the outer surface of the thylakoid membranes (Gantt and Conti 1966a, b), are composed of a core from which rods radiate. In most of the freshwater cyanobacteria the rods contain only phycocyanin (PC) while in many marine cyanobacteria phycoerythrin (PE) or phycoerythrocyanin (PEC) are found in the distal end of the rods. The major core phycobiliprotein is allophycocyanin (APC). Three other chromophorylated proteins ApcD, ApcF, and Lcm serve as terminal energy acceptors which transfer the harvest energy to the chlorophylls of photosystems (Gindt et al. 1994; Ashby and Mullineaux 1999; Dong et al. 2009). In addition the Lcm links the PB to the thylakoids.

Results revealing the existence of a blue-green light-induced photoprotective mechanism proposed to be associated with the phycobilisomes were first described in 2000 (El Bissati et al. 2000). Subsequently, spectral and kinetics data were presented confirming the existence of this mechanism and suggesting that it is induced by blue light-activated carotenoids (Rakhimberdieva et al. 2004). In 2000, photoinhibition and state 2 transition were the only two non-photochemical processes known to decrease fluorescence levels (non-photochemical quenching, NPQ). Under high light conditions (photoinhibition), the decrease of fluorescence is associated with the inactivation of PS II and damage to the D1 protein (for review, see Prasil et al. 1992; Aro et al. 1993; Vass and Aro 2007; Tyystjärvi 2008). Recovery of PS II activity and fluorescence requires the replacement of the damaged D1 and it does not occur in the presence of inhibitors of protein synthesis. In state transitions, the decrease of fluorescence (state 2 transition) is induced by the reduction of the PQ pool caused by a preferential illumination of PS II by orange light (or green when PE is present) principally absorbed by the phycobilisomes (for review, see Williams and Allen 1987; van Thor et al. 1998; Wollman 2001).

In contrast, blue-green light, principally absorbed by photosystem I (PS I), induces the oxidation of the PQ pool and an increase of fluorescence associated with state 1 transition. El Bissati et al. (2000) observed that exposure of *Synechocystis* PCC 6803 (hereafter called *Synechocystis*) cells to strong intensities of blue-green light induced PS II fluorescence quenching under conditions in which the plastoquinone (PQ) pool was largely oxidized and the oxygen evolving activity was not saturated. Moreover, the presence of DCMU, an inhibitor of electron transport between the QA and QB quinones, which prevents PQ pool reduction, did not inhibit the blue-green light-induced fluorescence quenching. In addition, fluorescence recovery occurs in the presence of inhibitors of translation (El Bissati et al. 2000) (Fig. 9.1). These results demonstrated that the blue-green light-induced quenching was not related to photoinhibition or state transitions. Since the induction of the fluorescence quenching was less affected by the lowering of the temperature or by the rigidity of the membranes than state transitions, the authors proposed that high light intensities induce phycobilisome fluorescence quenching accompanied by a decrease of the energy transfer from the phycobilisome to the photosystems. This proposition was later demonstrated. Rakhimberdieva et al. (2004) first showed that in a *Synechocystis* mutant lacking reaction center II and the Chl antennae CP43 and CP47, blue light induced a reversible quenching of the phycobilisome emission.

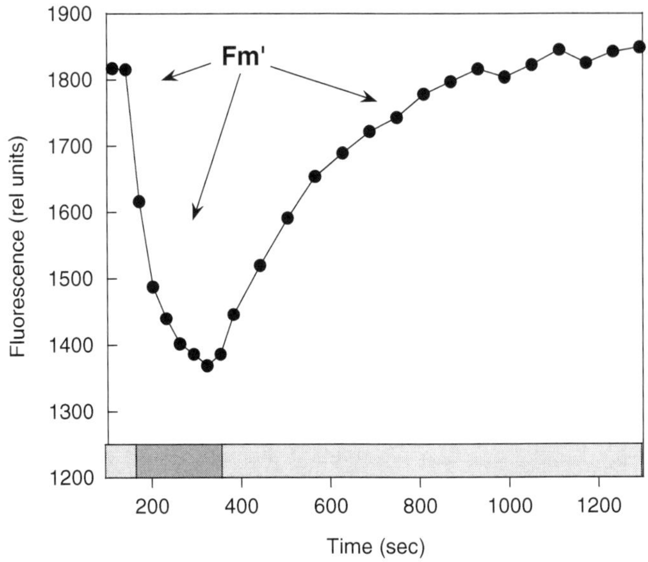

Fig. 9.1 The qE_{cya} mechanism followed by changes in fluorescence levels using a PAM fluorometer. When *Synechocystis* cells are illuminated with strong blue-green light a decrease of maximal fluorescence of PS II is observed. When the "quenched" cells are then illuminated with dim blue-green light fluorescence, they recover in the absence of protein synthesis. *Synechocystis* cells (3 μg Chl/ml) adapted to low intensities of blue-green light (400–550 nm, 80 μmol photons/m^2/s) were illuminated with high-intensity blue-green light (740 μmol photons/m^2/s), and then again with dim blue-green light. Saturating pulses were applied to measure maximal fluorescence levels (Fm'). Chloramphenicol (6 μg/ml) was present throughout the experiment

These results were confirmed by two other laboratories (Scott et al. 2006; Wilson et al. 2006).

In contrast, blue-green light was unable to induce fluorescence quenching in a phycobilisome-deficient mutant or in a mutant containing only the phycocyanin rods of the phycobilisome while the presence of only the core of the phycobilisome is sufficient to induce the quenching (Wilson et al. 2006). In WT *Synechocystis* cells, a specific decrease of the phycobilisome-related fluorescence was observed while there were no changes in the chlorophyll-related fluorescence emission (Scott et al. 2006; Wilson et al. 2006). Phycobilisomes are thus the essential components of the blue light-induced NPQ mechanism. Picosecond time-resolved fluorescence decay data were consistent with a fluorescence quenching at the phycobilisome core (Scott et al. 2006). It has also been proposed recently that in conditions in which all the PS II centers are closed, in the presence of DCMU + light, for example, chlorophyll excitation could also be quenched via the phycobilisome-quenching mechanism. In these conditions, excited chlorophyll molecules could be sufficiently long lived to allow the exciton to be transferred from PS II chlorophyll to the core of the phycobilisome and quenched (Rakhimberdieva et al. 2007a).

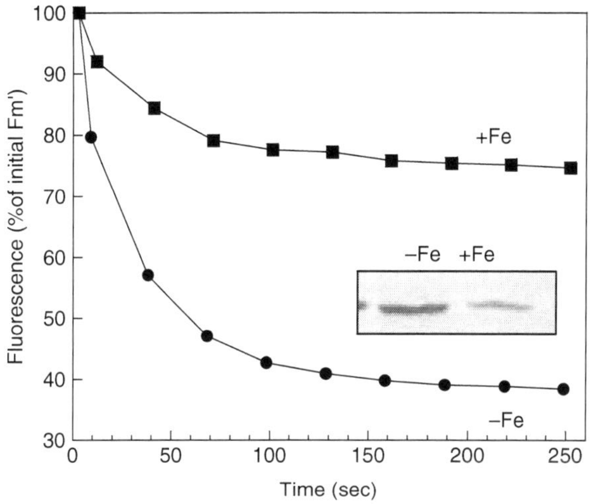

Fig. 9.2 Blue-green light-induced fluorescence quenching in iron-replete and iron-depleted *Synechocystis* cells. Blue-green light-induced decrease of the maximum fluorescence level (Fm') measured by saturating flashes with a PAM fluorometer in iron-containing (+Fe) and 10-day iron-starved (–Fe) *Synechocystis* cells. *Insert*: OCP immunoblot of MP fractions from these cells (each lane contained 1 μg of Chl)

Under stress conditions such as iron starvation (Cadoret et al. 2004; Joshua et al. 2005; Wilson et al. 2007; Rakhimberdieva et al. 2007a), but also in the absence of PS I (Kirilovsky, unpublished results), blue-green light induces a very large fluorescence quenching much larger than under complete medium conditions (Fig. 9.2). It was demonstrated that this light-induced fluorescence quenching is also associated with the phycobilisome-related photoprotective mechanism that becomes predominant under stress conditions (Cadoret et al. 2004; Joshua et al. 2005; Wilson et al. 2007; Rakhimberdieva et al. 2007a).

In plants and algae the equivalent mechanism converts the LHC, the chlorophyll membrane antenna of PS II, a very efficient collector of energy (under low-light conditions) into a very efficient dissipator of energy (Horton et al. 1996; Niyogi 1999; Pascal et al. 2005; Ruban et al. 2007). This energy dissipation is accompanied by a diminution of PS II-related fluorescence emission, known as non-photochemical-quenching fast (NPQ_f) or qE. This NPQ process is rapidly reversible (within seconds) in the dark in the absence of protein synthesis. In cyanobacteria the fluorescence changes associated with energy dissipation in the antenna are not faster than those associated with state transitions: both occur on a timescale of minutes. Thus, the name NPQ_{fast} is inappropriate for the cyanobacterial mechanism. qE is more appropriate since the fluorescence quenching is also associated with energy dissipation in cyanobacteria. However, the cyanobacteria and the plant qE are completely different mechanisms. To differentiate both mechanisms we will designate the cyanobacteria mechanism qE_{cya}.

9.3 The Orange Carotenoid Protein Is the Inducer of the qE_{cya} Mechanism

In higher plants, the antenna-associated qE or NPQ_f is induced by the ΔpH that is built up across the thylakoid membrane during electron transport. The drop of the lumen pH induces the conversion of LHC carotenoids by the xanthophyll cycle (violaxanthin to zeaxanthin) and the protonation of a PS II subunit (PsbS), a member of the LHC superfamily. Conformational changes in the LHCII, which are also induced by the acidification of the lumen, are crucial for this mechanism. These changes modify the interaction between chlorophylls and carotenoids and allow thermal dissipation (Horton, Ruban and Walters 1996; Niyogi 1999; Pascal et al. 2005; Ruban et al. 2007).

In contrast, the phycobilisome-associated qE_{cya} is not dependent on the presence of a *trans*-thylakoid ΔpH (Wilson et al. 2006). Moreover the xanthophyll cycle and the PsbS protein are absent from cyanobacteria cells. The induction of quenching also appears to be independent of excitation pressure on PS II or of changes in the redox state of the plastoquinone pool (El Bissati et al. 2000; Wilson et al. 2006). A specific characteristic of the qE_{cya} mechanism is that its induction depends on the quality of light; only blue-green light is able to induce it. The action spectrum for phycobilisome emission quenching in the PS II-deficient mutants of *Synechocystis* PCC 6803, showing peaks at 496, 470, and 430 nm, suggested that a carotenoid molecule could be involved in this process (Rakhimberdieva et al. 2004). Wilson et al. (2006) then demonstrated that indeed a carotenoid protein, the orange carotenoid protein (OCP), is specifically involved in the qE_{cya} mechanism. In the absence of OCP (in a ΔOCP *Synechocystis* mutant or in cyanobacteria strains lacking the OCP gene), fluorescence quenching induced by strong blue-green light in *Synechocystis* cells is completely inhibited and the cells are more sensitive to high light intensities; this is manifested as a faster decrease in PS II activity (Wilson et al. 2006; Boulay et al. 2008).

Under stress conditions (such as iron starvation), the larger fluorescence quenching observed correlates to a higher OCP concentration (relative to chlorophyll and phycobiliproteins) (Wilson et al. 2007; Boulay et al. 2008) showing the essential role of OCP in the acclimation of cyanobacteria to environmental changes (Fig. 9.2). The relationship between the concentration of the OCP and energy excitation energy dissipation was later confirmed using an overexpressing OCP *Synechocystis* mutant, in which large quantities of OCP were present and a very large fluorescence quenching was observed (Wilson et al. 2008).

OCP is specifically involved in the light-induced phycobilisome-associated qE_{cya} mechanism and not in other mechanisms affecting the levels of fluorescence-like state transitions. The absence of OCP does not affect the reorganization of the photosynthetic apparatus induced by changes in the redox state of the PQ pool: ΔOCP mutant cells were in state 2 (low fluorescence) in the dark and under orange light conditions (reduced PQ pool) and in state 1 (high fluorescence) under blue light conditions (oxidized PQ pool) (Wilson et al. 2006, 2007; Boulay et al. 2008).

The role of OCP suggested that it must interact with the phycobilisomes and/or the thylakoids. Indeed, studies using immunogold labeling and analysis by electron microscopy showed that OCP is present in the inter-thylakoid cytoplasmic region, on the phycobilisome side of the membrane (Wilson et al. 2006). The existence of an interaction between OCP and phycobilisomes and thylakoids was supported by the co-isolation of OCP with the phycobilisome-associated membrane fraction (MP) (Wilson et al. 2006, 2007; Boulay et al. 2008). The interaction of OCP with thylakoids is relatively strong. When the MP fraction was resuspended in a low-phosphate buffer to dissociate phycobilisomes from the membrane, most of the OCP remained attached. Moreover when *Synechocystis* cells were broken in MES buffer (40 mM, pH 6.8) the phycobiliproteins were completely dissociated from the membranes but OCP remained mostly linked (Wilson et al. 2006). In order to obtain OCP in the soluble fraction, the *Synechocystis* cells must be broken in a Tris–HCl buffer (100 mM, pH 8). In contrast, the interaction between OCP and the phycobilisomes is weak, at least in darkness or low light: OCP did not co-isolate with phycobilisomes when a high concentration of phosphate and Triton was used for phycobilisome isolation (Boulay et al. 2008). However, when isolated phycobilisomes were incubated with OCP or were reconstituted in its presence, it was possible to isolate a phycobilisome–OCP complex by co-migration in a sucrose gradient (Boulay et al. 2008).

9.4 The OCP Gene Is Present in Almost All Phycobilisome-Containing Cyanobacteria: The Few OCP-Lacking Strains Are More Sensitive to High Light and React Differently to Stress Conditions

In *Synechocystis*, OCP is encoded by the *slr1963* gene (Wu and Krogmann 1997) and it is constitutively expressed. It is present even in *Synechocystis* mutants lacking phycobilisomes (Wilson et al. 2007). Screening of the currently available cyanobacterial genomic databases showed that genes encoding homologs of *Synechocystis* OCP are present in most phycobilisome-containing cyanobacteria (Table 9.1). From the 39 already sequenced cyanobacterial genomes, 26 contain genes encoding homologs of OCP. Their sequence is highly conserved. The identity of OCP homologs relative to *Synechocystis* OCP varies between 50% (*Gloeobacter violaceus*) and 82% (*Arthrospira maxima*). OCP sequences from freshwater cyanobacteria present a higher similarity to *Synechocystis* OCP (66–82%) than those present in marine cyanobacteria (62–64%). The marine *Synechococcus* OCP sequences are very similar to each other, with identities ranging from 77 to 95%. The expression of full-length OCP was confirmed in seven strains where OCP was constitutively present (Boulay et al. 2008). All of these strains were able to perform blue-green light-induced fluorescence quenching substantiating the proposition that the OCP-related photoprotective mechanism is widespread in cyanobacteria and plays

Table 9.1 Occurrence of OCP-encoding genes and its shorter paralogs in cyanobacterial strains classified by decreasing percentage of the corresponding protein identity compared to *Synechocystis* OCP

Strain	Genes d'OCP	ID (%)	Genes OCP N-ter	Genes OCP C-ter
Synechocystis PCC 6803	*slr1963*	100	–	–
Arthrospira maxima CS-328	*OCP_SPIMA P83689*	82	–	–
Microcystis aeruginosa NIES-843	*MAE_18910*	81	–	–
M. aeruginosa PCC 7806	*IPF_5686*	81	–	–
Cyanothece sp. PCC 7424	*PCC7424_4467*	81	*PCC7424_1930*	*PCC7424_2720*
	PCC7424_5565	58	*PCC7424_1426*	
Lyngbya sp. PCC 8106	*L8106_2910*	80	*L8106_29395*	*L8106_29390*
			L8106_04666	
Cyanothece sp. CCY 0110	*CY0110_09677*	78	*CY0110_08696*	*CY0110_08806*
Cyanothece ATCC 51142	*cce_1649*	78	*cce_1537*	*cce_0742*
			cce_1991	
Anabaena sp. PCC 7120	*all3149*	75	*All1123*	*All4940*
			Alr4783	
			All4941	
			All3221	
Anabaena variabilis ATCC 29413	*Ava_3843*	75	*Ava_2052*	*Ava_2231*
			Ava_2230	
			Ava_4694	

Table 9.1 (continued)

Strain	Genes d'OCP	ID (%)	Genes OCP N-ter	Genes OCP C-ter
Microcoleus chthonoplastes PCC 7420	*MC7420_3617*	74	*MC7420_905*	*MC7420_6621*
Synechococcus sp. PCC 7335	*S7335_2791* *S7335_487* *S7335_655*	74 74 65	*S7335_4819* *S7335_2204*	*S7335_4977*
Synechococcus sp. PCC 7002	*SYNPCC7002_A28010*	72		
Synechococcus sp. WH 5701	*WH5701_04010* *WH5701_00210 (219aa)*	68 60		
Nostoc punctiforme PCC 73102	*Npun02003015* *Npun02000033*	66 55	*Npun0200500* *Npun02001317* *Npun02003030* *Npun02004998*	*Npun02003027*
Synechococcus sp. RCC307	*RCC307_1992*	65	–	–
Synechococcus sp. RS9917	*RS9917_00692*	64	–	–
Synechococcus sp. WH7803	*SynWH7803_0929*	64	–	–
Synechocccus sp. CC9311	*Sync_1803*	64	–	–
Synechococcus sp. BL107	*BL107_14105*	63	–	–

Table 9.1 (continued)

Strain	Genes d'OCP	ID (%)	Genes OCP N-ter	Genes OCP C-ter
Synechococcus sp. WH7805	*WH7805_01202*	63	–	–
Synechococcus sp. CC9902	*syncc9902_0973*	63	–	–
Synechococcus sp. WH8102	*SYNW1367*	62	–	–
Nodularia spumigena CCY9414	*N9414_13085*	56	*N9414_12098* *N9414_22258*	*N9414_22253*
'Nostoc azollae' 0708	*AazoDRAFT_1458*	54	*AazoDRAFT_3427* *AazoDRAFT_3422* *AazoDRAFT_5633*	*AazoDRAFT_3428*
Gloeobacter violaceus PCC 7421	*glr3935 (274aa)* *glr0050*	50 49	*gll0260 (217aa)* *gll0259*	*glr2503*
Cyanobium sp. PCC 7001	–	–	–	*CPCC7001_1352*
Cyanothece sp. PCC 7822	–	–	*Cyan7822DRAFT_2378* *Cyan7822DRAFT_2276* *Cyan7822DRAFT_4359*	*Cyan7822DRAFT_4358*
Nostoc sp. PCC 7425	–	–	*Cyan7425_1628* *Cyan7425_4810* *Cyan7425_0443*	*Cyan7425_1627*
Cyanothece sp. PCC 8801	–	–	*PCC8801_2551*	*PCC8801_2552*

Table 9.1 (continued)

Strain	Genes d'OCP	ID (%)	Genes OCP N-ter	Genes OCP C-ter
Cyanothece sp. PCC 8802	–	–	*Cyan8802DRAFT_1697*	*Cyan8802DRAFT_1698*
Crocosphaera watsonii WH8501	–	–	*CwatDRAFT_0985*	*CwatDRAFT_5349*
Thermosynechococcus elongatus BP-1	-	–	*tll1269*	*tll1268*

an important role in the physiology of these organisms (Boulay et al. 2008). Furthermore, in *A. maxima*, as in *Synechocystis*, iron starvation induces an increase in OCP concentration and a larger fluorescence quenching (Boulay et al. 2008) consistent with previous results (Hihara et al. 2001; Fulda et al. 2006; Wilson et al. 2007). The presence of a larger qE_{cya} protects the cells from "dangerous" functionally disconnected phycobilisomes under iron starvation conditions. The transfer of excess absorbed energy from phycobilisomes to thylakoids could induce oxidative damage (e.g., lipid peroxidation) if it is not thermally dissipated at the level of the phycobilisomes.

Synechocystis, Microcystis aeruginosa, A. maxima, Synechococcus 7002, and all marine *Synechococcus* strains contain only one copy of the OCP gene. Other cyanobacteria (mostly nitrogen fixing species) including *Lyngbya* sp. PCC 8106; *Cyanothece* PCC 7424, CCY0110, and ATCC51142; *Nostoc* sp. PCC 7120; *Nostoc azollae* 0708; *Anabaena variabilis*; *Nostoc punctiforme* PCC 73102; *Nodularia spumigena* CCY9414; and *G. violaceus* PCC 7421 have in addition multiple gene copies encoding the N-terminal domain located in disparate parts of the genome and a single copy of the gene coding for the C-terminal domain (Table 9.1). This gene is generally present near one of the N-terminal OCP-encoding genes. In *N. punctiforme* several of the N-terminal paralogs are known to be expressed (Anderson, Campbell and Meeks 2006). In *Cyanothece* ATCC51142, the C-terminal gene and one of the N-terminal genes are both expressed under light conditions but not in darkness (Stockel et al. 2008). The function of these fragments is unknown and their origin seems to be an ancient gene duplication.

Among PB-containing cyanobacteria, only a few strains (12 from 39) do not have an OCP gene homologue, including the freshwater *Synechococcus elongatus* PCC 7942 and PCC 6301; the thermophile *Thermosynechococcus elongatus*, *Synechococcus* sp. A, and *Synechococcus* sp. B′; the nitrogen fixing strains *Cyanothece* sp. PCC 7822, PCC 8801, and PCC 8802, and *Nostoc* sp. PCC 7425; and the marine *Synechococcus* sp. CC9605, *Cyanobium* PCC 7001, and *Crocosphaera watsonii* WH8501. However, *T. elongatus, Cyanothece* sp. PCC 7822, PCC 8801, and PCC 8802, and *Nostoc* sp. PCC 7425 contain separate but adjacent genes coding for the N- and C-terminal parts of the OCP (Table 9.1). In *C. watsonii* WH8501, they are in different parts of the genome. In the *S. elongatus* PCC 7942 and *T. elongatus* strains, which lack OCP, the qE_{cya} mechanism was absent confirming the correlation between the blue-green light-induced qE_{cya} and the presence of OCP-like genes (Boulay et al. 2008). The proteins encoded by the genes for the N- and C-terminal parts of OCP, if expressed, did not allow the induction of the qE_{cya} mechanism in *T. elongatus*. Strains which lack OCP are more sensitive to episodes of high light irradiance like the ΔOCP *Synechocystis* mutant (Boulay et al. 2008). Interestingly, *S. elongatus* PCC 7942 and *T. elongatus*, which lack the qE_{cya} mechanism, protect themselves by quickly decreasing the cellular content of phycobiliproteins to avoid the accumulation of potentially harmful functionally disconnected phycobilisomes under iron starvation conditions (Boulay et al. 2008).

9.5 OCP Is a Photoactive Protein: The Red Form Is the Active Form

OCPs isolated from *A. maxima* (Kerfeld et al. 2003) and *Synechocystis* (Kerfeld in preparation) consist of two domains: an all α-helical N-terminal domain and a mixed α/β sheet C-terminal one. The protein-embedded ketocarotenoid, 3′-hydroxyechinenone (hECN), which is composed of a conjugated chain of 11 carbon–carbon double bonds, is in an all-*trans*-configuration. The carotenoid spans both protein domains with its keto terminus in the hydrophobic pocket of the C-terminal domain. The hydroxyl terminus of the 3′-hydroxyechinenone is located in the N-terminal part. Tyr 44 and Trp110 which are strictly conserved make hydrophobic contacts with the hydroxy terminal end of the carotenoid whereas two other strictly conserved residues, Tyr203 and Trp290, form hydrogen bonds to the carbonyl moiety at the keto terminus of the pigment (more details in Kerfeld et al. 2003; Kerfeld et al. 2009).

Synechocystis OCP was isolated using Ni chromatography followed by an ion-exchange column from C-terminal His-tagged OCP mutants. In these mutants the *slr1963* gene containing the His-tag was expressed using its own promoter (His-tagged OCP mutant) or the *psbA2* promoter (overexpressing the His-tagged mutant) as the promoter (Wilson et al. 2008; Punginelli et al. 2009). In the latter case, large quantities of the OCP were present five to eight times more than in the wild-type (WT)] and a very large fluorescence quenching was observed. WT His-tagged OCP contained mostly hydroxyechinenone with only traces of echinenone (Wilson et al. 2008), while the overexpressed His-tagged OCP contained large quantities of echinenone (Fig. 9.3). In this strain, depending on the preparations, the echinenone

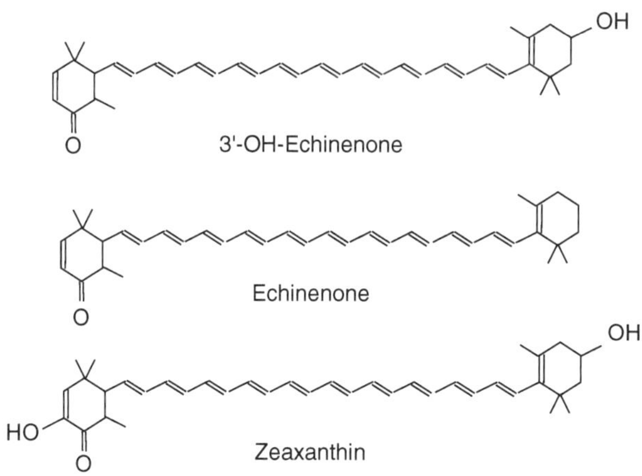

Fig. 9.3 Carotenoids in OCP. WT-OCP binds 3′-hydroxyechinenone (hECN) but it can also bind echinenone and zeaxanthin when hECN or echinenone are absent, or are not present in sufficient quantity

content varied from 53 to 82% of total carotenoid, while the hydroxyechinenone content varied from 11 to 35%. Zeaxanthin was also present in OCP preparations from an overexpressing OCP strain with a content varying from 6 to 14%. This is due to the fact that in *Synechocystis* cells there is only a limited quantity of hydroxyechinenone and this seems to be insufficient to bind to the large amount of OCP. OCP was also isolated from a double-mutant-overexpressing His-tagged OCP in the absence of the β-carotene monoketolase CrtO (Punginelli et al. 2009). In this mutant, echinenone and hECN are lacking and OCP binds mostly zeaxanthin (Fig. 9.3). However, the stability of the binding seems to be lower since less OCP protein was present in the ΔCrtO strains. The three carotenoids that were found bound to the OCP contain a central chain of 11 conjugated double bonds and differ in the oxygenated groups of their rings. Echinenone and hydroxyechinenone have a carbonyl group (site 2). In addition, hydroxyechinenone has a 3′-hydroxyl group. Zeaxanthin has two hydroxyl groups, one in each ring (sites 3 and 3′).

In darkness or dim light, the OCP isolated from the His-tagged OCP and the overexpressing His-tagged OCP strains appeared orange and their absorption spectra were similar (Punginelli et al. 2009). The OCP spectra of both strains presented a typical carotenoid shape with the 0–0 vibrational peak located at 496 nm (Kerfeld et al. 2003; Polivka et al. 2005; Wilson et al. 2008) (Fig. 9.4). In agreement with the crystallographic structure (Kerfeld et al. 2003), the spectrum corresponds to that of hECN (or echinenone) locked into an all-*trans* conformation by the surrounding protein (Polivka et al. 2005). The Raman spectra of the OCP from *A. maxima* and *Synechocystis* PCC 6803 also showed that the hECN is in an all-*trans* conformation (Wilson et al. 2008). OCP isolated from the mutant lacking CrtO is yellow (Punginelli et al. 2009). The resolution of the vibrational peaks was higher in the OCP isolated from this mutant and a blueshift (2–3 nm) of the maximum of the three vibrational bands was visible (Fig. 9.4).

Upon illumination with blue-green light (at 10°C), orange WT His-tagged OCP is completely photoconverted to a red form (Wilson et al. 2008) (Fig. 9.4a). The redshifted spectrum of the light-induced form with a maximum at 510 nm losses the resolution of the vibrational bands. The initial rate of light photoconversion strongly depended on the light intensity, reminiscent of the light dependency of the blue light-induced qE_{cya} in whole cells (Wilson et al. 2006). Light conversion occurs with a very low quantum yield (about 0.03); this can be explained by the fact that OCP is involved in a photoprotective mechanism that must be induced only under high irradiance (Wilson et al. 2008). In darkness, the red OCP spontaneously reverts to the orange form (Wilson et al. 2008). This step, which is not accelerated by illumination, shows marked temperature dependence. In vivo, the recovery qE_{cya} kinetics also showed a large temperature dependence; however, the kinetics were slower (El Bissati et al. 2000; Wilson et al. 2006; Rakhimberdieva et al. 2007b) than the red to orange dark conversion (Wilson et al. 2008), suggesting that the red OCP form is more stable in vivo than in vitro or that the fluorescence quenching remains longer than with the red form.

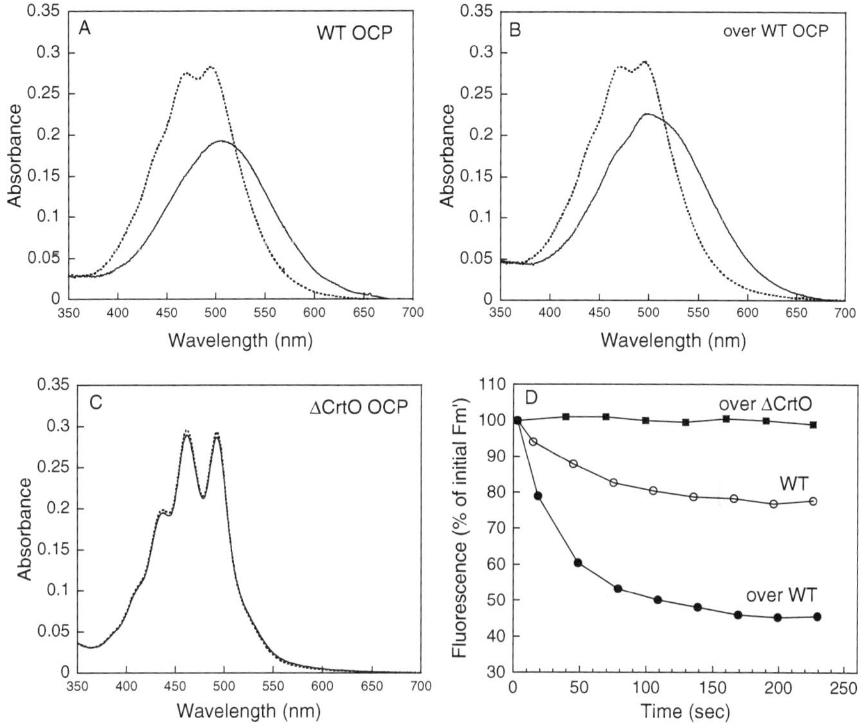

Fig. 9.4 Photoactivity of isolated OCP. The red form is the active form. Zeaxanthin cannot replace hECN (**a–c**): Absorbance spectra of the dark (*dotted line*) and light (*solid line*) forms of the OCP isolated from the His-tagged OCP strain (**a**), the overexpressing His-tagged OCP strain (**b**), and the ΔCrtO-overexpressing His-tagged OCP strain (**c**). To obtain the spectrum of the light form, the isolated OCP was illuminated with 1200 μmol photons/m^2/s, at 12°C, for 5 min. (**d**) Decrease of maximal fluorescence (Fm') induced by strong blue-green light in the overexpressing His-tagged OCP strain (*closed circles*), the His-tagged OCP strain (*open circles*), and the ΔCrtO-overexpressing His-tagged OCP strain

Illumination of OCP isolated from the ΔCrtO mutant did not change its color or the spectrum of the protein (Fig. 9.4c). Moreover, in strains containing zeaxanthin–OCP, light was unable to induce any fluorescence quenching although zeaxanthin–OCP absorbs blue-green light very well (Punginelli et al. 2009). His-tagged OCP isolated from the overexpressing strain was almost completely converted to the red form and the final spectrum was dependent on the zeaxanthin content but not on the different ratios of echinenone to hydroxyechinenone found in the OCP preparations (Fig. 9.4b). Furthermore, the kinetics of orange to red photoconversion at different light intensities, as well as red to orange dark conversion at different temperatures, were similar in WT-OCP and overexpressed OCP. Thus, echinenone can replace 3′-hydroxyechinenone in its function in the photoprotective mechanism since echinenone–OCP is photoactive and able to induce fluorescence quenching. In contrast, zeaxanthin–OCP is not photoactive, or at least it does not have a relatively stable light form, and it is unable to induce fluorescence quenching (Fig. 9.4d). Therefore, the presence of the carbonyl in echinenone and hECN is essential for the stabilization of the red form and the induction of fluorescence quenching. In contrast, the hydroxyl group is not required for the activity of OCP since echinenone behaves like hydroxyechinenone, and both can generate the red OCP form and induce fluorescence quenching (Punginelli et al. 2009).

These results also strongly suggested that the red OCP form is the active form. This was confirmed by the study of qE_{cya} in an overexpressing C-terminal His-tagged W110S OCP mutant. OCP carrying this point mutation is orange and contains mostly hECN. Although large quantities of OCP were present in this strain, almost no fluorescence quenching was induced by blue-green light and only a very small portion of this OCP variant was converted to the red form even after extended illumination with strong blue-green light at 10°C (Wilson et al. 2008; Punginelli et al. 2009). Therefore, the stabilization of the red form is essential for the induction of the fluorescence quenching. This was confirmed by the construction of other mutants that are unable to stabilize or to form the red OCP in which blue-green light does not induce the qE_{cya} mechanism (Wilson, Punginelli and Kirilovsky, in preparation). Moreover, absorbance spectra of *Synechocystis* cells overexpressing OCP measured before and after illumination with strong blue light strongly suggested that the red OCP form is accumulated in vivo under conditions which induce energy dissipation and fluorescence quenching (strong blue-green light) (Wilson et al. 2008).

9.6 Blue-Green Light Induces Changes in Both the Carotenoid and the Protein

Resonance Raman spectroscopy and light-induced Fourier transform infrared (FTIR) difference spectra demonstrated that light absorbance by OCP induces structural changes not only in the carotenoid but also in the protein (Wilson et al. 2008). Upon illumination of *A. maxima* and *Synechocystis* OCP, the apparent conjugation length of hECN increased by about one conjugated bond, and hECN reached a less distorted, more planar structure. These changes are in agreement with the observed redshift of the absorbance spectrum and strongly suggest that a *trans–cis* isomerization of the hECN does not occur upon illumination. It was concluded that hECN is still all-*trans* in the red OCP form. Although the specific changes in the carotenoid of the OCP remain to be elucidated they are clearly different from those occurring in retinal, a carotenoid derivative, which is the chromophore of rhodopsins (review about rhodopsins: Spudich et al. 2000). In rhodopsins, the retinal is covalently bound to the protein via a protonated Schiff-base linkage to a Lys residue and light induces an isomerization of the retinal initiating the photocycle. The photocycle involves proton transfer between the Schiff-base and a carboxylate residue. In OCP not only is no isomerization detected, the hECN is not covalently bound to the protein even though it is close to the absolutely conserved Arg155.

Large changes have also observed in the OCP FTIR spectra upon illumination, suggesting a major rearrangement of the protein backbone. The changes correspond to a less rigid helical structure in a significant part of the protein and a compaction of the β-sheet domain upon formation of the red form (Wilson et al. 2008). Further studies will be necessary to elucidate the specific nature of the changes in the carotenoid and the protein.

9.7 Conclusions and Perspectives

It is clear that OCP is a sensor of photoinhibitory conditions inducing a greater energy dissipation and fluorescence quenching (photoprotection) under increasing irradiance and lowering temperature (see model in Fig. 9.5). However, many questions about the qE_{cya} mechanism and the role of the OCP await elucidation; these include the following: Why is the red form the active form? How does it interact with the phycobilisome? Does OCP receive the energy from the phycobilisome and is it the energy quencher?

Based on the results already obtained, we can hypothesize that the red form of OCP is the energy and fluorescence quencher. We postulate that the light-induced modifications in the carotenoid and the protein are necessary to induce stronger interactions between OCP and the phycobilisomes and to allow energy transfer between the phycobilisome chromophores and the OCP. The redshift of the hECN (or echinenone) spectrum could be necessary to tune the optically forbidden S1 state of hECN to a level allowing energy transfer from the phycobilisomes to OCP.

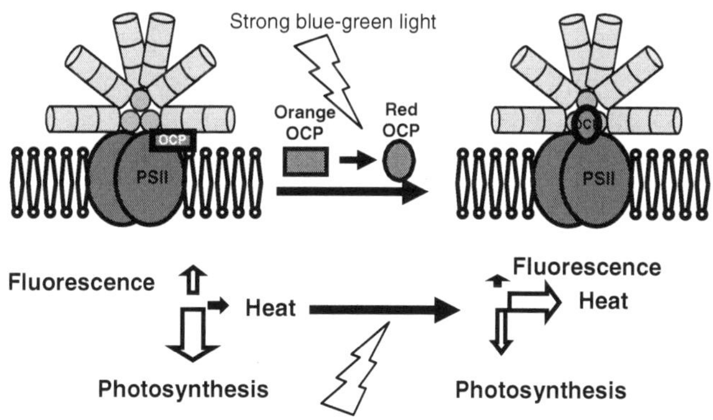

Fig. 9.5 The qE_{cya} mechanism and the role of OCP. The absorption of blue-green light by hECN induces conformational changes in the carotenoid and protein converting the orange OCP form into an active red form capable of inducing an increase of thermal excess energy dissipation as heat, leading to phycobilisome fluorescence quenching and to a decrease of energy arriving at the reaction centers. As a working hypothesis, we propose that the red OCP form is also the energy and fluorescence quencher capable of receiving the energy absorbed by the phycobilisome and of converting it into heat by interaction with the phycobilisome core

We propose that zeaxanthin–OCP and W110S OCP, due to the lack of a metastable "red" form, are unable to absorb the energy coming from the phycobilisome and thus they do not dissipate this energy. On the other hand, the conformational changes of the carotenoid and protein induced by blue-green light could alter the strength of the hydrogen bonds between the carbonyl and Tyr203 and/or Trp290, which could form a signal propagation pathway from the carotenoid to the surface of the protein, like in other photoactive proteins (PYP and LOV domains: Rubinstenn et al. 1998; Harper et al. 2003). These changes could also modify the interaction between the phycobilisome and the OCP (Kerfeld et al. 2003). We have hypothesized that the C-terminal domain of the OCP, by interacting with the center of an allophycocyanin (APC) trimer in the phycobilisome core, may bring the carotenoid into the proximity of the APC chromophores (Wilson et al. 2008). Maybe these interactions occur only with red OCP and they do not occur with zeaxanthin–OCP. The construction of OCP and phycobilisome mutants and the associated characterization of their phenotypes by spectroscopic and structural studies will be essential to elucidate which components of the PBS core are involved in the interaction with the OCP and in energy dissipation. Likewise, this will provide information about the role of specific amino acids that are essential for photoactivity and for energy dissipation, and it will help us to understand which side chains influence the properties of the carotenoid and the affinity of the carotenoid for the protein.

Acknowledgments Many thanks to my students Adjélé Wilson, Clémence Boulay, and Claire Punginelli who largely contributed to the results described in this work. Thanks to AW Rutherford for helpful discussions and critical reading of the manuscript. Thanks to Cheryl Kerfeld for helpful discussions and collaboration. The research of DK and her group is supported by the "Commissariat à l'Energie Atomique," the "Centre National de la Recherche Scientifique, and the "Agence Nationale de la Recherche" (project CAROPROTECT). The work was also partially supported by EU network INTRO2.

References

Adir N (2005) Elucidation of the molecular structures of components of the phycobilisome: reconstructing a giant. Photosynth Res 85:15–32

Anderson DC, Campbell EL, Meeks JC (2006) A soluble 3D LC/MS/MS proteome of the filamentous cyanobacterium *Nostoc punctiforme*. J Proteome Res 5:3096–3104

Aro EM, Virgin I, Andersson B (1993) Photoinhibition of photosystem II, inactivation, protein damage and turnover. Biochim Biophys Acta 1143:113–134

Ashby MK, Mullineaux CW (1999) The role of ApcD and ApcF in energy transfer from phycobilisomes to PSI and PSII in a cyanobacterium. Photosynth Res 61:169–179

Bailey S, Grossman A (2008) Photoprotection in cyanobacteria:regulation of light harvesting. Photochem Photobiol 84:1410–1420

Boulay C, Abasova L, Six C et al (2008) Occurrence and function of the orange carotenoid protein in photoprotective mechanisms in various cyanobacteria. Biochim Biophys Acta 1777: 1344–1354

Boulay C, Wilson A, Kirilovsky D (2008) Orange carotenoid protein (OCP) related NPQ in *Synechocystis* PCC 6803 OCP-phycobilisomes interactions. In: Photosynthesis 2007. Energy from the sun. Proceedings of the 14th International Congress on Photosynthesis. Springer, Heidelberg, pp 254–254

Cadoret JC, Demouliere R, Lavaud J et al (2004) Dissipation of excess energy triggered by blue light in cyanobacteria with CP43′ (IsiA). Biochim Biophys Acta 1659:100–104

Campbell D, Hurry V, Clarke AK et al (1998) Chlorophyll fluorescence analysis of cyanobacterial photosynthesis and acclimation. Microbiol Mol Biol Rev 62:667–683

Dong C, Tang A, Zhao J et al (2009) ApcD is necessary for efficient energy transfer from phycobilisomes to photosystem I and helps to prevent photoinhibition in the cyanobacterium *Synechococcus* sp. PCC 7002. Biochim Biophys Acta 1787:1122–1128

El Bissati K, Delphin E, Murata N et al (2000) Photosystem II fluorescence quenching in the cyanobacterium *Synechocystis* PCC 6803:involvement of two different mechanisms. Biochim Biophys Acta 1457:229–242

Fulda S, Mikkat S, Huang F et al (2006) Proteome analysis of salt stress response in the cyanobacterium *Synechocystis* sp. strain PCC 6803. Proteomics 6:2733–2745

Funk C, Vermaas W (1999) A cyanobacterial gene family coding for single-helix proteins resembling part of the light-harvesting proteins from higher plants. Biochemistry 38:9397–9404

Gantt E, Conti SF (1966a) Granules associated with chloroplast lamellae of *Porphyridium cruentum*. J Cell Biol 29:423–434

Gantt E, Conti SF (1966b) Phycobiliprotein localization in algae. Brookhaven Symp Biol 19: 393–405

Gindt YM, Zhou J, Bryant DA et al (1994) Spectroscopic studies of phycobilisome subcore preparations lacking key core chromophores: assignment of excited state energies to the L_{cm}, β^{18} and $\alpha^{AP\text{-}B}$ chromophores. Biochim Biophys Acta 1186:153–162

Glazer AN (1984) Phycobilisome – a macromolecular complex optimized for light energy-transfer. Biochim Biophys Acta 768:29–51

Grossman AR, Schaefer MR, Chiang GG et al (1993) The phycobilisome, a light-harvesting complex responsive to environmental-conditions. Microbiol Rev 57:725–749

Harper SM, Neil LC, Gardner KH (2003) Structural basis of a phototropin light switch. Science 301:1541–1544

Havaux M, Guedeney G, He Q et al (2003) Elimination of high-light-inducible polypeptides related to eukaryotic chlorophyll a/b-binding proteins results in aberrant photoacclimation in *Synechocystis* PCC6803. Biochim Biophys Acta 1557:21–33

He Q, Dolganov N, Bjorkman O et al (2001) The high light-inducible polypeptides in *Synechocystis* PCC6803. Expression and function in high light. J Biol Chem 276:306–314

Hihara Y, Kamei A, Kanehisa M et al (2001) DNA microarray analysis of cyanobacterial gene expression during acclimation to high light. Plant Cell 13:793–806

Holt TK, Krogmann DW (1981) A carotenoid-protein from cyanobacteria. Biochim Biophys Acta 637:408–414

Horton P, Ruban AV, Walters RG (1996) Regulation of light harvesting in green plants. Annu Rev Plant Physiol Plant Mol Biol 47:655–684

Ihalainen JA, D′Haene S, Yeremenko N et al (2005) Aggregates of the chlorophyll-binding protein IsiA (CP43′) dissipate energy in cyanobacteria. Biochemistry 44:10846–10853

Joshua S, Bailey S, Mann NH et al (2005) Involvement of phycobilisome diffusion in energy quenching in cyanobacteria. Plant Physiol 138:1577–1585

Karapetyan NV (2007) Non-photochemical quenching of fluorescence in cyanobacteria. Biochemistry (Mosc) 72:1127–1135

Kerfeld CA (2004a) Structure and function of the water-soluble carotenoid-binding proteins of cyanobacteria. Photosynth Res 81:215–225

Kerfeld CA (2004b) Water-soluble carotenoid proteins of cyanobacteria. Arch Biochem Biophys 430:2–9

Kerfeld CA, Alexandre M, Kirilovsky D (2009) The orange carotenoid protein of cyanobacteria. In: Landrum J (ed), Carotenoids: physical, chemical and biological functions and properties. Taylor and Francis Group, London

Kerfeld CA, Sawaya MR, Brahmandam V et al (2003) The crystal structure of a cyanobacterial water-soluble carotenoid binding protein. Structure 11:55–65

Kirilovsky D (2007) Photoprotection in cyanobacteria: the orange carotenoid protein (OCP)-related non-photochemical-quenching mechanism. Photosynth Res 93:7–16

MacColl R (1998) Cyanobacterial phycobilisomes. J Struct Biol 124: 311–334

Niyogi KK (1999) Photoprotection revisited: genetic and molecular approaches. Annu Rev Plant Physiol Plant Mol Biol 50:333–359

Pascal AA, Liu ZF, Broess K et al (2005) Molecular basis of photoprotection and control of photosynthetic light-harvesting. Nature 436:134–137

Polivka T, Kerfeld CA, Pascher T et al (2005) Spectroscopic properties of the carotenoid 3′-hydroxyechinenone in the orange carotenoid protein from the cyanobacterium *Arthrospira maxima*. Biochemistry 44:3994–4003

Prasil O, Adir N, Ohad I (1992) Dynamics of photosystem II:mechanism of photoinhibition and recovery processes. In: Barber J (ed) The photosystems:structure, function and molecular biology Elsevier Science, Amsterdam, pp 295–348

Punginelli C, Wilson A, Routaboul JM et al (2009) Influence of zeaxanthin and echinenone binding on the activity of the Orange Carotenoid Protein. Biochim Biophys Acta 1787:280–288

Rakhimberdieva MG, Bolychevtseva YV, Elanskaya IV et al (2007b) Protein-protein interactions in carotenoid triggered quenching of phycobilisome fluorescence in *Synechocystis* sp. PCC 6803. FEBS Lett 581:2429–2433

Rakhimberdieva MG, Stadnichuk IN, Elanskaya IV et al (2004) Carotenoid-induced quenching of the phycobilisome fluorescence in photosystem II-deficient mutant of *Synechocystis* sp. FEBS Lett 574:85–88

Rakhimberdieva MG, Vavilin DV, Vermaas WF et al (2007a) Phycobilin/chlorophyll excitation equilibration upon carotenoid-induced non-photochemical fluorescence quenching in phycobilisomes of the cyanobacterium *Synechocystis* sp. PCC 6803. Biochim Biophys Acta 1767:757–765

Ruban AV, Berera R, Ilioaia C et al (2007) Identification of a mechanism of photoprotective energy dissipation in higher plants. Nature 450:575–U522

Rubinstenn G, Vuister GW, Mulder FA et al (1998) Structural and dynamic changes of photoactive yellow protein during its photocycle in solution. Nat Struct Biol 5:568–570

Scott M, McCollum C, Vasil′ev S et al (2006) Mechanism of the down regulation of photosynthesis by blue light in the Cyanobacterium *Synechocystis* sp. PCC 6803. Biochemistry 45:8952–8958

Spudich JL, Yang CS, Jung KH et al (2000) Retinylidene proteins:structures and functions from archaea to humans. Annu Rev Cell Dev Biol 16:365–392

Stockel J, Welsh EA, Liberton M et al (2008) Global transcriptomic analysis of *Cyanothece* 51142 reveals robust diurnal oscillation of central metabolic processes. Proc Natl Acad Sci USA 105:6156–6161

Tandeau de Marsac N (2003) Phycobiliproteins and phycobilisomes:the early observations. Photosynth Res 76:197–205

Tyystjärvi E (2008) Photoinhibition of photosystem II and photodamage of the oxygen evolving manganese cluster. Coordination Chem Rev 252:361–376

van Thor JJ, Mullineaux CW, Matthijs HCP et al (1998) Light harvesting and state transitions in cyanobacteria. Botanica Acta 111:430–443

Vass I, Aro EM (2007) Photoinhibition of photosystem II electron transport. In:Renger G, (ed) Primary processes of photosynthesis:basic principles and apparatus. Royal Society of Chemistry, Cambridge, pp 393–411

Williams WP, Allen JF (1987) State-1/State-2 changes in higher-plants and algae. Photosynth Res 13:19–45

Wilson A, Ajlani G, Verbavatz JM et al (2006) A soluble carotenoid protein involved in phycobilisome-related energy dissipation in cyanobacteria. Plant Cell 18:992–1007

Wilson A, Boulay C, Wilde A et al (2007) Light-induced energy dissipation in iron-starved cyanobacteria:roles of OCP and IsiA proteins. Plant Cell 19:656–672

Wilson A, Punginelli C, Gall A et al (2008) A photoactive carotenoid protein acting as light intensity sensor. Proc Natl Acad Sci USA 105:12075–12080

Wollman FA (2001) State transitions reveal the dynamics and flexibility of the photosynthetic apparatus. Embo J 20:3623–3630

Wu YP, Krogmann DW (1997) The orange carotenoid protein of *Synechocystis* PCC 6803. Biochim Biophys Acta 1322:1–7

Yeremenko N, Kouril R, Ihalainen JA et al (2004) Supramolecular organization and dual function of the IsiA chlorophyll-binding protein in cyanobacteria. Biochemistry 43:10308–10313

Chapter 10
Proteomic Analysis of the Developing Intracytoplasmic Membrane in *Rhodobacter sphaeroides* During Adaptation to Low Light Intensity

Kamil Woronowicz and Robert A. Niederman

Abstract Although the primary photochemical events in the facultative photoheterotrophic purple bacterium *Rhodobacter sphaeroides* are now well understood, comparatively little is known about how their photosynthetic apparatus is assembled. Here we present a proteomic analysis of the intracytoplasmic membrane (ICM) assembly process during adaptation to lowered light intensity, in which the size of the photosynthetic units is greatly expanded by addition of the light-harvesting 2 (LH2) peripheral antenna complex. When the isolated ICM-derived chromatophore vesicles were subjected to clear native gel electrophoresis (CNE), four pigmented bands appeared; the top and bottom bands contained the reaction center – light-harvesting 1 (RC–LH1) core complex and LH2 peripheral antenna, respectively, while the two bands of intermediate migration contained associations of the LH2 and core complexes. Proteomic analysis revealed a large array of other proteins associated with the CNE gel bands – in particular, several F_1F_O-ATP synthase subunits gave unexpectedly high spectral counts, given the inability to detect this coupling factor, as well as the more abundant cytochrome bc_1 complex, by atomic force microscopy. Significant levels of general membrane assembly factors were also found, as well as numerous proteins of unknown function including high counts for RSP6124 that were correlated with LH2 levels. When combined with further AFM and spectroscopic studies, these proteomic approaches are expected to provide a much-improved understanding of the overall assembly process.

10.1 Introduction

The facultative photoheterotrophic bacterium *Rhodobacter sphaeroides* provides a unique combination of accessible molecular genetics with an intracytoplasmic photosynthetic membrane (ICM) that is amenable to an unparalleled variety of

R.A. Niederman (✉)
Department of Molecular Biology and Biochemistry, Rutgers University, Piscataway, NJ, 08854, USA
e-mail: rniederm@rci.rutgers.edu

P.C. Hallenbeck (ed.), *Recent Advances in Phototrophic Prokaryotes*, Advances in Experimental Medicine and Biology 675, DOI 10.1007/978-1-4419-1528-3_10,

biochemical, spectroscopic, and ultrastructural probes. As a result, *R.. sphaeroides* has served as a simple, dynamic, and experimentally accessible model system that has greatly enhanced the understanding of primary photochemical events and energy transduction processes, and of how the levels of the participating protein components are regulated (Hunter et al. 2008). In contrast, much less is known about the mechanisms involved in their assembly, how their pattern of localization is established within the cell membranes, or how numerous assembly factors cooperate with the apoprotein components to form functional photosynthetic complexes in the growing ICM.

When *R. sphaeroides* is grown photoheterotropically (under anaerobic conditions in the light), an ICM system is formed that houses photosynthetic units consisting of rows of dimeric light-harvesting 1 (LH1)–reaction center (RC) core structures, interspersed with narrow lanes of the peripheral LH2 antenna complex, as demonstrated by the AFM topographs shown in Fig. 10.1. The radiant energy harvested by LH2 is transferred to LH1, which funnels these excitations to the RC–BChl special pair, where they are transduced into a transmembrane charge separation. This initiates a cycle of electron transfer reactions between the primary iron–quinone acceptor (Q_A) and the ubiquinol–cytochrome c_2 oxidoreductase (cytochrome bc_1) complex and cytochrome c_2, resulting in the formation of electrochemical proton gradient coupled to the synthesis of ATP.

During photoheterotrophic growth, cellular ICM levels, together with those of LH2 relative to the RC–LH1 core complexes, are related inversely to light intensity, thereby permitting examination of the differential biosynthesis of photosynthetic complexes during adaptation to lower levels of illumination. Although ICM formation is repressed by high oxygen tensions during chemoheterotrophic growth, lowering the oxygen partial pressure initiates a gratuitous induction of ICM assembly in the dark by invagination of the CM, together with the synthesis and assembly of the LH and RC complexes (Koblízek et al. 2005). These sites of CM invagination can be isolated as an upper-pigmented band (UPB), sedimenting more

→

Fig. 10.1 (**a**) AFM topograph of *R. sphaeroides* ICM patch showing LH1–RC core complex arrays surrounded by LH2 complexes (Bahatyrova et al. 2004). The central protruding feature within the cores corresponds to RC-H subunit, indicating that the cytoplasmic face of the ICM is uniformly exposed at the surface. *Inset* at bottom: Existing structural data were used to model the region denoted by the central *dashed box*; a typical LH1–RC dimer is delineated in both images by a *red outline* and a representative LH2 complex by a *green circle*. (**b**) High-resolution AFM topograph, which shows a 3-D representation of a small ICM region revealing LH1–RC core complexes with associated LH2. *Green arrows* indicate contact points for energy transfer between LH2 and LH1–RC complexes. *Asterisks* denote nine-unit structure of αβ-heterodimers visible in LH2 rings. Encircled LH2 complex is sandwiched between two LH1–RC complexes. (**c**) View of membrane patches showing two types of arrangements of photosynthetic complexes. The *dashed red box* demarcates a membrane patch consisting of LH2 complexes clustered together at different levels, which form an LH2-only domain lacking the RC–LH1 core complexes; the *red arrow denotes* the typical LH2 ring within the LH2-only domain. RC–LH1 core complexes predominate in the other patches

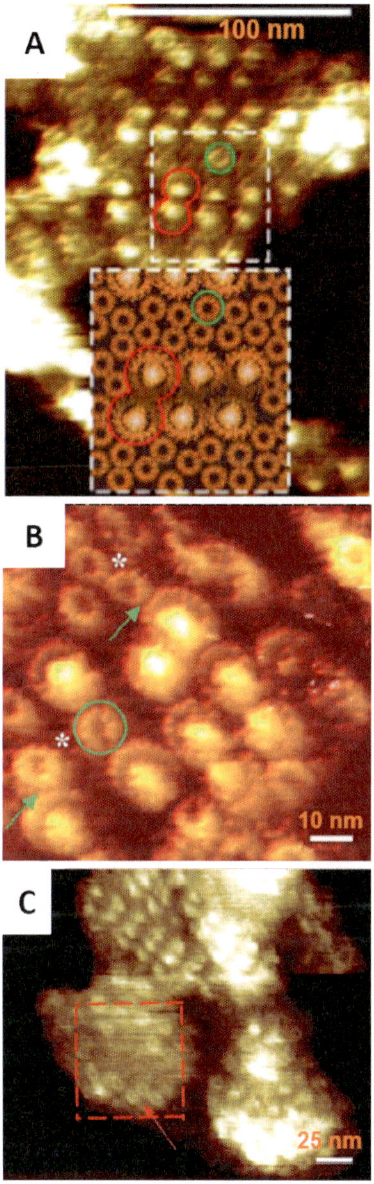

Fig. 10.1 (continued)

slowly than the ICM-derived chromatophore vesicle fraction during sucrose density gradient centrifugation (Niederman et al. 1979). Extensive biochemical, biophysical, and ultrastructural characterization of these membrane fractions (reviewed by Niederman 2006) has demonstrated that photosynthetic units are assembled sequentially, with RC–LH1 core structures inserted initially into the CM in a form that is largely inactive in forward electron transfer. This is followed by the activation of functional electron flow, together with the addition of LH2, resulting in further invagination and vesicularization of the membrane to form the ICM. During the developmental process, LH2 is thought to pack initially between linear arrays of dimeric core complexes, as shown in Fig. 10.1, and when these regions become filled, the LH2 complex ultimately forms a light-responsive peripheral antenna complement by clustering into LH2-only domains (Fig. 10.1c, Hunter et al. 2005).

In an effort to further characterize the ICM assembly process, we have initiated a detailed structural and functional proteomic analysis of the developing membrane fractions. These proteomics approaches are focusing upon the identification of proteins that are temporally expressed during ICM development and spatially localized in membrane domains isolated in both the UPB and chromatophores. These membrane fractions have been further purified by a two-phase partitioning procedure and subjected to clear native electrophoresis (CNE) (Wittig et al. 2007). Here we report the first results, in which the proteome of the CNE gel bands obtained from the chromatophore fraction has been examined.

10.2 Materials and Methods

10.2.1 Membrane Preparation

For membrane preparation, *R. sphaeroides* NCIB 8253 cells were grown as previously described (Holmes et al. 1980) at high light intensity of 1,100 W/m^2 until they reached $OD_{680\ nm}$ of 0.2 and then transferred to low-light illumination of 100 W/m^2 for 2, 4, 6, or 24 h. Light intensities were monitored with YSI-Kettering radiometer. Following growth, cells were cooled to 4°C, washed, and resuspended in 1 mM Tris, pH 7.5; a few crystals of DNAse I were added and passed twice through a French press. Debris and unbroken cells were removed by centrifugation at 12,000×*g*. The supernatant was layered onto a 5–35% (w/v) sucrose gradient and subjected to rate-zone ultracentrifugation for 3 h at 28,000 rpm in Beckman SW-28 rotor. Two bands were collected: the UPB, containing developing photosynthetic membranes, and main pigmented band, containing mature chromatophores.

10.2.2 Purification of Membranes by Aqueous Partitioning in a Two-Phase Dextran–Polyethylene Glycol (PEG) System

All steps were performed at 4°C. Dextran and PEG phases were collected, after solution containing 5.7% (w/v) dextran T-500 (average 425,000–575,000 mol. wt., Sigma), 5.7% (w/v) PEG (20,000 mol. wt., Sigma), 10 mM sodium phosphate, 5 mM Tris buffer, pH 7.5, and 10 mM EDTA was allowed to separate overnight. To extract UPB, 5 mL of each UPB, dextran, and PEG phases were mixed for 20 min in a 15 mL centrifuge tube, then centrifuged 20 min at 1000×*g*. At this time the upper phase containing UPB was removed and re-extracted with additional 7 mL of dextran phase. Extraction was repeated a third time, and the PEG layer was removed along with purified UPB. The purity was examined by comparing absorption peaks at UV (260–280 nm) and near-IR (850–875 nm). Centrifugation for 90 min at 55,000 rpm in a Beckman 60-Ti rotor further removed PEG, and UPB was resuspended in 1 mM Tris, pH 7.5. Chromatophores were also subjected to this procedure for further purification.

10.2.3 Clear Native Electrophoresis

Chromatophore and UPB samples were solubilized in either 15 mM *n*-octyl β-D-glucopyranoside, 15 mM deoxycholate detergent, or 2 g/g digitonin/total protein in sample and applied to a 3–12% gradient clear native gel as described recently by Wittig et al. (2007). The gel was run in a Vertical Slab Unit Model SE-400 (Hoeffer Scientific Instruments) at 10 mA, until current was no longer measurable. Up to four bands were identified without staining, and gels were also scanned using a visible light scanner (Canon) and Typhoon imager (GE Healthcare) set in a fluorescence mode with blue laser excitation source (488 nm) and emission filter of 670 nm. Each band was excised and absorption spectra were obtained in a DU-640 Spectrophotometer (Beckman), using gel containing no sample as a blank.

10.2.4 Proteomic Analysis

Excised bands from CNE were fixed for 30 min using 40% methanol, 10% acetic acid solution and were subjected to in-gel digestion with trypsin or chymotrypsin, followed by LC–MS/MS using Thermo LTQ and Dionex U-3000 systems operated in the nanoLC mode to obtain sensitivity at the sub-femtomole level.

10.3 Results

10.3.1 Establishing a Protocol for ICM Assembly Studies During Low-Light Adaptation

Figure 10.2 shows the near-IR absorption spectra of membrane fractions from cells harvested at various stages over a 24-h period during a shift from high (1,100 W/m^2)- to low-intensity illumination (100 W/m^2). Only low levels of LH2 are present (LH2/LH1 molar ratio = 0.14) in the UPB isolated from the initial exponential-phase high-light cells (time = 0), with the predominance of the RC–LH1 core complex indicated by the LH1 absorption maximum at 875 nm and the redshift and broadening on the red side of the 800-nm band, indicating the presence of the accessory RC–BChl band (maximum at 804 nm). This is supported further by the presence of a small band near 740 nm in the early fractions (see also

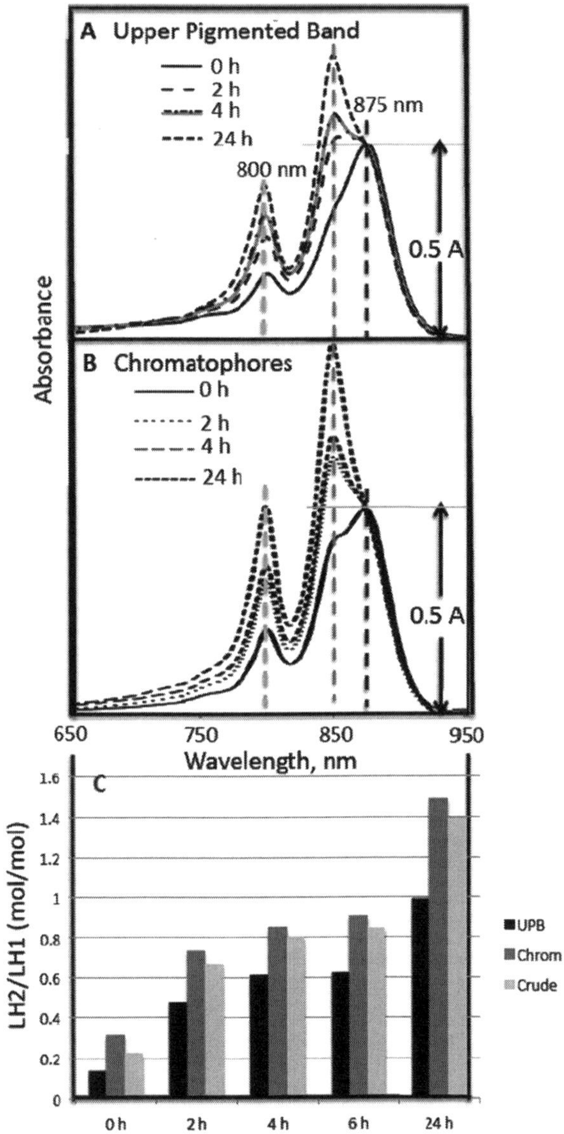

Fig. 10.2 Near-IR absorption spectra of the UPB (**a**) and chromatophore (**b**) fractions isolated by rate-zone sedimentation on sucrose density gradients from cells undergoing adaptation to low light intensity. Over the 24-h time course, the LH1/LH2 molar ratio was maintained at 1.5- to 2.3-fold higher levels in the UPB in comparison to chromatophores, indicating that these putative CM invagination sites act as hotspots for initial assembly of LH1–RC cores. (**c**) Spectral deconvolution (Sturgis et al. 1988) revealed that the molar ratios of LH2/LH1 gradually increased, reaching a 4.7-fold higher level at 24 h in chromatophores, as compared to 7.2-fold in the UPB

Fig. 10.3d), attributable to the bacteriopheophytin component of the RC. Overall, the dramatic increase in LH2 levels demonstrated by these spectral results and their analysis (Fig. 10.2) indicate that the essential membrane fractions can be isolated during the light intensity downshift to critically test the proposed mode of expansion of the LH apparatus during adaptation to lowered light levels.

The isolated membrane fractions have been further purified by polymer two-phase partitioning to facilitate proteomic and AFM analyses, which require highly

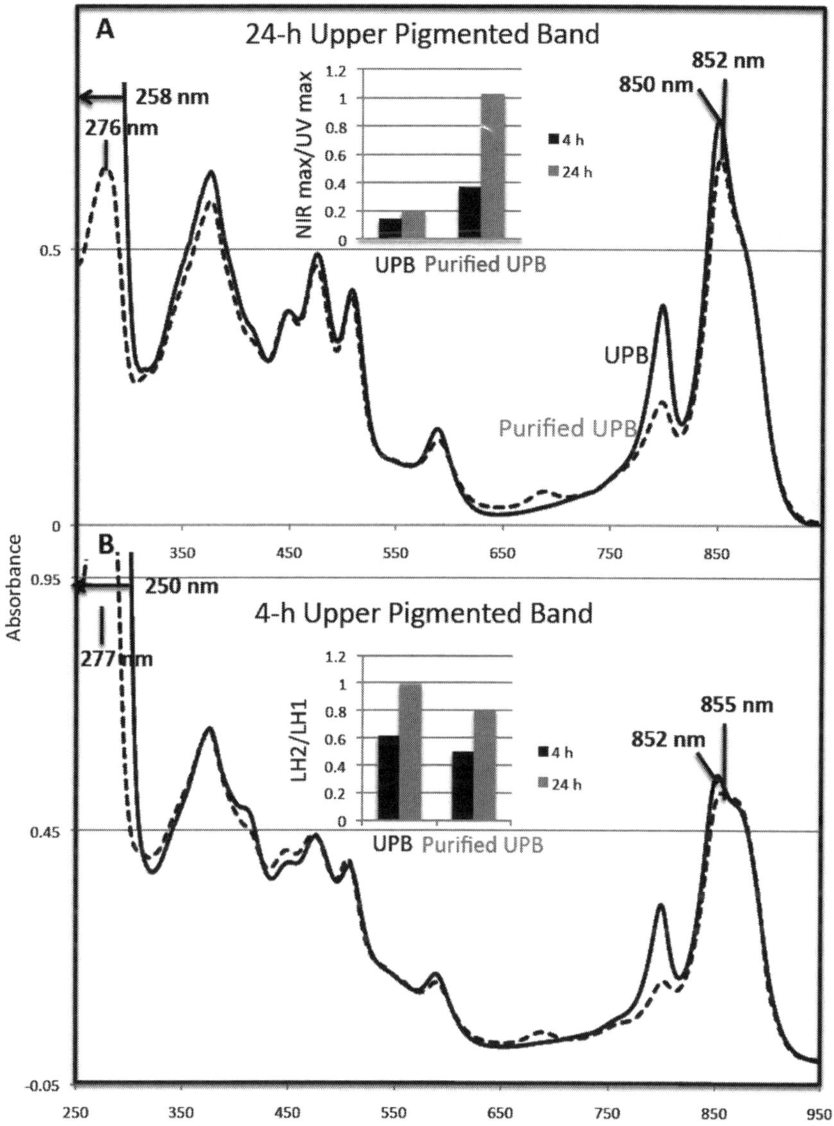

Fig. 10.3 Absorption spectra of UPB fractions purified by polymer two-phase partitioning. The UPB fraction was enriched in pigmented membranes after partitioning into the polyethylene glycol-rich upper phase, while ribosomes are found in the dextran-containing lower phase, along with some CM and cell wall material. When the purified UPB is subjected to sodium dodecyl sulfate polyacrylamide gel electrophoresis (SDS-PAGE), RC and LH bands are found largely superimposed upon co-migrating CM polypeptide bands, as expected of a site of membrane invagination. Contaminating material is also removed by partitioning the sucrose gradient purified chromatophore fraction into the upper phase (not shown). *Insets*: (**a**) ratios of the maximum absorbance of the near-IR/UV peaks of purified vs. unpurified samples; (**b**) overall LH2/LH1 molar ratios of purified vs. unpurified samples

purified material in an unaggregated state (Fig. 10.3). The high level of purification of the resulting unaggregated UPB fractions is demonstrated by the marked reduction in material absorbing in the UV near 260 nm (mainly contaminating ribosomes), which is replaced by a new band of much lower absorption with the maximum shifted to ~280 nm. This was reflected by comparison of the ratios of the maximum absorbance of the near-IR/UV peaks of the purified vs. unpurified samples (Fig. 10.3a, inset), which reached 2.6- and 5.1-fold increases in the 4 and 24 h preparations, respectively; however, the LH2 B800 and B850 bands were diminished upon purification and decreases of nearly 20% were observed in the overall LH2/LH1 molar ratios in each of the respective samples (Fig. 10.3b, inset). It is also noteworthy that the coincidence of the paired spectra throughout the visible and UV regions indicates that no aggregation of the samples has occurred during their purification.

10.3.2 Establishing a Clear Native Polyacrylamide Gel Electrophoresis Procedure for Separating Bacteriochlorophyll–Protein Complexes

In order to critically assess the assembly of the LH2 and RC–LH1 complexes within the developing ICM, we have established a procedure in which the highly purified membranes are subjected to CNE (Fig. 10.4). This method has been adapted from the high-resolution CNE procedure developed recently by Wittig et al. (2007) for direct in-gel functional and fluorescence assays of well-separated mitochondrial inner membrane protein complexes. The use of Coomassie blue, which interferes with catalytic activity and fluorescence determinations in blue-native electrophoresis (Schägger and von Jagow 1991), is circumvented with non-colored mixtures of anionic and neutral detergents in the cathode buffer, which enhance the solubility of membrane proteins during electrophoresis and, like the Coomassie dye, impose a charge shift that augments their anodic migration. Protein aggregation and band broadening are also prevented, which results in superior resolution, while facilitating the in-gel functional assays.

Panels a–c of Fig. 10.4 show the results of the application of this procedure to the chromatophores isolated from cells undergoing low-light adaptation, after solubilization with an *n*-octyl β-D-glucopyranoside/deoxycholate (BOG/DOC) detergent mixture. The accompanying absorption spectra, obtained directly with gel slices, show a marked enrichment of the RC–LH1 core and LH2 complexes in the upper and lower bands (Fig. 10.3d, e, respectively). Moreover, bands of intermediate migration are also seen; SDS-PAGE (Fig. 10.4f) shows that they contain associations of these two complexes such that the upper intermediate band is enriched in LH2, while the lower intermediate band represents an LH1-enriched mixture. These intermediate complexes may represent detergent–protein micelles arising from ICM regions seen in AFM where the LH2 and LH1 complexes are associated (Fig. 10.1a, b). Importantly, this new clear gel procedure has provided the basis for a

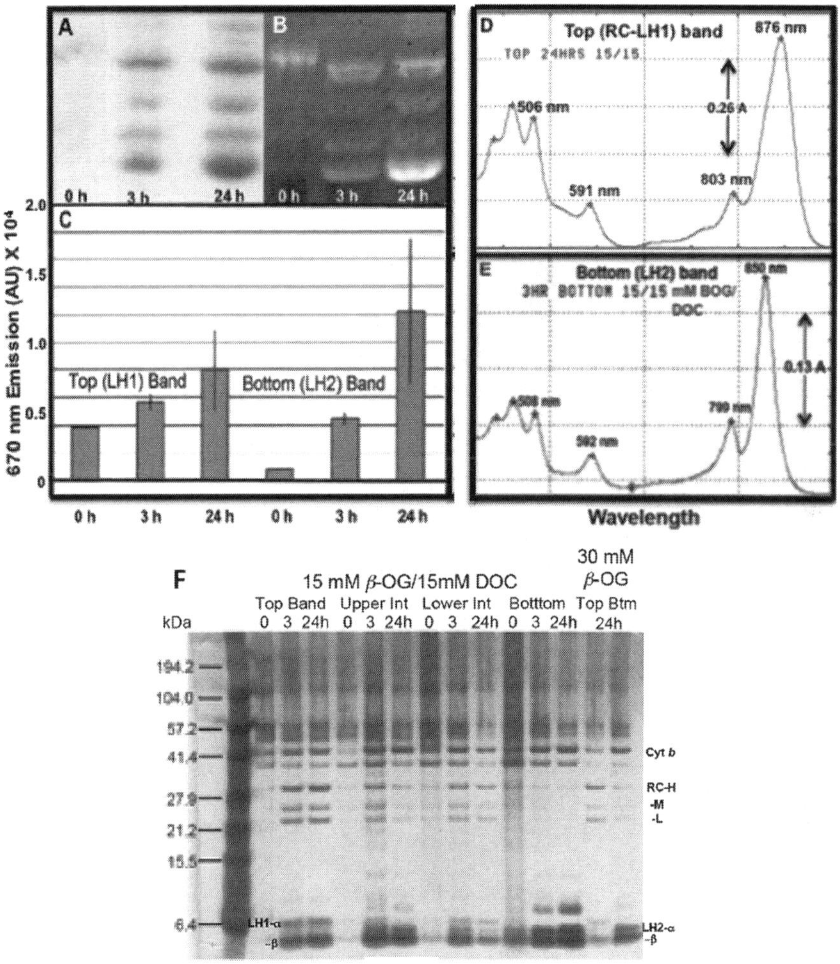

Fig. 10.4 Separation of intact pigment–protein complexes by CNE of BOG–DOC solubilized chromatophores from cells undergoing low-light adaptation. (**a**) Scan of unstained gel. (**b**) Intensity profile of 670-nm fluorescence emission from unstained gel obtained with a Typhoon imager; excitation at 488 nm. The emission apparently arises largely from the blue tail of the emission from the LH2 complex (**c**) quantification of fluorescence emission. (**d**, **e**) Absorption spectra obtained directly on indicated gel slices. Note the high level of spectral purity of the LH1 and RC complexes in the top band (absorption bands at 876 and 803 nm, respectively) and the LH2 complex in the bottom band (maxima at 850 and 799 nm). Also noteworthy are the differences in carotenoid contents of the complexes, reflecting the BChl:carotenoid molar ratios near 1.0 reported for LH1 and ~2.0 for LH2, as well as the blueshift in the position of the absorption maximum of the redmost carotenoid absorption band (Hunter et al. 1988). (**f**) Silver-stained gel of CNE bands subjected directly to SDS-PAGE, showing that apoprotein distributions essentially confirm observed absorption spectra. The last two lanes are from chromatophores solubilized with 30 mM BOG alone. Note the respective well-separated α- and β-polypeptides, with no significant cross contamination. No LH bands of intermediate migration appeared. See text for tentative identification of some of the additional gel bands in the various gel lanes

critical proteomic analysis of the assembly of the pigment–protein complexes. Some of the other major bands observed in the SDS gel have been tentatively identified and are discussed below.

10.3.3 Proteomic Analysis of Developing Photosynthetic Membranes

A detailed analysis of relevant portions of the *R. sphaeroides* proteome is essential for filling in gaps in our understanding of the ICM assembly process. The separation of intact pigment–protein complexes by CNE has permitted in-gel trypsin digestion of the various pigmented gel bands shown in Fig. 10.4 which were then subjected to LC–MS/MS as described in Section 10.2 (Fig. 10.5). It should be emphasized that the proteomics data at this point are expressed in spectral counts, which only reflect the ability of trypsin to act at possible cleavage sites and are therefore semi-quantitative in nature. This means that proteins that are not highly abundant with a large number of sites give higher counts than more abundant proteins with far fewer sites; in extreme cases where no sites are available, no counts are found in the resulting pivot table, viz., PufB (LH1-β-polypeptide) and PucA (LH2-α); however, the latter polypeptide could be detected after chymotrypsin cleavage. On the other hand, valid comparisons can be made between the same components in different gel bands and the designated clusters of orthologous groups into which they are classified (Fig. 10.5).

In general, the spectral counts arising from the proteomics analysis of the LH and RC complexes in the gel bands confirmed their distributions as determined in near-IR absorption spectra (Fig. 10.4d, e) and are also in line with the levels of their apoprotein components found in SDS gels (Fig. 10.4f). Some differences were found for the purified 3-h chromatophore samples, including diminished levels of contaminating outer membrane proteins in several preparations and a marked enrichment in ATP synthase in the LH2-rich upper intermediate band, accounting for a total of 45% of the spectral counts. The highest spectral counts were detected for PuhA (RC-H subunit), and surprisingly, Puc2A (LH2-α), encoded by the *puc2BA*operon, was second highest in abundance, despite the report by

Fig. 10.5 Proteomic analysis of photosynthetic complexes isolated from developing ICM. Spectral count distributions after in-gel trypsin digestion of the four CNE gel bands obtained from chromatophores. These chromatophore preparations are from a different experiment (isolated from cell samples taken 3 and 24 h after the shift from high to low light intensity, along with the 3-h preparation after purification by the two-phase partitioning). (**a**) Top (RC–LH1) bands; (**b**) upper intermediate bands (LH2-enriched associations of LH1 and LH2); (**c**) lower intermediate bands (LH1-enriched associations of LH1 and LH2); (**d**) bottom bands (LH2). The distributions shown are for clusters of orthologous groups (Tatusov et al. 1997) that account for patterns of sequence similarities, in which orthologs typically have the same function. The usual energy production and conversion category have been broken down into subgroups to account for the distinct metabolic capabilities unique to photoheterotrophically grown *R. sphaeroides*

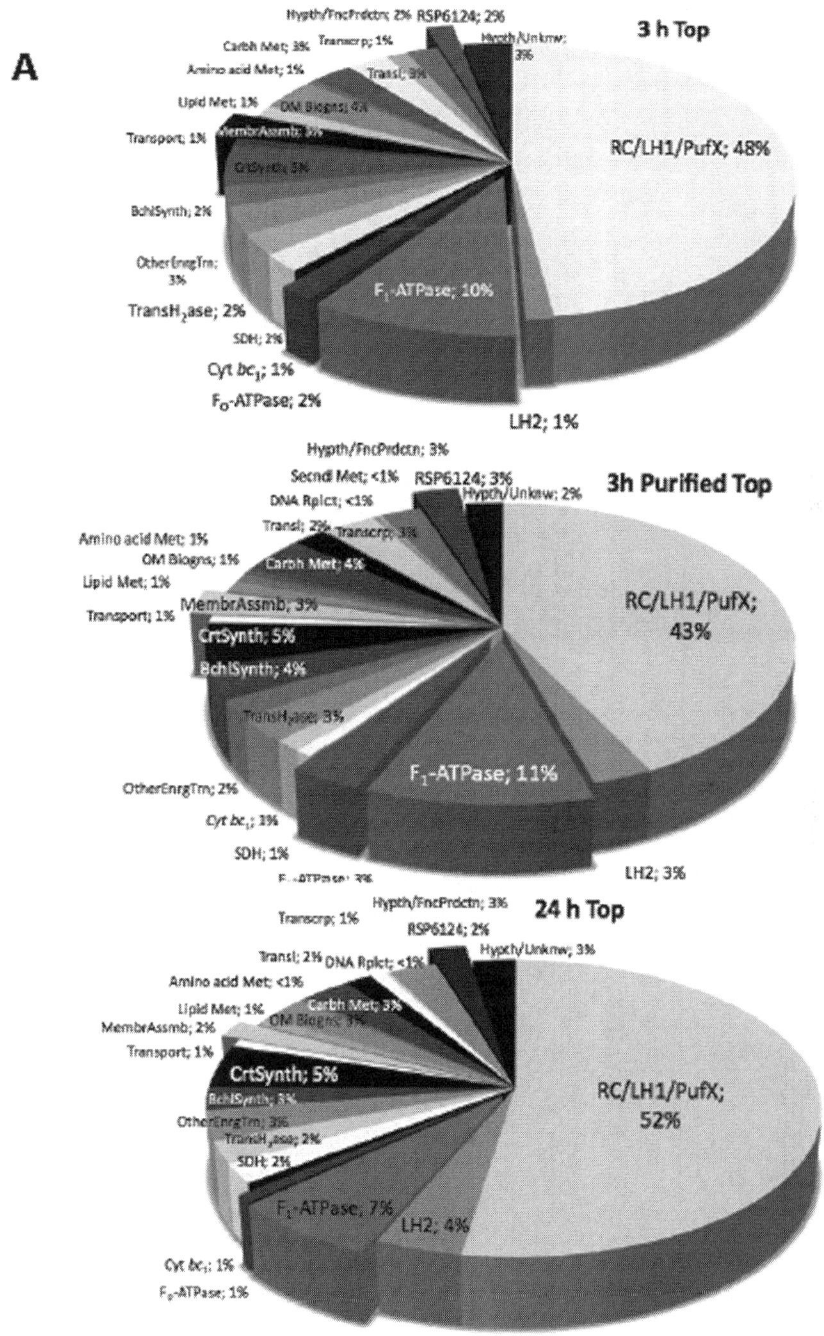

Fig. 10.5 (continued)

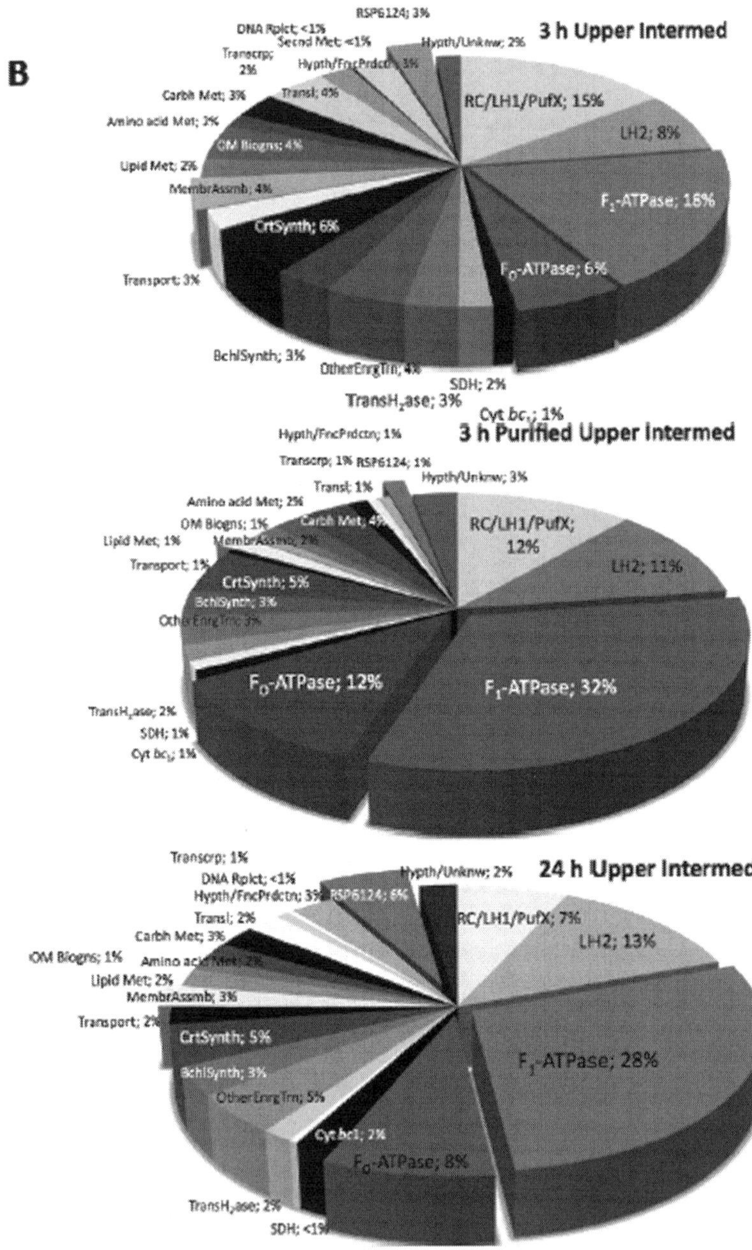

Fig. 10.5 (continued)

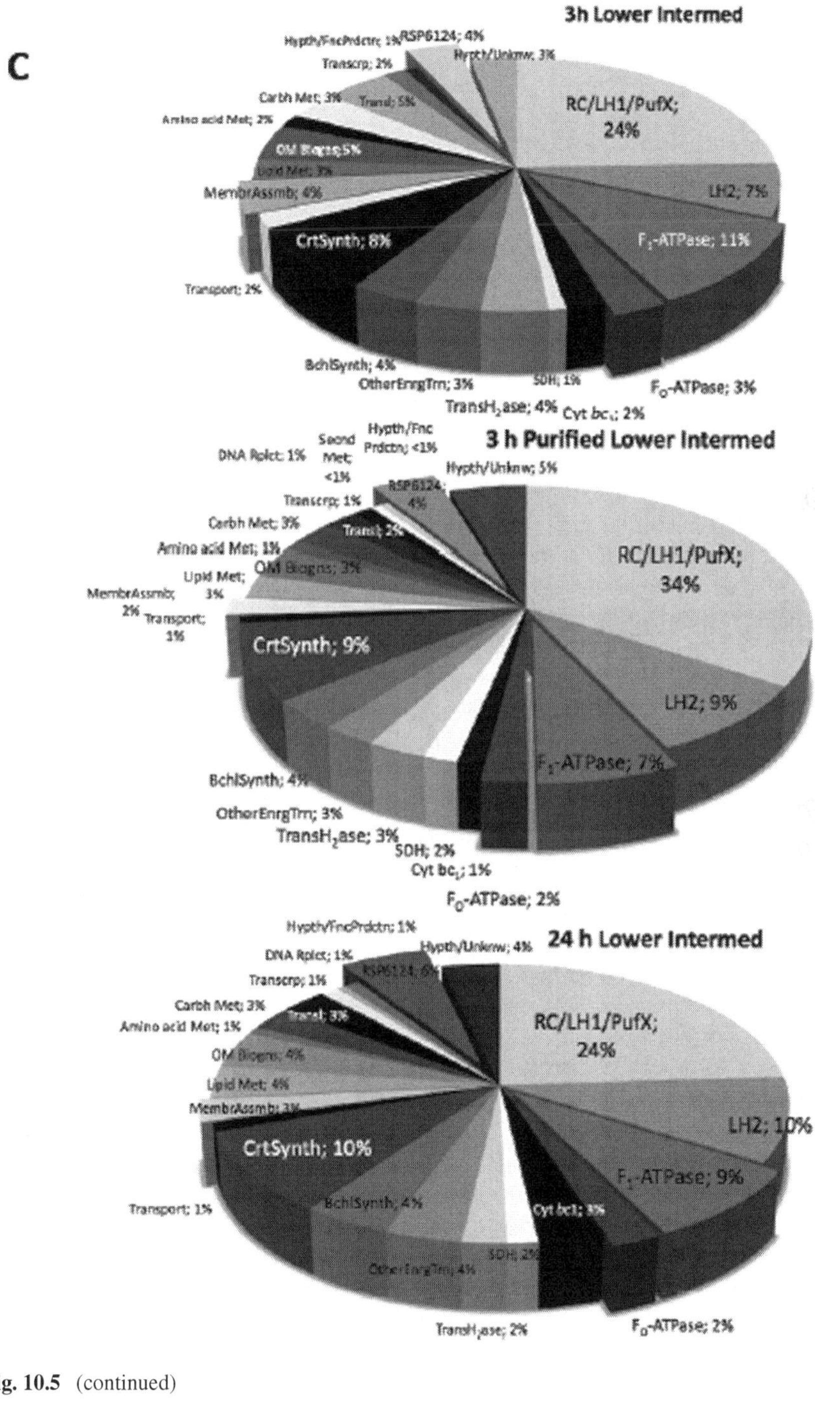

Fig. 10.5 (continued)

Zeng et al. (2003) that this polypeptide is not incorporated into LH2 complexes. PufA (LH1-α) was third in abundance, followed by the β-subunit of F_1-ATP synthase (*atpD*), while lower counts were found for PucB (LH2-β) and Puc2B (LH2-β). The unexpectedly high spectral counts for the F_1F_O-ATPase subunits, as noted earlier, were particularly intriguing, given the reported low abundance of the ATPase in chromatophores (Feniouk et al. 2001; see, however, Gubellini et al. 2007; Sturgis et al. 2009), and the inability to detect this coupling factor as well as the more abundant cytochrome bc_1 complex in AFM (Bahatyrova et al. 2004). In this connection, the high levels of these ATPase subunits together with those of the bc_1 complex apparently associated with the LH2-enriched fractions are consistent with the possibility that these components are localized at ICM vesicle edges, thought to contain LH2-only domains that are outside the flat ICM vesicle regions imaged by AFM (Fig. 10.1).

Significant levels of a large number of other proteins, including membrane assembly factors, were also observed, most notably in 3-h upper gel bands. These included the preprotein translocase YidC homolog, along with lesser amounts of other general membrane assembly factors such as TatA, a subunit of the twin arg translocation system, the SecY preprotein translocase subunit, bacterial type 1 signal peptidase, and the putative preprotein translocase subunit YajC. While the response regulators RegA and RegB were detected, no complex-specific assembly factors (Young and Beatty 2003) were found in any of the fractions. Moreover, about 45 proteins of unknown function were found associated with the gel bands. One hypothetical protein, RSP6124, which ranked fifth in abundance in spectral counts is of particular interest, as it seems to be associated with LH2-rich bands in the clear gels. This component reached levels of 10% in the 3-h LH2 bottom band (~1/2 was removed by the purification procedure) and 12% in the 24-h LH2 band. Zeng et al. (2007) have also observed RSP6124 recently, along with RSP1760 and RSP3246, in their proteomic analysis of the ICM. Interestingly, RSP1760 ranked second during low-light growth (3 W/m^2) in their studies, while RSP6124 ranked fifth under those conditions. In our study, RSP1760 ranked only 43rd, and RSP3246 was not observed. Zeng et al. (2007) also found significant levels of RSP6124 in the cytoplasm and topological predictions suggested that this protein lacks transmembrane domains; however, the high levels found in association with LH2-enriched CNE bands suggest RSP6124 may play at least a transient role in the assembly or function of the LH2 antenna.

10.4 Discussion

10.4.1 Clear Native Polyacrylamide Gel Electrophoresis

A native PAGE procedure was previously developed in which membranes were solubilized with lithium dodecyl sulfate at 4°C and subjected to electrophoresis in the presence of this detergent (Broglie et al. 1980). Up to 11 pigment–protein complexes were resolved, including LH2 which banded near the top of the gel and a faster migrating band (band 10), containing spectrally homogeneous LH1. Bands of intermediate migration were also resolved that consisted of LH1/LH2 mixtures in various proportions, and it was suggested that they represented in vivo associations existing in the membrane. Subsequent studies with a mutant lacking LH2 (Westerhuis et al. 2002) revealed a very regular banding pattern, indicative of an oligomeric series of LH1 complexes that banded in positions identical to those of the wild-type intermediate complexes, thus indicating that the latter may represent fortuitously co-migrating LH1 and LH2. Thus, in addition to giving rise to LH2 and RC–LH1 preparations, the CNE procedure developed here greatly simplifies analysis of putative inter-LH complex associations, since only two complexes of intermediate migration are isolated, an upper one enriched in LH2 and a lower one enriched in LH1 (Fig. 10.4).

Regarding the tentative identification of the polypeptide bands in the analytical SDS gel of the pigmented CNE bands that migrated to positions distinct from those of the BChl-associated apoproteins (Fig. 10.4f), it seems reasonable to point out that major bands were consistently observed near expected mass positions deduced from the *R. sphaeroides* genome sequence (Mackenzie et al. 2001). These components consisted of the α-, β-, and γ-polypeptides of the F_1-ATPase (55.2, 50.4, and 31.2 kDa, respectively), well resolved in the LH2-enriched CNE bands, cytochrome *b* (50.1 kDa), migrating to the same position as in Western blots (Sturgis et al. 2009), RSP6124 (11.4 kDa), prominent in 3- and 24-h bottom bands. Verification awaits further mutant studies and Western blotting analyses. It is important to note that each of these putative polypeptide components gave rise to high to reasonable numbers of spectral counts in the proteomic analysis.

It is possible that the CNE procedure may provide additional experimental approaches that could have bearing on the authenticity and physiological relevance of intercomplex associations observed between LH1 and LH2 and those between the LH complexes and non-pigmented energy-transducing complexes in the various pigmented gel bands. Accordingly, as shown recently by Wittig et al. (2007), the mild detergent digitonin can be used for inner mitochondrial membrane solubilization during the isolation of labile protein assemblies or the preservation of supramolecular associations of multiprotein complexes. This obviates the necessity of chemical cross-linking for the identification of physiological protein–protein interactions.

Since we were unable to resolve the pigmented bands when the UPB fraction was solubilized with BOG/DOC mixtures, digitonin was used instead, which gave rise to the respective top and bottom RC–LH1 and LH2 bands and a lower intermediate band, as well as all four bands when applied to chromatopores (not shown). As this mild procedure is expected to preserve authentic physiological associations, these results would suggest that the LH1–LH2 associations in the intermediate complexes represent hydrophobic protein–protein interactions that exist within the membrane, but definitive proof awaits measurements of the energy transfer efficiencies between the LH2 and LH1 complexes within these gel bands. In addition, we are in the process of applying in-gel activity assays to localize NADH: Ubiquinone oxidoreductase, succinate:ubiquinone oxidoreductase, the cytochrome bc_1 complex, cytochrome *c* oxidase, and the ATP synthase, which will ultimately have bearing on possible associations with specific pigmented gel bands.

10.4.2 Proteome of the Clear Native Electrophoresis Gel Bands

R. sphaeroides is genetically well characterized and the availability of the annotated genome (http://genome.jgi-psf.org/finished_microbes/rhosp/rhosp.home.html) has formed the basis for DNA microarray analysis of the effects of oxygen tension and light intensity on transcriptome expression (Roh et al. 2004). In addition to the expression of the well-described, traditional pattern of photosynthesis-associated genes, a further complement of genes responding to oxygen and light regulation was identified, covering 25–30% of the genome and representing genes involved in physiological functions with significant regulatory potential. The transcriptional activities of several key regulatory elements and membrane assembly factors were also revealed. However, definitive physiological extrapolation of these gene expression profiles required a detailed proteomic analysis of relevant gene products, given the probable interplay of posttranscriptional control in the ultimate phenotypic expression and metabolic outcome of the effects of the initial regulatory events exerted at the transcriptional level.

Accordingly, Zeng et al. (2007) have recently reported a detailed proteomic characterization of the ICM and other subcellular fractions obtained from *R. sphaeroides*. In addition to the protein components of the well-studied complexes involved in the primary photochemical and energy conservation events, numerous other ICM-localized proteins were encountered including a number of abundant newly discovered proteins, as noted above. Altered photosynthetic growth phenotypes were observed when genes encoding the RSP1760 and RSP1467 components were mutated, and it was suggested that they may be involved in ICM formation. Of further significance to the work described here was the demonstration that BChl and carotenoid biosynthetic enzymes are enriched in the CM and the detection of additional chaperonin and preprotein translocases including GroEL, a protein transporter of the Tim44 family, and protein-export membrane protein secD in both the CM and the ICM. Importantly, the application of these highly informative proteomic methodologies on both whole membrane fractions and their protein complexes, in conjunction with structural and functional analyses of the membrane assembly, will ultimately reveal how the biological potential encoded in the *R. sphaeroides* genome controls and integrates events at the protein level that drive the formation of the specialized energy-transducing membrane housing the photosynthetic apparatus.

At present, the work reported here has formed a baseline for future studies in which the assembly of BChl-protein and other energy-transducing complexes will be perturbed by blocking the energy currency needed to drive the membrane translocation of their nascent polypeptide components, or for the examination of mutant strains in which conversion of the invaginating membrane into mature ICM structures is arrested. Under both of these circumstances, it is expected that additional assembly factors will accumulate, such as the group of complex-specific assembly factors (Young and Beatty 2003), not yet detected in the *R. sphaeroides* CNE band-associated proteome.

Acknowledgments This work was supported by Supported by US Department of Energy Grant No. DE-FG02-08ER15957 from the Chemical Sciences, Geosciences and Biosciences Division, Office of Basic Energy Sciences, Office of Science. We thank Oluwatobi B. Olubanjo for assistance with membrane purification procedures and Prof. Peter Lobel and Dr. Haiyan Zheng of the Center for Advanced Biotechnology and Medicine, University of Medicine and Dentistry of New Jersey, for conducting the proteomics analysis.

References

Bahatyrova S, Frese RN, Siebert CA, van der Werf KO, van Grondelle R, Niederman RA, Bullough PA, Otto C, Olsen JD, Hunter CN (2004) The native architecture of a photosynthetic membrane. Nature 430:1058–1062

Broglie RM, Hunter CN, Delepelaire P, Niederman RA, Chua N-H, Clayton RK (1980) Isolation and characterization of pigment-protein complexes of *Rhodopseudomonas sphaeroides* by lithium dodecyl sulfate/polyacrylamide gel electrophoresis. Proc Natl Acad Sci USA 77:87–91

Feniouk BA, Cherepanov DA, Voskoboynikova NE, Mulkidjanian AY, Junge W (2002) Chromatophore vesicles of *Rhodobacter capsulatus* contain on average one F_OF_1-ATP synthase each. Biophys J 82:1115–1122

Gubellini F, Francia F, Turina P, Lévy D, Venturoli G, Melandri BA (2007) Heterogeneity of photosynthetic membranes from *Rhodobacter capsulatus*: Size dispersion and ATP synthase distribution. Biochim Biophys Acta 1767:1340–1352

Holmes NG, Hunter CN, Niederman RA, Crofts AR (1980) Identification of the pigment pool responsible for the flash-induced carotenoid band shift in *Rhodopseudomonas sphaeroides*. FEBS Lett 115:43–48

Hunter CN, Daldal F, Thurnauer MC, Beatty JT (eds) (2008) The purple phototrophic bacteria. Advances in photosynthesis and respiration, vol 28. Springer, Dordrecht

Hunter CN, Pennoyer JD, Sturgis JN, Farrelly D, Niederman RA (1988) Oligomerization states and associations of light-harvesting pigment-protein complexes of *Rhodobacter sphaeroides* as analyzed by lithium dodecyl sulfate-polyacrylamide gel electrophoresis. Biochemistry 27:3459–3467

Hunter CN, Tucker JD, Niederman RA (2005) The assembly and organisation of photosynthetic membranes in *Rhodobacter sphaeroides*. Photochem Photobiol Sci 4:1023–1027

Koblízek M, Shih JD, Breitbart SI, Ratcliffe EC, Kolber ZS, Hunter CN, Niederman RA (2005) Sequential assembly of photosynthetic units in *Rhodobacter sphaeroides* as revealed by fast repetition rate analysis of variable bacteriochlorophyll a fluorescence. Biochim Biophys Acta 1706:220–231

Mackenzie C, Choudhary M, Larimer FW, Predki PF, Stilwagen S, Armitage JP, Barber RD, Donohue TJ, Hosler JP, Newman JE, Shapleigh JP, Sockett RE, Zeilstra-Ryalls J, Kaplan S (2001) The home stretch, a first analysis of the nearly completed genome of *Rhodobacter sphaeroides* 2.4.1. Photosynth Res 70:19–41

Niederman RA, Mallon DE, Parks LC (1979) Membranes of *Rhodopseudomonas sphaeroides*. VI. Isolation of a fraction enriched in newly synthesized bacteriochlorophyll *a*-protein complexes. Biochim Biophys Acta 555:210–220

Niederman RA (2006) Structure, function and formation of bacterial intracytoplasmic membranes. In: Shively JM (ed) Complex intracellular structures in prokaryotes, microbiology monographs, vol 2. Springer, Berlin, Heidelberg, pp 193–227

Roh JH, Smith WE, Kaplan S (2004) Effects of oxygen and light intensity on transcriptome expression in *Rhodobacter sphaeroides* 2.4.1. Redox active gene expression profile. J Biol Chem 279:9146–9155

Schägger H, von Jagow G (1991) Blue native electrophoresis for isolation of membrane protein complexes in enzymatically active form. Anal Biochem 199:223–231

Sturgis JN, Hunter CN, Niederman RA (1988) Spectra and extinction coefficients of near-infrared absorption bands in membranes of *Rhodobacter sphaeroides* mutants lacking light-harvesting and reaction center complexes. Photochem Photobio l 48:243–247

Sturgis JN, Tucker JD, Olsen JD, Hunter CN, Niederman RA (2009) Atomic force microscopy studies of native photosynthetic membranes. Biochemistry 48:3679–3698

Tatusov RL, Koonin EV, Lipman DJ (1997) A genomic perspective on protein families. Science 278:631–637

Westerhuis WHJ, Sturgis JN, Ratcliffe EC, Hunter CN, Niederman RA (2002) Isolation, size estimates and spectral heterogeneity of an oligomeric series of light-harvesting 1 complexes from *Rhodobacter sphaeroides*. Biochemistry 41:8698–8707

Wittig I, Karas M, Schägger H (2007) High resolution clear native electrophoresis for in-gel functional assays and fluorescence studies of membrane protein complexes. Mol Cell Proteomics 6:1215–1225

Young CS, Beatty JT (2003) Multi-level regulation of purple bacterial light-harvesting complexes. In: Green BR, Parson WW (ed) Light-harvesting antennas in photosynthesis advances in photosynthesis and respiration, vol 13. Kluwer Academic, Dordrecht, pp 449–470

Zeng X, Choudhary M, Kaplan S (2003) A second and unusual *pucBA*operon of *Rhodobacter sphaeroides* 2.4.1: Genetics and function of the encoded polypeptides. J Bacteriol 185: 6171–6184

Zeng X, Roh JH, Callister SJ, Tavano CL, Donohue TJ, Lipton MS, Kaplan S (2007) Proteomic characterization of the *Rhodobacter sphaeroides* 2.4.1 photosynthetic membrane: identification of new proteins. J Bacteriol 189:7464–7474

Chapter 11
A Glimpse into the Proteome of Phototrophic Bacterium *Rhodobacter capsulatus*

Ozlem Onder, Semra Aygun-Sunar, Nur Selamoglu, and Fevzi Daldal

Abstract A first glimpse into the proteome of *Rhodobacter capsulatus* revealed more than 450 (with over 210 cytoplasmic and 185 extracytoplasmic known as well as 55 unknown) proteins that are identified with high degree of confidence using nLC–MS/MS analyses. The accumulated data provide a solid platform for ongoing efforts to establish the proteome of this species and the cellular locations of its constituents. They also indicate that at least 40 of the identified proteins, which were annotated in genome databases as unknown hypothetical proteins, correspond to predicted translation products that are indeed present in cells under the growth conditions used in this work. In addition, matching the identification labels of the proteins reported between the two available *R. capsulatus* genome databases (ERGO-light with *RRCxxxxx* and NT05 with *NT05RCxxxx* numbers) indicated that 11 such proteins are listed only in the latter database.

11.1 Introduction

The Gram negative, purple non-sulfur facultative phototrophic bacterium *Rhodobacter capsulatus* is a model organism that is intensely studied for various aspects of major metabolic pathways such as photosynthesis and respiration (Zannoni 1995; Hunter et al. 2009). Using this species, which exhibits highly versatile growth modes including anoxygenic light (i.e., photosynthesis), anoxygenic dark (i.e., anaerobic respiration), and oxygenic (i.e., light-independent aerobic respiration) metabolisms, physiologically relevant cellular responses to the availability of light or oxygen can be examined at the molecular level (Hunter et al. 2009). Indeed, a complete definition of the presence, regulation, and biogenesis of various cellular components in response to the changing environmental conditions greatly benefits from global analyses approaches, including transcriptomic and proteomic studies

F. Daldal (✉)
Department of Biology, University of Pennsylvania, Philadelphia, PA 19104, USA
e-mail: fdaldal@sas.upenn.edu

P.C. Hallenbeck (ed.), *Recent Advances in Phototrophic Prokaryotes*, Advances in Experimental Medicine and Biology 675, DOI 10.1007/978-1-4419-1528-3_11,

(Park et al. 2005; Wasinger 2006). Thus, along with the transcriptional studies, availability of qualitative and quantitative description of *R. capsulatus* proteomes under defined growth conditions is extremely invaluable. Toward this end, we have initiated an effort to define the complete proteome of this species (Onder et al. 2008), and we present here a first glimpse to this developing protein identification dataset.

11.2 Materials and Methods

Wild-type *R. capsulatus* strain MT1131 was grown under respiratory growth condition in enriched MPYE (mineral–peptone–yeast extracts) medium (Daldal et al. 1986) under standard culture conditions (1 l medium in 2 l flask, shaken at approximately 150 rpm at 35°C in dark for approximately 36 h) (Myllykallio et al. 1997). Cells were harvested by centrifugation at 4°C (5000×*g*, 20 min), washed with ice-cold 20 mM Tris–HCl, pH 8.0, and gently resuspended in cold extraction buffer consisting of 1 mg/ml of polymyxin B sulfate, 20 mM Tris–HCl, 250 mM NaCl (pH 8.0) (5 ml/g of wet cells) (Ren and Thony-Meyer 2001; Onder et al. 2008). The suspension was gently stirred for 1 h at 4°C and centrifuged at 10,000×*g* for 20 min at 4°C. The supernatant was carefully transferred into a clean tube, re-centrifuged at 150,000×*g* for 2 h at 4°C, and saved as periplasmic fractions at –20°C. Polymyxin-treated cell pellets were then resuspended in 50 mM MOPS, 100 mM KCl (pH 8.0) buffer, and processed using a French pressure cell, as described earlier (Onder et al. 2008). After centrifugation for 2 h at 45,000 rpm, supernatants (i.e., cytoplasmic fractions) and pellets (i.e., membrane fractions) were processed separately for analyses of their protein contents.

Soluble or periplasmic proteins were precipitated with trichloroacetic (TCA)/acetone (20% w/v), washed twice with ice-cold acetone to remove residual TCA, and dried under vacuum. Pellets were resuspended in two-dimensional gel electrophoresis (2D-GE) sample solubilization buffer (SSB: 8 M urea, 4% CHAPS, 40 mM Tris, 0.2% Bio-Lyte-pH 3–10, 65 mM DTT) at room temperature until complete solubilization. For 2D-GE, samples containing 300 μg of solubilized proteins were applied to 18 cm, pH 4–7 immobilized pH gradient (IPG) strips (Bio-Rad), and following a 12 h passive rehydration, isoelectrofocusing (IEF) was carried out by using PROTEAN IEF Cell (Bio-Rad) at 20°C at a maximum of 7000 V for 15–18 h, and the strips thus prepared were kept frozen at –20°C until use. For the second dimension sodium dodecyl sulfate-polyacrylamide gel electrophoresis (SDS-PAGE), IPG strips were reduced with 1% w/v dithiothreitol (DTT) and alkylated with 2.5% w/v iodoacetamide at room temperature, both prepared in equilibration buffer consisting of 50 mM Tris–HCl, pH 8.8, 6 M urea, 30% v/v glycerol, 2% SDS, and 0.02% bromophenol blue. After equilibration, the IPG strips were layered on top of the second dimension resolving gel slabs and overlaid with a solution of molten 0.5% agarose in SDS electrophoresis buffer. The second dimension Laemmli-type SDS-PAGE was carried out using 11% gels without any stacking at 40 mA per gel in a Protean II XL cell (Bio-Rad), and gels were stained with colloidal coomassie brilliant blue (Neuhoff et al. 1990). Spots or bands were manually excised from gels and subjected to in-gel trypsin digestions (Onder et al. 2006). Tryptic peptide extracts were analyzed using a nanoLC–MS/MS setup (LCQ Deca XP Plus mass spectrometer from Thermo Scientific coupled to an Ultimate Nano liquid chromatography system from DIONEX). Tryptic peptide mixtures were first loaded onto a μ-precolumn (C18, 5 μm, 100 Å, 300 μm i.d. × 5 mm) (DIONEX) and washed for 4 min at a flow rate of 0.25 μl/min with the LC buffer A, then transferred onto an analytical C18-nanocapillary HPLC column [DIONEX, Acclaim PepMap100, C18 column (75 μm i.d. by 150 mm)] with a 3 μm particle size and a 100 Å pore diameter for fractionation. A fused silica tip with 8-μm aperture (New Objective, Woburn, MA) was used for nanospray ionization of peptides eluting from the column. Mass spectra were analyzed using the DTA generation and SEQUEST search algorithms within Bioworks 3.3 software (Thermo Scientific) and *R. capsulatus* protein databases (ERGO-light from http://www.ergo-light.com or in-house available NT05 data bases). The following parameters were selected as proteolytic enzymes, allowing trypsin cleavage only after arginine and lysine and with the maximal number of internal (missed) cleavage set to 2–4. Mass tolerance for precursor and fragment ions was 2.0–2.5 and 1.0, respectively. Methionine oxidation and cysteine carbamidomethylation were set as variable modification with maximum modification set to 2, and default setting was used for all other variables. Matching peptides were filtered according to correlation scores (X_{Corr} at least 1.5, 2.0, and 2.5 for +1, +2, and +3 charged peptides, respectively, and $\Delta C_N > 0.1$) to give high confidence of protein identification, and proteins were considered as significant only when at least two peptides were identified with the SEQUEST filter settings mentioned above. The *R. capsulatus* proteins thus identified were examined with bioinformatic tools. Prediction software SignalP (version 3.0) (Bendtsen et al. 2004) (http://www.cbs.dtu.dk/services/SignalP/) and PSORTb (version 2.0) (Gardy et al. 2005) (http://www.psort.org/psortb/) were used to predict the likely sub-cellular localization of identified proteins.

11.3 Results and Discussion

A long-term objective of this study is the establishment of a comprehensive proteomic reference database for *R. capsulatus* under different growth conditions. This physiological proteomics will then establish an important step to enable detailed analyses of major metabolic pathways of this organism extending from photosynthesis to respiration. To this end, we use modern proteomic approaches involving 1D- and 2D-GE combined with liquid chromatography and tandem mass spectrometry techniques. Representative 1D and 2D analyses are shown in Fig. 11.1. In addition, as comprehensive proteome analyses are often limited by physiological constraints and technical issues, to maximize the number of proteins identified, to reduce the sample complexity, and also to detect low abundance

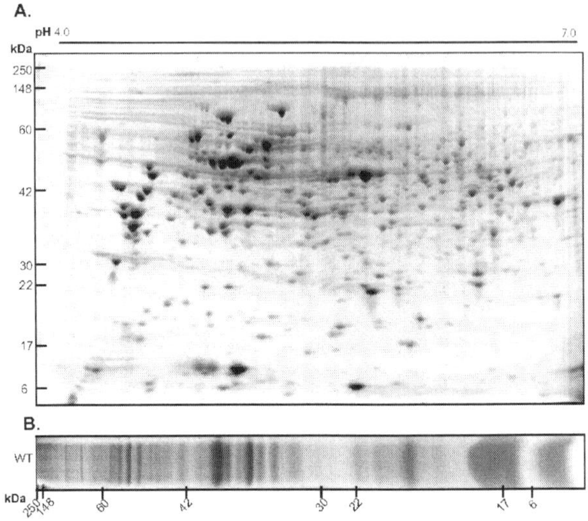

Fig. 11.1 Representative 1D and 2D gel analyses of *R. capsulatus* protein samples. (**a**) Periplasmic fraction proteins of *R. capsulatus* were separated by 2D-GE, with pH 4–7 IPG and SDS-PAGE, as first and second dimensions. (**b**) Chromatophore membranes of *R. capsulatus* were separated on 1D-GE using SDS-PAGE. In both cases, protein samples were prepared as described in Section 11.2, gels were stained with colloidal coomassie blue and protein spots or bands were excised and trypsin digested for identification by nano-LC–MS/MS

proteins, we prepare overlapping subproteomes of periplasmic, cytoplasmic, and membrane fractions of *R. capsulatus* cells grown under different conditions and analyze these samples as described in Section 11.2. These analyses identified so far more than over 450 *R. capsulatus* proteins that are grouped into different cellular function categories listed in Table 11.1. Tentative sub-cellular localizations of these proteins are also indicated using the PSORTb v 2.0 prediction software trained on bacterial protein sub-cellular localization prediction (Gardy et al. 2005). Of these proteins, 218 are predicted to be cytoplasmic and 187 are assigned to be exported proteins. Among the latter extracytoplasmic proteins, 52 and 94 of them are considered to be periplasmic and integral membrane proteins, respectively. Clearly, the predictive analysis is highly successful as for only 53 out of the 450 identified proteins a cellular localization could not be attributed reliably using PSORTb program. Moreover, global distributions of the identified proteins into different cellular function categories are also analyzed, and the results are presented as a pie chart (Fig. 11.2). Even though the available data set is not yet exhaustive, it appears that "energy metabolism" and "transport and binding proteins" categories are among the most populated groups of proteins with 82 and 79 members, respectively. Remarkably, while a large fraction of the identified proteins correspond to proteins of known functions, 49 of them have undefined

Table 11.1 Complete list of *R. capsulatus* proteins identified by nLC–MS/MS analyses. Identified proteins are categorized according to their predominant function as reported in the current literature (ref). Protein names, enzyme commission (EC) numbers, gene acronyms, and ERGO-light (*RRCxxxxx*) and NT05 (*NT05RCxxxx*) identification numbers are indicated. Sub-cellular localization of each protein was predicted by PSORTb and SignalP analyses, and C, EC, P, IM, OM, and M were used to designate cytoplasmic, extracytoplasmic, periplasmic, inner membrane, outer membrane, and membrane proteins, respectively

Protein name	EC Number	*Gene name*	ERGO*	NT05*	Location
AMINO ACID BIOSYNTHESIS					
Aromatic amino acid family					
Ornithine cyclodeaminase, Ocd	4.3.1.12	*arcB*	RRC00632	NT05RC2206	C
Tryptophan synthase beta subunit, TrpB	4.2.1.20	*trpB*	RRC01000	NT05RC2658	C
Tryptophan synthase alpha subunit, TrpA	4.2.1.20	*trpA*	RRC00192	NT05RC3179	C
Aromatic amino acid aminotransferase, Arat (aroat)	2.6.1.57	*tyrB*	RRC01763	NT05RC3527	C
Aspartate family					
Dihydrodipicolinate synthase	4.2.1.52	*dapA*	RRC02939	NT05RC0396	C
Aspartokinase (aspartate kinase)	2.7.2.4	*lysC*	RRC02984	NT05RC0439	C
5,10-Methylenetetrahydrofolate reductase	1.5.1.20	*metF*	RRC03981	NT05RC0510	C
5-Methyltetrahydrofolate homocysteine methyltransferase/ methionine synthase I	2.1.1.13	*metH*	RRC01233	NT05RC2273	C
Threonine synthase, ThrC	4.2.99.2	*thrC*	RRC00242	NT05RC3231	C
Glutamate family					
Glutamate synthase NADPH large chain, NADPH-Gogat	1.4.1.13	*gltA*	RRC03219	NT05RC0170	C
Glutamate synthase NADPH small chain, NADPH-Gogat	1.4.1.13	*gltD*	RRC03172	NT05RC0172	C
Argininosuccinate synthase	6.3.4.5	*argG*	RRC04654	NT05RC0219	C
Glutamate *N*-acetyltransferase/arginine biosynthesis bifunctional protein, ArgJ	2.3.1.35/2.3.1.1	*argJ*	RRC03917	NT05RC0229	C
N-Acetyl-gamma-glutamyl-phosphate reductase, ArgC	1.2.1.38	*argC*	RRC04153	NT05RC0575	C
Pyrroline-5-carboxylate reductase	1.5.1.2	*proC*	RRC03526	NT05RC1196	C
Glutamine synthetase, GlnA	6.3.1.2	*glnA*	RRC01355	NT05RC1748	C
Acetyltransferase/gnat family	2.3.1.–			NT05RC2400	

Table 11.1 (continued)

Protein name	EC Number	*Gene name*	ERGO*	NT05*	Location
Histidine family			RRC04158		
Imidazole glycerol phosphate synthase, cyclase subunit, HisF	4.1.3.–	*hisF*	RRC04159	NT05RC1201	C
Phosphoribosylformimino-5-aminoimidazole carboxamide ribotide isomerase	5.3.1.16	*hisA*	RRC04493	NT05RC1202	C
Imidazole glycerol phosphate synthase, glutamine amidotransferase subunit, HisH	2.4.2.–	*hisH*	RRC01022	NT05RC1232	C
Histidinol dehydrogenase, HisD	1.1.1.23	*hisD*		NT05RC2681	C
Pyruvate family			RRC03293		
3-Isopropylmalate dehydrogenase	1.1.1.85	*leuB*	RRC02733	NT05RC0087	C
Ketol-acid reductoisomerase/2-dehydropantoate 2-reductase	1.1.1.86	*ilvC*		NT05RC1266	C
Serine family			RRC00998		
Cysteine synthase A, CysK	2.5.1.47	*cysK*		NT05RC2657	
O-Acetylhomoserine thiol-lyase (*O*-acetylhomoserinesulfhydrylase)/homocysteine synthase	2.5.1.49/4.2.99.8	*metY*	RRC00762	NT05RC2915	C
Cysteine synthase (*O*-acetylserine sulfhydrylase)/*O*-acetylserine thiol-lyase, CSase	2.5.1.47/4.2.99.8		RRC00087	NT05RC3089	
Phosphoserine aminotransferase	2.6.1.52	*serC*	RRC05753	NT05RC3593	CM
BIOSYNTHESIS OF COFACTORS, PROSTHETIC GROUPS, AND CARRIERS					
Chlorophyll and bacteriochlorphyll					
Light-independent protochlorophyllide reductase, iron–sulfur ATP-binding protein	1.18.–.–	*bchL*	RRC02850	NT05RC0691	C
Bacteriochlorophyll synthase 33 kDa chain/geranylgeranyl bacteriochlorophyllide synthase	2.5.1.–	*bchG*	RRC02841	NT05RC0700	CM
Bacteriochlorophyll-geranylgeranyl reductase	1.3.1.–	*bchP*	RRC02839	NT05RC0702	CM
Magnesium chelatase ATPase subunit I	6.6.1.1	*bchI*	RRC02835	NT05RC0706	C

Table 11.1 (continued)

Protein name	EC Number	*Gene name*	ERGO*	NT05*	Location
2-Desacetyl-2-hydroxyethyl bacteriochlorophyllide *a* dehydrogenase, BchC	1.–.–.–	*bchC*	RRC02826	NT05RC0715	CM
Chlorophyllide reductase 35 kDa protein, BchA (XYZ)	1.18.–.–	*bchA*	RRC02825	NT05RC0716	C
Folic acid					
MoxR-like ATPase/AAA_3 ATPase			RRC04552	NT05RC0619	C
5-Methyltetrahydrofolate corrinoid iron–sulfur protein methyltransferase/dihydropteroate synthase, Dhps	1.2.99.2	*acsE*	RRC03586	NT05RC1686	C
Para-aminobenzoate synthetase component I, PabB	4.1.3.–/6.3.5.8	*pabB*	RRC01115	NT05RC2770	C
Methylenetetrahydrofolate dehydrogenase (NADP+)/methenyltetrahydrofolate cyclohydrolase/bifunctional protein, FolD	1.5.1.5/3.5.4.9	*folD*	RRC00338	NT05RC3327	C
Glutathione and analogs					
Glutathione synthetase	6.3.2.3	*gshB*	RRC04009	NT05RC0483	C
Heme, porphyrin, and cobalamin					
Ferrochelatase	4.99.1.1	*hemH*	RRC04666	NT05RC0206	C
Cobaltochelatase subunit, CobS	6.6.1.2	*cobS*	RRC01954	NT05RC0874	C
Uroporphyrinogen decarboxylase	4.1.1.37	*hemE*	RRC04134	NT05RC1222	CM
Porphobilinogen deaminase	2.5.1.61	*hemC*	RRC04136	NT05RC1223	C
5-Aminolevulinic acid synthase	2.3.1.37	*hemA*	RRC04400	NT05RC1506	CM
Delta-aminolevulinic acid dehydratase	4.2.1.24	*hemB*	RRC06033	NT05RC1890	
Precorrin-8X methylmutase/precorrin isomerase, HBAsynthase	5.4.1.2	*cobH*	RRC00650	NT05RC2138	C
Cobalamin biosynthesis protein, CobD		*cobD*	RRC00706	NT05RC2147	CM
Nicotinate-nucleotide-dimethylbenzimidazole phosphoribosyltransferase, CobT	2.4.2.21	*cobT*	RRC00590	NT05RC2151	C
Menaquinone and ubiquinone					
3-Demethylubiquinone 3-methyltransferase	2.1.1.64	*ubiG*	RRC03170	NT05RC0221	C

Table 11.1 (continued)

Protein name	EC Number	*Gene name*	ERGO*	NT05*	Location
Ubiquinone biosynthesis protein AARF/2-polyprenylphenol 6-hydroxylase		*ubiB*	RRC03386	NT05RC3680	CM
Pantothenate and coenzyme A					
Pantoate-beta-alanine ligase	6.3.2.1	*panC*	RRC02990	NT05RC0445	C
Pyridine nucleotides					
UDP-sugar diphosphatase/5′-nucleotidase	3.6.1.45/3.1.3.5	*ushA*	RRC01089	NT05RC2743	P
Riboflavin, FMN, and FAD					
Riboflavin biosynthesis protein, RibF	2.7.1.26/2.7.7.2	*ribF*	RRC00565	NT05RC2178	C
Riboflavin biosynthesis protein, RibAB/GTP cyclohydrolase II/3,4-dihydroxy-2-butanone-4-phosphate synthase	3.5.4.25/4.1.2.–	*ribAB*	RRC00220	NT05RC3210	C
Thiamine					
Thiamine-phosphate pyrophosphorylase	2.5.1.3	*thiE*	RRC06023	NT05RC1674	C
Other					
Iron–sulfur cluster assembly ATP-dependent transporter, SufC		*sufC*	RRC01541	NT05RC1963	C, CM
CELL ENVELOPE					
Biosynthesis and degradation of murein sacculus and peptidoglycan					
Glucose-1-phosphate thymidylyltransferase	2.7.7.24	*rfbA*	RRC03209	NT05RC0184	C
dTDP-4-dehydrorhamnose reductase	1.1.1.133	*rfbD*	RRC03208	NT05RC0185	EC
dTDP-glucose-4,6-dehydratase	4.2.1.46	*rfbB*	RRC03206	NT05RC0187	EC
dTDP-4-dehydrorhamnose 3,5-epimerase (reductase)	5.1.3.13	*rfbC*	RRC03205	NT05RC0188	EC
Glucans biosynthesis protein, MdoG		*mdoG*	RRC04027	NT05RC0391	P
Soluble lytic murein transglycosylase	3.2.1.–	*slt*	RRC02941	NT05RC0397	P
Cell elongation-specific D,D-transpeptidase/peptidoglycan glycosyltransferase	2.4.1.129	*ftsI*	RRC04232	NT05RC0836	C, CM

Table 11.1 (continued)

Protein name	EC Number	*Gene name*	ERGO*	NT05*	Location
UDP-*N*-acetylglucosamine-*N*-acetylmuramyl-pentapeptide-pyrophosphoryl-undecaprenol *N*-acetylglucosamine transferase	2.4.1.227	*murG*	RRC04656	NT05RC0856	CM
Peptidoglycan-binding domain 1 protein			RRC02722	NT05RC1256	EC
D-Alanyl-D-alanine serine-type carboxypeptidase	3.4.16.4		RRC02699	NT05RC1306	CM
UDP-*N*-acetylglucosamine diphosphorylase/glucosamine-1-phosphate *N*-acetyltransferase, GlmU	2.7.7.23/2.3.1.157	*glmU*	RRC04347	NT05RC1565	C
Alanine racemase, Alr	5.1.1.1	*alr*	RRC03627	NT05RC1640	C
Succinoglycan biosynthesis transport protein, ExoP	2.7.1.112	*exoP*	RRC01472	NT05RC2037	CM
D-Alanyl-D-alanine carboxypeptidase (D-alanyl-D-alanine-endopeptidase)/penicillin-binding protein 4, PBP-4	*dacB*	RRC01098	NT05RC2752	C	
Peptidoglycan-associated lipoprotein/OmpA-MotB family protein			RRC04613	NT05RC3334	OM
Membrane-bound lytic murein transglycosylase B	3.2.1.–	*mltB*	RRC05698	NT05RC3538	CM, P
Other					
Competence lipoprotein/tetratricopeptide TPR_2 repeat protein		*comL*	RRC01947	NT05RC0865	M
CELLULAR PROCESSES					
Adaptations to atypical conditions					
Transcriptional regulator/methionine-*R*-sulfoxide reductase	1.8.4.–	*msrB*	RRC00410	NT05RC3399	EC
Cell division					
Cell division ATP-binding protein, FtsE	3.6.3.–	*ftsE*	RRC00317	NT05RC3307	CM
Chemotaxis and motility					
Flagellar hook protein, FlgE		*flgE*	RRC00713	NT05RC0007	OM
Flagellar hook-associated protein 1, FlgK		*flgK*	RRC00714	NT05RC0008	OM
Flagellar hook-associated protein 3, FlgL		*flgL*	RRC00715	NT05RC0009	OM
Flagellar biosynthetic protein, FlhA		*flhA*	RRC03323	NT05RC0057	CM

Table 11.1 (continued)

Protein name	EC Number	*Gene name*	ERGO*	NT05*	Location
Methyl-accepting chemotaxis sensory transducer			RRC04109	NT05RC1237	CM
Methyl-accepting chemotaxis protein, McpA		*mcpA*	RRC01082	NT05RC2737	CM
Flagellin protein		*fliC*	RRC03417	NT05RC3674	P
Detoxification					
Dihaem cytochrome c551 peroxidase	1.11.1.5	*ccpA*	RRC00728	NT05RC0020	P
Glutathione peroxidase	1.11.1.9	*gpx*	RRC02811	NT05RC0730	P
Phosphinothricin *N*-acetyltransferase/gnat family	2.3.1.–	*pat*	RRC04226	NT05RC0842	
NAD-dependent aldehyde dehydrogenase	1.1.1.1	*adh*	RRC04460	NT05RC1433	C
Catalase/peroxidase, HPI	1.11.1.6/1.11.1.7	*katG*	RRC03460	NT05RC1815	
Beta-lactamase/hypothetical protein	3.5.2.6		RRC00641	NT05RC2200	CM
Beta-lactamase family protein/putative beta-lactamase	3.5.2.6	*bla*	RRC00550	NT05RC2365	P
Superoxide dismutase (Fe)	1.15.1.1	*sodB*	RRC01365	NT05RC2413	
Toluene tolerance protein, Ttg2D		*ttg2D*	RRC01770	NT05RC3534	
Peroxiredoxin, AhpC/glutaredoxin domain protein	1.11.1.15	*ahpC*	RRC05718	NT05RC3559	
Pathogenesis					
Virulence-associated protein E			RRC00816	NT05RC2862	
Iron-regulated protein frpC/rhizobiocin, RzcA		*frpC/rzcA*	RRC00341	NT05RC3330	EC
Toxin production and resistance					
Bicyclomycin resistance protein/drug resistance transporter, Bcr			RRC02414	NT05RC1893	M
Iron-regulated protein, FrpC/bifunctional hemolysin-adenylate cyclase (cyclolysin), AC-HLY		*frpC*	RRC01135	NT05RC2792	EC
CENTRAL INTERMEDIARY METABOLISM					
Amino sugars					
N-Acetylglucosamine-6-phosphate deacetylase	3.5.1.25	*nagA*	RRC05751	NT05RC3592	

Table 11.1 (continued)

Protein name	EC Number	*Gene name*	ERGO*	NT05*	Location
Nitrogen fixation					
Nitrogenase iron protein	1.18.6.1	*nifH*	RRC03785	NT05RC0610	C
Iron–sulfur cluster assembly NifU domain protein	1.18.6.1	*nifU*	RRC00380	NT05RC3369	C
Nitrogenase cofactor biosynthesis protein, NifB	1.18.6.1	*nifB*	RRC03905	NT05RC3406	C
Nitrogenase MoFe cofactor biosynthesis protein, NifE	1.18.6.1	*nifE*	RRC00431	NT05RC3420	C
Nitrogen metabolism					
Nitrogen regulatory protein P-II		*glnB*	RRC03527	NT05RC1747	C
Polyamine biosynthesis					
Agmatinase	3.5.3.11	*speB*	RRC03474	NT05RC1799	C
Arginase, RocF	3.5.3.1	*rocF*	RRC00631	NT05RC2207	C
One-carbon metabolism					
Trimethylamine methyltransferase family protein, MttB		*mttB*	RRC02448	NT05RC1856	CM
2-Hydroxy-6-oxo-6-phenylhexa-2,4-dienoate hydrolase or acyltransferase	3.7.1.–	*bphD*	RRC01136	NT05RC2793	C
Phosphorus compounds					
Bacterial phosphonate metabolism protein, PhnI		*phnI*	RRC04121	NT05RC1247	C
Sulfur metabolism					
Nitrilotriacetate monooxygenase component B	1.14.13.–	*ntaB*	RRC00711	NT05RC0005	C
3′(2′),5′-Bisphosphate nucleotidase	3.1.3.7	*cysQ*	RRC05208	NT05RC0131	C, CM
Other					
Succinate-semialdehyde dehydrogenase [NADP+], Ssdh	1.2.1.16	*gabD*	RRC02976	NT05RC0431	C
Methylmalonyl-CoA mutase alpha subunit	5.4.99.2	*mutB*	RRC02025	NT05RC0954	
S-Adenosylmethionine synthetase	2.5.1.6	*metK*	RRC01795	NT05RC1167	C
Gamma-aminobutyrate aminotransferase/aminotransferase class-III	2.6.1.19/2.6.1.18		RRC01241	NT05RC2282	C

Table 11.1 (continued)

Protein name	EC Number	*Gene name*	ERGO*	NT05*	Location
Exopolyphosphatase/pyrophosphate phosphohydrolase family II	3.6.1.11	*ppx*	RRC00932	NT05RC2590	C
Acetate-CoA ligase, AcsA	6.2.1.1	*acsA*	RRC01093	NT05RC2746	CM
DNA METABOLISM					
DNA replication, recombination, and repair					
DNA polymerase III beta chain	2.7.7.7	*dnaN*	RRC00708	NT05RC0002	C
ATP-dependent DNA helicase II, PcrA	3.6.1.–	*pcrA*	RRC04357	NT05RC1555	
Replicative DNA helicase	3.6.1.–	*dnaB*	RRC03626	NT05RC1641	C
NAD-dependent DNA ligase, LigA	6.5.1.2	*ligA*	RRC03535	NT05RC1737	C
Single-stranded DNA-binding protein, Ssb/helix-destabilizing protein		*ssb*	RRC02420	NT05RC1887	C
DNA polymerase I, Pol I	2.7.7.7	*pol1*	RRC00937	NT05RC2594	C
DNA primase	2.7.7.–	*dnaG*	RRC00201	NT05RC3187	C
Primosomal protein N′		*priA*	RRC00361	NT05RC3351	C
Exodeoxyribonuclease III	3.1.11.2	*exoDNase_III*	RRC00493	NT05RC3477	
DNA mismatch repair protein, MutS		*mutS*	RRC05797	NT05RC3643	CM
ENERGY METABOLISM					
Amino acids and amines					
Adenosylhomocysteinase	3.3.1.1	*ahcY*	RRC03330	NT05RC0050	C
Serine hydroxymethyl transferase	2.1.2.1	*glyA*	RRC03001	NT05RC0457	C
Alanine dehydrogenase	1.4.1.1	*ald*	RRC03987	NT05RC0504	C
Tyrosine phenol-lyase/beta-tyrosinase	4.1.99.2	*tlp*	RRC03952	NT05RC0536	CM
Sarcosine oxidase beta subunit/FAD-dependent oxidoreductase	1.5.3.1	*soxB*	RRC03939	NT05RC0550	CM
Aminomethyltransferase/glycine cleavage system T protein	2.1.2.10	*gcvT*	RRC03644	NT05RC1190	C
Methylmalonate-semialdehyde dehydrogenase (acylating)	1.2.1.27	*mmsA*	RRC03635	NT05RC1631	C
Propanediol utilization protein, PduB		*pduB*	RRC01267	NT05RC2309	
Tyrosine phenol-lyase/beta-tyrosinase	4.1.99.2	*tpl*	RRC01364	NT05RC2411	C
Sarcosine oxidase gamma subunit	1.5.3.1	*soxG*	RRC01367	NT05RC2415	
Sarcosine oxidase alpha subunit	1.5.3.1	*soxA*	RRC01368	NT05RC2416	C

Table 11.1 (continued)

Protein name	EC Number	*Gene name*	ERGO*	NT05*	Location
Sarcosine oxidase beta subunit	1.5.3.1	*soxB*	RRC01370	NT05RC2418	C
Aminotransferase, DegT/DnrJ/EryC1/StrS family			RRC00767	NT05RC2910	
LMW–protein–tyrosine–phosphatase, PTPase/small, acidic phosphotyrosine protein phosphatase (PYprotein phosphatase)	3.1.3.48	*ptpA*	RRC00274	NT05RC3261	
D-Alanine aminotransferase (D-aspartateaminotransferase)/D-amino acid aminotransferase/D-amino acid transaminase, Daat	2.6.1.21	*daat*	RRC01700	NT05RC3299	C
Ornithine cyclodeaminase	4.3.1.12	*arcB*	RRC05768	NT05RC3610	C
Anaerobic					
Trimethylamine-*N*-oxide reductase 1, TMAO reductase 1	1.7.2.3		RRC05612	NT05RC2976	P
ATP-proton motive force interconversion					
ATP synthase subunit I b chain	3.6.3.14	*ATP_synt_b*	RRC04287	NT05RC0776	EC
ATP synthase F1 beta subunit, AtpD	3.6.1.34	*atpD*	RRC00103	NT05RC3105	C
ATP synthase F1 alpha subunit, AtpA	3.6.1.34	*atpA*	RRC00106	NT05RC3107	C
ATP synthase F1 delta subunit, AtpH	3.6.1.34	*atpH*	RRC00107	NT05RC3108	C
Biosynthesis and degradation of polysaccharides					
Anhydromuramoyl-peptide exo-beta-*N*-acetylglucosaminidase	3.2.1.–	*nagZ*	RRC01974	NT05RC0895	C
Periplasmic beta-glucosidase (Gentiobiase) (Cellobiase)/beta-D-glucoside glucohydrolase	3.2.1.21	*bglX*	RRC01186	NT05RC2528	P
Electron transport					
Glutaryl-CoA dehydrogenase, Gcd	1.3.99.7	*gcd*	RRC04664	NT05RC0023	C
Electron transfer flavoprotein alpha or large subunit (alpha-Etf or Etfls)		*alpha-etf*	RRC03239	NT05RC0149	
Electron transfer flavoprotein beta or small subunit (beta-Etf or Etfss)		*beta-etf*	RRC03238	NT05RC0150	
Oxidoreductase, FAD-binding/homologous to D-amino acid dehydrogenase/glycine oxidase family			RRC02945	NT05RC0400	

Table 11.1 (continued)

Protein name	EC Number	*Gene name*	ERGO*	NT05*	Location
Electron transfer flavoprotein-ubiquinone oxidoreductase. EFT-Qo (electron-transferring-flavoprotein dehydrogenase)	1.5.5.1	*EFT-Qo*	RRC03955	NT05RC0533	
Sulfide-quinone reductase, Sqr	1.8.5.–	*sqr*	RRC04245	NT05RC0820	CM, P
Cytochrome *c* oxidase, cbb3-type, subunit I	1.9.3.1	*ccoN*	RRC04162	NT05RC1206	CM
Cytochrome *c* oxidase, cbb3-type, subunit II monoheme subunit	1.9.3.1	*ccoO*	RRC04163	NT05RC1207	CM
Cytochrome *c* oxidase, cbb3-type, subunit III, diheme subunit	1.9.3.1	*ccoP*	RRC04165	NT05RC1209	CM
Cytochrome *c*2		*cycA*	RRC02712	NT05RC1293	P
NADH-quinone oxidoreductase chain D	1.6.5.3	*nuoD*	RRC02454	NT05RC1585	C
NADH-quinone oxidoreductase chain E	1.6.5.3	*nuoE*	RRC02455	NT05RC1586	CM
Acyl-CoA dehydrogenase family member 8, Acad-8/isobutyryl-CoA dehydrogenase/activator-recruited cofactor 42 kDa component, Arc42	1.3.99.2		RRC02499	NT05RC1630	C
Ferredoxin-NADP reductase	1.18.1.2	*fpr*	RRC03609	NT05RC1659	C
N-Ethylmaleimide reductase/xenobiotic reductase, XenB	1.7.1.–	*xenB*	RRC03600	NT05RC1668	
Cytochrome *c*551 peroxidase/cytochrome *c* peroxidase, Ccp	1.11.1.5	*ccpA*	RRC03475	NT05RC1798	P
Thiol-disulfide isomerase/thioredoxin, redoxin domain protein			RRC02397	NT05RC1909	
Cytochrome P450-like protein			RRC01166	NT05RC2544	C
Cytochrome *c*-type cyt cy		*cycY*	RRC00961	NT05RC2616	CM, P
Ubiquinol-cytochrome *c* reductase iron–sulfur subunit, PetA	1.10.2.2	*petA*	RRC00780	NT05RC2896	CM
Ubiquinol-cytochrome *c* reductase cytochrome *b* subunit, PetB	1.10.2.2	*petB*	RRC00781	NT05RC2897	CM
Ubiquinol-cytochrome *c* reductase cytochrome *c*1 subunit, PetC	1.10.2.2	*petC*	RRC00779	NT05RC2898	CM
Ferredoxin I, FdI		*fd1*	RRC00758	NT05RC2920	C
Cytochrome *c*-type biogenesis protein CcmH (Ccl2)		*ccl2/ccmH*	RRC00287	NT05RC3273	CM, P
Cytochrome P450 family protein	1.14.15.3		RRC00405	NT05RC3394	C
Sulfide dehydrogenase (flavocytochrome *c*)/FAD-dependent pyridine nucleotide-disulfide oxidoreductase	1.8.2.–	*fccB*	RRC00443	NT05RC3435	EC
Fermentation					
Phosphate acetyltransferase/phosphate butyryltransferase	2.3.1.8/2.3.1.19	*pta*	RC01951	NT05RC0870	C

Table 11.1 (continued)

Protein name	EC Number	*Gene name*	ERGO*	NT05*	Location
Glycolysis/gluconeogenesis					
Phosphoenolpyruvate-protein phosphotransferase, PtsP	2.7.3.9	*ptsP*	RRC02985	NT05RC0440	C
Triosephosphate isomerase	5.3.1.1	*tpiA*	RRC03602	NT05RC1666	C
Fructose-bisphosphate aldolase type I	4.1.2.13	*fba*	RRC03492	NT05RC1781	C
Phosphoglycerate kinase	2.7.2.3	*pgk*	RRC03491	NT05RC1782	C
Enolase/phosphopyruvate hydratase	4.2.1.11	*eno*	RRC03484	NT05RC1789	C
Fructose-bisphosphate aldolase, class II, Calvin cycle subtype	4.1.2.13	*fba*	RRC02394	NT05RC1912	C
Glyceraldehyde 3-phosphate dehydrogenase B	1.2.1.12	*gap*	RRC02393	NT05RC1913	C
Glyceraldehyde-3-phosphate dehydrogenase, type I Gap	1.2.1.12	*gap*	RRC04329	NT05RC2258	C
Fructose-bisphosphate aldolase, putative aldolase, YneB	4.1.2.13	*yneB*	RRC00008	NT05RC3011	CM
L-Lactate dehydrogenase (FMN-linked) [cytochrome]/(*S*)-2-hydroxy-acid oxidase	1.1.2.3/1.1.3.15	*lldD*	RRC00194	NT05RC3180	
Pentose phosphate pathway					
Ribose 5-phosphate isomerase A	5.3.1.6	*rpiA*	RRC01215	NT05RC2502	C
Ribulose-phosphate 3-epimerase/pentose-5-phosphate 3-epimerase	5.1.3.1	*rpe*	RRC00013	NT05RC3016	CM
Fructose-1,6-bisphosphatase, class II GlpX	3.1.3.11	*glpX*	RRC00208	NT05RC3195	CM
Transaldolase, putative	2.2.1.2		RRC00360	NT05RC3349	C
Photosynthesis					
Light-harvesting protein B-870, alpha chain		*pufA*	RRC02820	NT05RC0721	P
Light-harvesting protein B-800/850, alpha chain		*pucA*	RRC02342	NT05RC2651	P
Light-harvesting protein B-800/850, gamma chain		*pucE*	RRC00994	NT05RC2653	CM
Photosynthetic reaction center H subunit		*puhA*	RRC02853	NT05RC0688	CM
Photosynthetic reaction center L subunit		*pufL*	RRC02819	NT05RC0722	CM
Photosynthetic reaction center M subunit		*pufM*	RRC02818	NT05RC0723	CM
Sugars					
Xylose isomerase	5.3.1.5	*xylA*	RRC00720	NT05RC0014	C
Xylulose kinase	2.7.1.17	*xylB*	RRC00721	NT05RC0015	C

Table 11.1 (continued)

Protein name	EC Number	*Gene name*	ERGO*	NT05*	Location
Nucleoside-diphosphate-sugar epimerase			RRC03125	NT05RC0264	CM
Phosphoglycolate phosphatase, Gph	3.1.3.18	*gph*	RRC02398	NT05RC1908	
Phosphoribulokinase 1	2.7.1.19	*cbbP*	RRC02391	NT05RC1915	C
Ribulose bisphosphate carboxylase large chain 2	4.1.1.39	*rbcL*	RRC02395	NT05RC1911	C
Phosphoglucomutase/glucose phosphomutase, Pgm	5.4.2.2	*pgm*	RRC02387	NT05RC1919	C
TCA cycle					
Malate dehydrogenase, NAD dependent	1.1.1.37	*mdh*	RRC02795	NT05RC0747	C
Succinyl-CoA synthetase beta chain	6.2.1.5	*sucC*	RRC02793	NT05RC0750	C
Succinyl-CoA synthetase alpha chain	6.2.1.5	*sucD*	RRC02792	NT05RC0751	C
Isocitrate dehydrogenase, NADP dependent	1.1.1.42	*idh2*	RRC01533	NT05RC1971	C
Aconitate hydratase	4.2.1.3	*acnA*	RRC04337	NT05RC2248	C
FATTY ACID AND PHOSPHOLIPID METABOLISM					
Biosynthesis					
CDP-diacylglycerol-glycerol-3-phosphate 3-phosphatidyltransferase	2.7.8.5	*pgsA*	RRC02959	NT05RC0413	CM
MaoC domain protein dehydratase		*maoC*	RRC02782	NT05RC0761	C
Dehydrogenase with MaoC-like domain			RRC02783	NT05RC0761	C
Putative phosphatidylserine decarboxylase homolog	4.1.1.65	*psd*	RRC04174	NT05RC0957	CM
Acyl-CoA thioesterase I precurser/lipolytic enzyme, G-D-S-L family	3.1.2–	*acot1*	RRC04139	NT05RC1226	P
Malonyl-CoA-[acyl-carrier-protein] transacylase, FabD	2.3.1.39	*fabD*	RRC03525	NT05RC1749	
3-Oxoacyl-[acyl-carrier protein] synthase, FabF	2.3.1.179	*fabF*	RRC03521	NT05RC1754	C
Enoyl-[acyl-carrier protein] reductase (NADH)	1.3.1.9	*fabI*	RRC01406	NT05RC2453	C
3-Oxoacyl-(acyl-carrier protein) reductase/short-chain dehydrogenase/reductase, SDR	1.1.1.100	*fabG*	RRC01189	NT05RC2525	C
Enoyl-[acyl-carrier-protein] reductase [NADH] (NADH-dependent enoyl-ACP reductase)	1.3.1.9		RRC05546	NT05RC2794	C
3-Oxoacyl-[acyl-carrier-protein] synthase II, FabF		*fabF*	RRC01138	NT05RC2795	C
Beta-hydroxyacyl-(acyl-carrier-protein) dehydratase, FabA	4.2.1.60	*fabA*	RRC01139	NT05RC2796	C

Table 11.1 (continued)

Protein name	EC Number	*Gene name*	ERGO*	NT05*	Location
3-Hydroxyacyl-CoA dehydrogenase type-2, type II HADH/3-hydroxy-2-methylbutyryl-CoA dehydrogenase	1.1.1.35	*hadh*	RRC00129	NT05RC3126	C
Degradation					
Succinyl-CoA:3-ketoacid-coenzyme A transferase subunit b, Oxct b	2.8.3.5	*scoB*	RRC00248	NT05RC3237	C
Other					
Acetyl-CoA acetyltransferase/acetoacetyl-CoA thiolase	2.3.1.9	*Acat*	RRC00329	NT05RC3317	C
PROTEIN FATE					
Degradation of proteins, peptides, and glycopeptides					
Secreted protease C precursor	3.4.24.–		RRC03223	NT05RC0167	EC
Peptidase T, PepT	3.4.11.–	*pepT*	—	NT05RC0370	C
Peptidase, M20/M25/M40 family	3.5.1.14		RRC04401	NT05RC1505	C
N-Acyl-L-amino acid amidohydrolase/Hippuricase	3.5.1.14		RRC02444	NT05RC1861	C
ATP-dependent protease La/ATP-dependent endopeptidase, Lon	3.4.21.53		RRC01049	NT05RC2706	C
Leucyl/phenylalanyl-tRNA-protein transferase	2.3.2.–	*aat*	RRC01073	NT05RC2731	CM
ATP-dependent endopeptidase clp proteolytic subunit, ClpP	3.4.21.92	*clpP*	RRC01078	NT05RC2735	CM
Oligoendopeptidase F, putative	3.4.24.–	*pepF*	RRC00748	NT05RC2930	C
Protein and peptide secretion and trafficking					
Protein-export chaperone, SecB		*secB*	RRC00736	NT05RC0029	C
Protein secretion ABC efflux permease and ATP-binding protein/type I secretion processing peptidase			RRC03199	NT05RC0194	CM
Preprotein translocase, SecA subunit		*secA*	RRC03160	NT05RC0231	C
Peptidase M16 domain protein/zinc protease	3.4.99.–		RRC03108	NT05RC0280	CM
Peptidase M16 domain protein/insulinase family zinc protease	3.4.99.–		RRC03107	NT05RC0281	EC
60 kDa inner membrane insertion protein, YidC		*yidC*	RRC03045	NT05RC0360	CM
General secretion pathway protein D/type II and III secretion system protein		*gspD*	RRC03968	NT05RC0522	OM

Table 11.1 (continued)

Protein name	EC Number	*Gene name*	ERGO*	NT05*	Location
SecY stabilizing integral membrane protein		*secY*	RRC01986	NT05RC0911	CM
Hemin import ATP-binding protein, HmuV/transporter		*hmuV*	RRC04502	NT05RC1071	C, CM
Preprotein translocase, YajC subunit		*yajC*	RRC02443	NT05RC1862	CM
Survival protein, SurA/PpiC-type peptidyl-prolyl *cis–trans* isomerase	5.2.1.8	*surA*	—	NT05RC2819	EC
Microcin-processing peptidase 2 protein, TldD		*tldD*	RRC00243	NT05RC3232	M
Tol system periplasmic component, YbgF/tetratricopeptide TPR_2		*ybgF*	RRC05669	NT05RC3333	EC
Outer membrane lipoprotein-carrier protein, LolA		*lolA*	RRC00485	NT05RC3469	EC
Type I secretion adaptor protein (HlyD family)/RTX secretion protein D		*type_I_hlyD*	RRC05697	NT05RC3537	C, CM
Protein folding and stabilization					
PpiC-type peptidyl-prolyl *cis–trans* isomerase	5.2.1.8		—	NT05RC0230	OM
Chaperone protein, DnaK		*dnaK*	RRC03154	NT05RC0235	C
Thiol:disulfide interchange protein DsbA (DsbA oxidoreductase)		*dsbA*	RRC03149	NT05RC0240	EC
Peptidyl-prolyl *cis–trans* isomerase, cyclophilin type			RRC03490	NT05RC1783	P
Peptidyl-prolyl *cis–trans* isomerase, PPIase/potamase/cyclophilin/cyclosporin A-binding protein, CYP B	5.2.1.8	*cypB*	RRC03489	NT05RC1784	CM
Outer membrane protein thiol:disulfide interchange protein DsbA like		*dsbA*	RRC03455	NT05RC1820	EC
Trigger factor, PPIase, Tig	5.2.1.8	*tig*	RRC00692	NT05RC2098	C
Putative peptidyl-prolyl *cis–trans* isomerase	5.2.1.8		RRC01410	NT05RC2457	
Endopeptidase, DegP/periplasmic serine protease	3.4.21.–	*degP*	RRC01208	NT05RC2508	P
10 kDa chaperonin, GroES		*groS*	RRC00934	NT05RC2591	C
60 kDa chaperonin, GroEL		*groL*	RRC00935	NT05RC2592	C
ATP-dependent chaperone, ClpB		*clpB*	RRC05711	NT05RC3551	C
Co-chaperone, GrpE		*grpE*	RRC05796	NT05RC3642	C
Protein modification and repair					
Carboxyl-processing protease (tail-specific protease)	3.4.21.102		RRC03302	NT05RC0077	CM
Universal stress family protein		*uspA*	RRC02802	NT05RC0740	C

Table 11.1 (continued)

Protein name	EC Number	*Gene name*	ERGO*	NT05*	Location
Universal stress protein UspA and related nucleotide-binding domain protein		*uspA*	RRC04161	NT05RC1205	C
Nitrilase/cyanide hydratase and apolipoprotein *N*-acyltransferase			RRC01014	NT05RC2673	
Leucyl aminopeptidase	3.4.11.1	*pepA*	RRC04532	NT05RC2823	C
Protein-L-isoaspartate *O*-methyltransferase	2.1.1.77	*Pcmt1*	RRC00264	NT05RC3250	C
Other					
von Willebrand factor type A domain protein/CBBO			RRC03776	NT05RC0600	C
PROTEIN SYNTHESIS					
Ribosomal proteins: synthesis and modification					
LSU ribosomal protein, L12 (L7/L12)		*rplL*	RRC03083	NT05RC0305	C
SSU ribosomal protein, S3		*rpsC*	RRC03069	NT05RC0320	C
LSU ribosomal protein, L14		*rplN*	RRC03064	NT05RC0325	C
LSU ribosomal protein, L6		*rpl6P*	RRC03059	NT05RC0330	C
SSU ribosomal protein, S5		*rpsE*	RRC03057	NT05RC0332	C
SSU ribosomal protein, S13		*rpsM*	RRC03051	NT05RC0339	C
SSU ribosomal protein, S11		*rpsK*	RRC03050	NT05RC0340	C
LSU ribosomal protein, L19 (or L31)		*rpmE*	RRC03026	NT05RC0377	C
SSU ribosomal protein, S1		*rpsA*	RRC01790	NT05RC1173	C
SSU ribosomal protein, S2		*rpsB*	RRC02403	NT05RC1903	C
LSU ribosomal protein, L9		*rplI*	RRC00691	NT05RC2101	C
Translation factors					
Translation elongation factor Tu		*tuf*	RRC03808	NT05RC0153	C
Protein translation elongation factor G (EF-G)		*fusA*	RRC02509	NT05RC0311	C
Protein translation elongation factor Tu (EF-TU)		*tuf*	RRC03235	NT05RC0312	C
Protein translation elongation factor Tu (EF-TU)		*tuf*	RRC03808	NT05RC0312	C
Translation elongation and release factor G (GTPase)		*GTPase*	RRC04353	NT05RC1559	C
Peptide chain release factor 2, RF-2		*prfB*	RRC02481	NT05RC1613	C

Table 11.1 (continued)

Protein name	EC Number	*Gene name*	ERGO*	NT05*	Location
Protein translation elongation factor Ts (EF-Ts)		*tsf*	RRC02402	NT05RC1904	C
Bacterial peptide chain release factor 3 (RF-3), PrfC		*prfC*	RRC06068	NT05RC2425	C
tRNA and rRNA base modification					
Aspartyl/glutamyl-tRNA amidotransferase subunit A	6.3.5.–	*gatA*	RRC00798	NT05RC2879	CM
Aspartyl/glutamyl-tRNA amidotransferase subunit B (Asp/Glu-ADT subunit B)	6.3.5.–	*gatB*	RRC01959	NT05RC0880	C
D-Tyrosyl-tRNA deacylase, Dtd	3.1.–.–	*dtd*	RRC00975	NT05RC2630	C
tRNA aminoacylation					
Tryptophanyl-tRNA synthetase	6.1.1.2	*trpS*	RRC04002	NT05RC0490	C
Aspartyl-tRNA synthetase	6.1.1.12	*aspS*	RRC03937	NT05RC0553	C
Seryl-tRNA synthetase	6.1.1.11	*serS*	RRC04397	NT05RC1510	C, CM
Prolyl-tRNA synthetase		*proRS*	—	NT05RC2575	C
Phenylalanyl-tRNA synthetase, beta subunit	6.1.1.20	*pheT*	RRC06104	NT05RC3289	C
PURINES, PYRIMIDINES, NUCLEOSIDES, AND NUCLEOTIDES					
Nucleotide and nucleoside interconversions					
Adenylate kinase (ATP–AMP transphosphorylase)/nucleoside-diphosphate kinase	2.7.4.3/2.7.4.6	*adk*	RRC03052	NT05RC0338	C
Nucleoside diphosphate kinase, Ndk/NDP kinase/nucleoside-2-P kinase	2.7.4.6	*ndk*	RRC03479	NT05RC1794	C
Purine ribonucleotide biosynthesis					
GMP synthase (glutamine-hydrolyzing)	6.3.5.2	*guaA*	RRC00895	NT05RC1941	C
Formyltetrahydrofolate deformylase	3.5.1.10	*purU*	RRC01205	NT05RC2511	
Pyrimidine ribonucleotide biosynthesis					
Carbamoyl-phosphate synthase, large subunit	6.3.5.5	*carB*	RRC03938	NT05RC0552	CM, P

Table 11.1 (continued)

Protein name	EC Number	*Gene name*	ERGO*	NT05*	Location
Orotate phosphoribosyltransferase	2.4.2.10	*pyrE*	RRC03625	NT05RC1642	C
Dihydropyrimidine dehydrogenase [NADP+] beta subunit/dihydroorotate dehydrogenase family protein	1.3.1.2	*DPYD*	RRC00991	NT05RC2646	C
Salvage of nucleosides and nucleotides					
Adenine phosphoribosyltransferase, Apt	2.4.2.7	*apt*	RRC01698	NT05RC3227	C
Purine nucleotide phosphorylase	2.4.2.1		RRC00387	NT05RC3376	C
2′-Deoxyribonucleotide metabolism					
Pyrimidine-specific ribonucleoside hydrolase, RihA/cytidine/uridine-specific hydrolase	3.2.–.–	*rihA*	RRC03481	NT05RC1792	C
Other					
Nucleoside-binding protein/Bmp family protein			RRC04244	NT05RC0821	EC
Xanthine dehydrogenase, small subunit	1.17.1.4	*xdhA*	RRC04239	NT05RC0827	C
REGULATORY FUNCTIONS					
DNA interactions					
Transcriptional regulator, LuxR family		*luxR*	RRC03806	NT05RC0343	C
Cell cycle transcriptional regulator, CtrA/response regulator SokA		*ctrA*	RRC03536	NT05RC1736	C
Transcriptional regulatory protein BaeR, alkaline phosphatase synthesis		*baeR*	RRC00026	NT05RC3031	C
Protein interactions					
Regulatory protein, SenC (PrrC)		*senC/prrC/scoI*	RRC03336	NT05RC0044	CM
Phosphohistidine phosphatase, SixA/phosphoglycerate mutase family protein	3.1.3.–	*sixA*	RRC04024	NT05RC0354	
Protease activity modulator		*hflK*	RRC01212	NT05RC2505	

Table 11.1 (continued)

Protein name	EC Number	*Gene name*	ERGO*	NT05*	Location
Small-molecule interactions					
Diguanylate cyclase/phosphodiesterase with PAS/PAC and GAF sensor	2.7.7.65		RRC00149	NT05RC3141	C
Two-component systems					
Two-component response regulator (winged helix family)			—	NT05RC0276	C
Other					
Response regulator/ggdef domain protein			RRC06082	NT05RC2762	
Phosphate transport system regulatory protein, PhoU		*phoU*	RRC05801	NT05RC3647	C
SIGNAL TRANSDUCTION					
Two-component systems					
Methyl-accepting chemotaxis sensory transducer, McpA		*mcpA*	RRC04620	NT05RC1409	CM
Sensory transduction histidine protein kinase	2.7.3.–		RRC05028	NT05RC1570	CM
Methyl-accepting chemotaxis protein, McpC		*mcpC*	RRC03579	NT05RC1694	CM
Sensory transduction protein kinase	2.7.3.–		RRC01105	NT05RC2760	
Two-component response sensory transduction histidine kinase, potential phosphate regulatory protein			RRC06094	NT05RC2997	C
Two-component system integral membrane signal transduction histidine kinase	2.7.3.–		RRC00027	NT05RC3032	CM
TRANSCRIPTION					
Degradation of RNA					
Polyribonucleotide nucleotidyltransferase/polynucleotidephosphorylase, PNPase	2.7.7.8		RRC03090	NT05RC0297	C
Ribonuclease E, RNase E	3.1.4.–	*RNaseEG*	RRC05484	NT05RC2263	C
Translation initiation inhibitor/endoribonuclease L-PSP			RRC05722	NT05RC3562	CM

Table 11.1 (continued)

Protein name	EC Number	*Gene name*	ERGO*	NT05*	Location
DNA-dependent RNA polymerase					
DNA-directed RNA polymerase beta subunit	2.7.7.6	*rpoB*	RRC03082	NT05RC0306	C
DNA-directed RNA polymerase beta subunit	2.7.7.6	*rpoC*	RRC03080	NT05RC0307	C
DNA-directed RNA polymerase, alpha subunit	2.7.7.6	*rpoA*	RRC03049	NT05RC0341	C
Transcription factors					
Sigma 54 modulation protein/ribosomal protein S30EA or factor Y		*rpoX*	RRC03265	NT05RC0121	C
Transcription elongation factor, GreA		*greA*	RRC03954	NT05RC0534	C
Transcriptional regulator, LysR family		*lysR*	RRC03919	NT05RC0572	C
Transcription elongation factor, NusA/N utilization substance protein A		*nusA*	RRC03167	NT05RC224	C
Other					
Putative AraC family transcriptional regulator			RRC04375	NT05RC1537	CM
TRANSPORT AND BINDING PROTEINS					
Amino acids, peptides, and amines					
Transporter/import inner membrane translocase, subunit Tim44 homolog		*Sec?*	RRC00738	NT05RC0031	CM
Oligopeptide-binding protein, OppA/extracellular solute-binding protein, family 5		*oppA*	RRC03230	NT05RC0161	P
Glutamate/glutamine/aspartate/asparagine-binding protein, BztA		*bztA*	RRC03799	NT05RC0350	P
Glutamine-binding protein/extracellular solute-binding protein, family 3			RRC04018	NT05RC0474	P
Type II secretion system protein E			RRC03965	NT05RC0525	CM, P
Putative lysine–arginine–ornithine or histidine-binding periplasmic protein			RRC02872	NT05RC0650	P
Oligopeptide-binding protein OppA/extracellular solute-binding protein family		*oppA*	RRC02807	NT05RC0735	CM, P
Dipeptide-binding protein/extracellular solute-binding protein, family			RRC01964	NT05RC0885	P

Table 11.1 (continued)

Protein name	EC Number	*Gene name*	ERGO*	NT05*	Location
Dipeptide transport ATP-binding protein, DppD/oligopeptide transport ATP-binding protein, OppD		*oppD*	RRC01967	NT05RC0888	C, CM
Glycine betaine/L-proline transport system-binding protein, ProX		*proX*	RRC02010	NT05RC0936	
Spermidine/putrescine ABC transporter ATP-binding subunit			RRC02709	NT05RC1297	C, CM
Putative polyamine ABC transporter, periplasmic polyamine-binding protein			RRC02708	NT05RC1298	EC
Polyamine ABC transporter, permease protein			RRC02707	NT05RC1299	CM
Probable phosphatidylinositol-4-phosphate-5-kinase			RRC04081	NT05RC1419	EC
ABC polyamine/opine/phosphonate transporter, periplasmic ligand-binding protein			RRC04469	NT05RC1423	EC
ABC transporter ATP-binding protein/methionine import ATP-binding protein, MetN		*metN*	RRC03629	NT05RC1638	CM
Leucine-, isoleucine-, valine-, threonine-, and alanine-binding protein/extracellular ligand-binding receptor	*braC*	RRC00901	NT05RC1935	P	
Putrescine transport ATP-binding protein, PotG		*potG*	RRC01528	NT05RC1977	C
Putrescine transport system permease protein, PotI		*potI*	RRC01527	NT05RC1978	CM
Spermidine/putrescine transport system permease protein, PotB/binding-protein-dependent inner membrane transport component	*potB*	RRC01526	NT05RC1979	CM	
Spermidine/putrescine-binding protein/extracellular solute-binding protein, family 1			RRC01245	NT05RC2286	P
Taurine ABC transporter, periplasmic taurine-binding protein			RRC01299	NT05RC2343	EC
Spermidine/putrescine-binding periplasmic protein			RRC01326	NT05RC2370	P
Oligopeptide-binding protein, OppA		*oppA*	RRC01333	NT05RC2377	P
Branched-chain amino acid ABC transporter, ATP-binding protein, LivF		*livF*	RRC01157	NT05RC2553	C, CM
His/Glu/Gln/Arg/opine family ABC transporter/periplasmic His/Glu/Gln/Arg/opine family-binding protein		RRC00775	NT05RC2902	P	

Table 11.1 (continued)

Protein name	EC Number	*Gene name*	ERGO*	NT05*	Location
Spermidine/putrescine-binding protein, ABC polyamine transporter, periplasmic substrate-binding protein		RRC00091	NT05RC3093	P	
Branched-chain amino acid aminotransferase	2.6.1.42		RRC00377	NT05RC3367	C
ABC transporter, Cyd cysteine exporter family permease/ATP-binding protein, CydC		*cydC*	RRC00404	NT05RC3393	CM
ABC branched amino acid transporter family, periplasmic substrate-binding subunit			RRC05738	NT05RC3577	EC
ABC branched amino acid transporter family, ATPase subunit			RRC05739	NT05RC3578	C, CM
Anions					
Molybdate ABC transporter, periplasmic molybdate-binding protein		*modA*	RRC03762	NT05RC0585	P
Molybdate ABC transporter, ATP-binding protein, ModC		*modC*	RRC03907	NT05RC0587	C, CM
Thiosulfate-binding protein			—	NT05RC2870	P
Membrane protein MosC, major facilitator superfamily MFS_1		*mosC*	RRC00195	NT05RC3181	CM
Phosphate transport ATP-binding protein, PstB		*pstB*	RRC05802	NT05RC3648	C, CM
Phosphate ABC transporter, permease protein, PstA		*pstA*	RRC05803	NT05RC3649	CM
ABC phosphate transporter, inner membrane subunit, PstC		*pstC*	RRC03394	NT05RC3650	CM
Phosphate-binding periplasmic protein			RRC03395	NT05RC3651	P
Carbohydrates, organic alcohols, and acids					
D-Xylose ABC transporter, D-xylose-binding protein		*xylF*	RRC00724	NT05RC0018	P
Mannitol-binding protein/Trap dicarboxylate transporter, Dctp subunit			RRC04462	NT05RC1431	EC
Alpha-glucoside-binding protein/ABC alpha-glucoside transporter, periplasmic substrate-binding protein		*algE*	RRC03427	NT05RC1851	EC
Alpha-glucoside transport ATP-binding protein, AglK		*algK*	RRC03422	NT05RC1855	C, CM
Fructose-binding periplasmic protein, FrcB		*frcB*	—	NT05RC2111	P
Sorbitol-binding protein/potential maltose-binding protein			RRC00677	NT05RC2114	EC
ABC sugar transporter, periplasmic-binding protein/D-galactose-binding/ribitol-binding protein			RRC02337	NT05RC2476	P
Mannitol-binding protein			RRC01191	NT05RC2524	EC

Table 11.1 (continued)

Protein name	EC Number	*Gene name*	ERGO*	NT05*	Location
Ribose ABC transporter, D-ribose-binding protein			RRC01127	NT05RC2784	P
Ribose transport ATP-binding protein RbsA/ABC transporter, carbohydrate uptake transporter-2 (CUT2) family	*rbsA*	RRC01129	NT05RC2786	C, CM	
C4-dicarboxylate-binding periplasmic protein			RRC00169	NT05RC3156	P
Cations and iron carrying compounds					
Manganese transport protein, MntH		*mntH*	RRC03871	NT05RC0626	CM
Iron (III) dicitrate-binding periplasmic protein		*fecB*	RRC04484	NT05RC1088	EC
Outer membrane siderophore receptor/TonB-dependent siderophore receptor		*fiu*	RRC04402	NT05RC1504	OM
Sensor protein, KdpD	2.7.3.–	*kdpD*	RRC01313	NT05RC2357	CM
Transporter/TrkA-C domain protein			—	NT05RC2388	CM
TonB-system energizer ExbB/TolQ protein		*exbB*	RRC01434	NT05RC2479	CM
2,3-Dihydro-2,3-dihydroxybenzoate dehydrogenase/short-chain dehydrogenase/reductase, SDR	1.3.1.28	*entA*	RRC01183	NT05RC2531	C
25 kD outer membrane immunogenic protein/rhizobiocin, RzcA		*rzcA*	RRC01603	NT05RC2572	OM
Iron-regulated protein, FrpC		*frpC*	RRC01604	NT05RC2573	EC
Iron-regulated protein FrpC /poly(beta-D-mannuronate) C5 epimerase 3 (mannuronanepimerase 3)		*frpC*	RRC00309	NT05RC3297	EC
Porins					
Porin			RRC03116	NT05RC0272	OM
Other					
Putative membrane protein			RRC03140	NT05RC0250	CM
Acetyltransferase/MobC protein	2.3.1.–	*mobC*	RRC02973	NT05RC0428	
Acriflavin resistance plasma membrane protein B		*acrB*	RRC02995	NT05RC0450	CM
Acriflavin resistance periplasmic protein			RRC02881	NT05RC0641	P
Periplasmic component of efflux system/HlyD family protein			RRC02879	NT05RC0643	CM, P

Table 11.1 (continued)

Protein name	EC Number	*Gene name*	ERGO*	NT05*	Location
ABC transporter, substrate-binding protein, aliphatic sulfonates/taurine-binding periplasmic protein			RRC06012	NT05RC1459	EC
ABC transporter ATP-binding protein			RRC04420	NT05RC1484	CM
Transporter			RRC04392	NT05RC1517	
ABC transporter, fused ATPase, and inner membrane subunit			RRC01360	NT05RC2407	CM
Putative transport protein, YidE/integral membrane protein with TrkA domains		*yidE*	RRC01177	NT05RC2535	CM
Oligopeptide-binding protein/extracellular solute-binding protein, family 5			RRC00960	NT05RC2615	EC
ABC-type nitrate/sulfonate/bicarbonate transport systems, periplasmic component, TauA		*tauA*	RRC00984	NT05RC2639	EC
Iron(III)-binding periplasmic protein			RRC01044	NT05RC2700	P
Thiamine/thiamine pyrophosphate ABC transporter, thiamine/thiamine pyrophospate-binding protein, ThiB		*thiB*	RRC01667	NT05RC2825	P
Acriflavin resistance protein E/efflux transporter, RND family, MFP subunit			RRC00805	NT05RC2872	CM, P
Acriflavin resistance plasma membrane protein			RRC00803	NT05RC2873	CM
TolB protein		*tolB*	RRC00346	NT05RC3335	P
ABC transporter, periplasmic substrate-binding protein			RRC00383	NT05RC3372	P
UNCLASSIFIED					
Role category not yet assigned					
Myo-inositol-1(or 4)-monophosphatase (I-1-Pase)	3.1.3.25		RRC03984	NT05RC0507	C
Myo-inositol-1(or 4)-monophosphatase family protein	3.1.3.25		RRC01989	NT05RC0915	
Lipoprotein, putative			RRC01112	NT05RC2767	EC
Inositol monophosphatase/ADP-ribosylglycohydrolase, CysQ protein	3.1.3.25		RRC00266	NT05RC3253	
UNKNOWN FUNCTION					
General					
Filament-A precursor (peptidase M23B)		*filA*	RRC03301	NT05RC0078	C, CM

Table 11.1 (continued)

Protein name	EC Number	*Gene name*	ERGO*	NT05*	Location
Putative iojap protein family homolog			RRC03298	NT05RC0081	C
Pentapeptide repeat family protein			RRC04445	NT05RC1451	EC
Outer membrane protein E/membrane protein involved in aromatic hydrocarbon degradation			RRC01502	NT05RC2007	OM
Hypothetical proteins					
Conserved domain proteins					
ErfK/YbiS/YcfS/YnhG family protein			RRC02993	NT05RC0448	
Conserved hypothetical protein			RRC03927	NT05RC0563	
Conserved hypothetical protein			RRC03926	NT05RC0564	
Conserved hypothetical membrane-spanning protein			RRC02854	NT05RC0687	CM
Conserved hypothetical protein			RRC02456	NT05RC1587	
Hypothetical cytosolic protein of unknown function			RRC02483	NT05RC1615	C
Hypothetical protein/GfdT protein			RRC03541	NT05RC1732	C
Conserved hypothetical membrane-associated protein			RRC03520	NT05RC1755	C, CM
Hypothetical membrane-associated protein/invasion-associated family protein			RRC00884	NT05RC1952	EC
Hypothetical protein			RRC01494	NT05RC2016	
Conserved hypothetical cytosolic protein			RRC00597	NT05RC2240	C
Hypothetical exported protein			RRC01358	NT05RC2405	EC
Hypothetical membrane-associated protein/pyrrolo-quinoline quinine			RRC01359	NT05RC2406	CM
Hypothetical protein			—	NT05RC2634	
Hypothetical protein/lipoprotein, putative			RRC01128	NT05RC2785	
Hypothetical membrane-spanning protein of unknown function			RRC06091	NT05RC2994	CM
Conserved hypothetical protein			RRC06095	NT05RC3026	
Conserved hypothetical protein, phage minor tail protein			RRC02354	NT05RC3046	CM
Hypothetical protein			RRC05643	NT05RC3064	C
Hypothetical cytosolic protein, prophage Lp2 protein			RRC00072	NT05RC3071	CM
Conserved hypothetical protein			—	NT05RC3075	

Table 11.1 (continued)

Protein name	EC Number	*Gene name*	ERGO*	NT05*	Location
Conserved hypothetical protein			RRC00313	NT05RC3301	C
Conserved hypothetical protein/permease			RRC00389	NT05RC3378	EC
Conserved hypothetical protein			RRC00393	NT05RC3382	EC
Conserved hypothetical cytosolic protein			RRC00413	NT05RC3402	C
Conserved hypothetical exported protein			RRC01759	NT05RC3524	CM, P
Conserved hypothetical protein			RRC03402	NT05RC3658	P
Conserved hypothetical proteins					
Conserved hypothetical (transglutaminase-like) protein			RRC03327	NT05RC0053	
Conserved hypothetical cytosolic protein			RRC03249	NT05RC0139	C
Conserved hypothetical protein/glucose-1-phosphate adenyltransferase	2.7.7.27		—	NT05RC0925	
OmpA family hypothetical protein			RRC03674	NT05RC0952	EC
Conserved hypothetical protein/aminomethyltransferase	2.1.2.10		RRC03662	NT05RC0993	C
Conserved hypothetical protein			RRC01788	NT05RC1175	
Hypothetical cytosolic protein			RRC03645	NT05RC1186	
Hypothetical protein			RRC02694	NT05RC1315	
Conserved hypothetical protein			RRC02685	NT05RC1327	
Conserved hypothetical protein			RRC06004	NT05RC1351	EC
Outer membrane protein/inner membrane lipoprotein YiaD		*yiaD*	RRC00951	NT05RC2605	OM
Hypothetical membrane-spanning protein			RRC01102	NT05RC2757	CM
Conserved hypothetical exported protein (*mucoidy inhibitor A*)			RRC01654	NT05RC2839	EC
Conserved hypothetical exported protein			RRC03610	NT05RC2917	CM, P
GTP-binding protein/conserved hypothetical protein			RRC00370	NT05RC3360	C

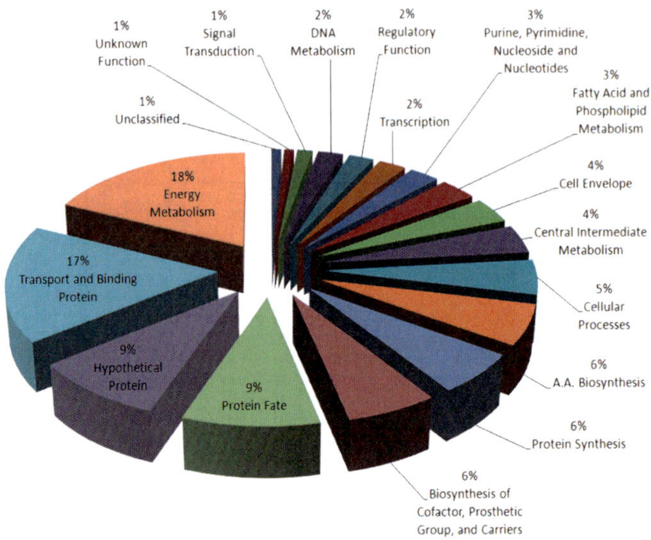

Fig. 11.2 Functional category distribution of identified proteins

roles, and among the latter group 41 of them were initially annotated as hypothetical proteins in either or both (ERGO-light with *RRCxxxxx* and NT05 with *NT05RCxxxx* numbers) of *R. capsulatus* genome databases (Table 11.1). Our data establish that these translation products are indeed real, and searches for defining their functions can now be initiated on firm grounds. We believe that the data set that is under construction is of great value for rapid progress of current and future studies focused on *Rhodobacter* species (Du et al. 2008). This ever-growing data set is providing a platform onto which we can build future qualitative and quantitative comparisons for various cellular components under defined physiological conditions. A current example to the point is illustrated by one of our recent studies where a portion of our accumulated proteomics data of *R. capsulatus*, once combined with standard in-depth biochemical and molecular genetic approaches, yielded exquisite understanding in detailed molecular terms of an unusual physiological situation simply detected by a growth phenotype (Onder et al. 2008). We believe that establishment of a comprehensive proteomic data for *R. capsulatus* species, like in many other cases, will be invaluable to provide much needed impetus for understanding the global biology of this organism with a "systems" level organization.

Acknowledgments This work is supported by grants from DOE ER9120053 and NIH GM38237 to F.D.

References

Bendtsen JD, Nielsen H, von Heijne G et al (2004) Improved prediction of signal peptides: SignalP 3.0. J Mol Biol 340:783–795

Daldal F, Cheng S, Applebaum J et al (1986) Cytochrome *c* (2) is not essential for photosynthetic growth of *Rhodopseudomonas capsulata*. Proc Natl Acad Sci USA 83:2012–2016

Du X, Callister SJ, Manes NP et al (2008) A computational strategy to analyze label-free temporal bottom-up proteomics data. J Proteome Res 7:2595–604

Gardy JL, Laird MR, Chen F et al (2005) PSORTb v.2.0: expanded prediction of bacterial protein subcellular localization and insights gained from comparative proteome analysis. Bioinformatics 21:617–623

Hunter NC, Daldal F, Thurnauer MC, Beatty JT (eds) (2009) The purple phototrophic bacteria. Springer, Dordrecht

Myllykallio H, Jenney FE Jr, Moomaw CR et al (1997) Cytochrome c(y) of *Rhodobacter capsulatus* is attached to the cytoplasmic membrane by an uncleaved signal sequence-like anchor. J Bacteriol 179:2623–2631

Neuhoff V, Stamm R, Pardowitz I et al (1990) Essential problems in quantification of proteins following colloidal staining with coomassie brilliant blue dyes in polyacrylamide gels, and their solution. Electrophoresis 11:101–117

Onder O, Turkarslan S, Sun D et al (2008) Overproduction or absence of the periplasmic protease DegP severly compromises bacterial growth in the absence of the dithiol:disulfide oxidoreductase DsbA. Mol Cell Proteomics 7:875–890

Onder O, Yoon H, Naumann B et al (2006) Modifications of the lipoamide-containing mitochondrial subproteome in a yeast mutant defective in cysteine desulfurase. Mol Cell Proteomics 5:1426–1436

Park SJ, Lee SY, Cho J et al (2005) Global physiological understanding and metabolic engineering of microorganisms based on omics studies. Appl Microb Biotechnol 68: 567–79

Ren Q, Thony-Meyer L (2001) Physical interaction of CcmC with heme and the heme chaperone CcmE during cytochrome c maturation. J Biol Chem 276:32591–32596

Wasinger V (2006) Holistic biology of microorganisms: genomics, transcriptomics, and proteomics. Methods Biochem Anal 49:3–14

Zannoni D (1995) Aerobic and anaerobic electron transport chains in anoxygenic phototrophic bacteria. In: Blankenship RE, Madigan MT, Bauer CE (eds) Anoxygenic photosynthetic bacteria. Kluwer Academic, Dordrecht, pp 949–971

Chapter 12
Phycobiliprotein Biosynthesis in Cyanobacteria: Structure and Function of Enzymes Involved in Post-translational Modification

Wendy M. Schluchter, Gaozhong Shen, Richard M. Alvey, Avijit Biswas, Nicolle A. Saunée, Shervonda R. Williams, Crystal A. Mille, and Donald A. Bryant

Abstract Cyanobacterial phycobiliproteins are brilliantly colored due to the presence of covalently attached chromophores called bilins, linear tetrapyrroles derived from heme. For most phycobiliproteins, these post-translational modifications are catalyzed by enzymes called bilin lyases; these enzymes ensure that the appropriate bilins are attached to the correct cysteine residues with the proper stereochemistry on each phycobiliprotein subunit. Phycobiliproteins also contain a unique, post-translational modification, the methylation of a conserved asparagine (Asn) present at β-72, which occurs on the β-subunits of all phycobiliproteins. We have identified and characterized several new families of bilin lyases, which are responsible for attaching PCB to phycobiliproteins as well as the Asn methyl transferase for β-subunits in *Synechococcus* sp. PCC 7002 and *Synechocystis* sp. PCC 6803. All of the enzymes responsible for synthesis of holo-phycobiliproteins are now known for this cyanobacterium, and a brief discussion of each enzyme family and its role in the biosynthesis of phycobiliproteins is presented here. In addition, the first structure of a bilin lyase has recently been solved (PDB ID: 3BDR). This structure shows that the bilin lyases are most similar to the lipocalin protein structural family, which also includes the bilin-binding protein found in some butterflies.

12.1 Introduction

The brilliantly colored phycobiliproteins (PBPs), major components of the phycobilisome (PBS) light-harvesting complex, are responsible for the characteristic colors of cyanobacteria, because these proteins can constitute up to 40–50% of the total proteins in the cell when cyanobacteria are cultured under low-light conditions (Glazer 1989). Their spectroscopic properties are primarily due to covalently attached chromophores called bilins, and the prosthetic groups allow these proteins

W.M. Schluchter (✉)
Department of Biological Sciences, University of New Orleans, New Orleans, LA 70148, USA
e-mail: wschluch@uno.edu

P.C. Hallenbeck (ed.), *Recent Advances in Phototrophic Prokaryotes*, Advances in Experimental Medicine and Biology 675, DOI 10.1007/978-1-4419-1528-3_12,

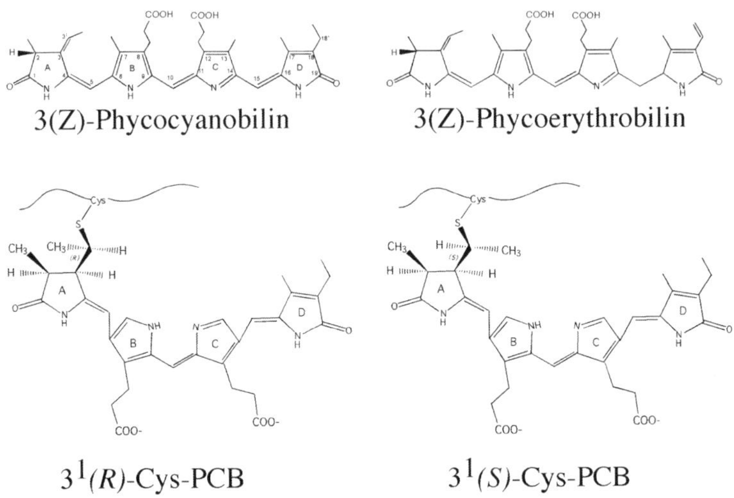

Fig. 12.1 Structures of free bilins phycocyanobilin and phycoerythrobilin as synthesized in cyanobacteria (*upper*) and structures of two stereoisomers of phycocyanobilin as attached to β-phycocyanin (*lower*)

to absorb light in the visible region where chlorophyll *a* has minimal absorption. Each PBP is made up of two different polypeptides, denoted α and β, which are similar in sequence and derived from ancient gene duplication events. Each α or β-subunit carries at least one, and as many as three, bilin(s), which are covalently attached to the protein via thioether linkages at specific Cys residues (see Fig. 12.1). The subunit structure for PBPs, the core of which is structurally related to members of the globin family, consists primarily of α-helices (Schirmer et al. 1985; Betz 1997).

PBPs are usually isolated as trimers $(\alpha\beta)_3$ or as hexamers $(\alpha\beta)_6$, and specific linker proteins mediate the association of one hexameric disc to another and modulate the spectroscopic properties of PBPs (Glazer 1984) (see Fig. 12.2). There are three major classes of PBPs: phycoerythrin (PE; $\lambda_{max} \sim 565$ nm), phycocyanin (PC; $\lambda_{max} \sim 620$ nm), and allophycocyanin (AP; $\lambda_{max} \sim 650$ nm). AP and related minor PBPs form the core of the PBS; six to eight rods composed of PC radiate out from

Fig. 12.2 Model of the events required for the biosynthesis of phycocyanin and allophycocyanin subunits and of the rods and cores in *Synechococcus* sp. PCC 7002. It is unknown whether αβ monomer assembly occurs before or after these post-translational events, although apo-αβ-subunits could be recognized by CpcT and CpcSU bilin lyases and the CpcM methyltransferase in in vitro reactions. The only bilin lyase that can remove bilins is CpcEF as indicated by a *double-sided arrow*. Once the proteins are fully modified, oligomers of these proteins form with the aid of various linker proteins as reviewed elsewhere (Gingrich et al. 1983; Glazer et al. 1983, 1985)

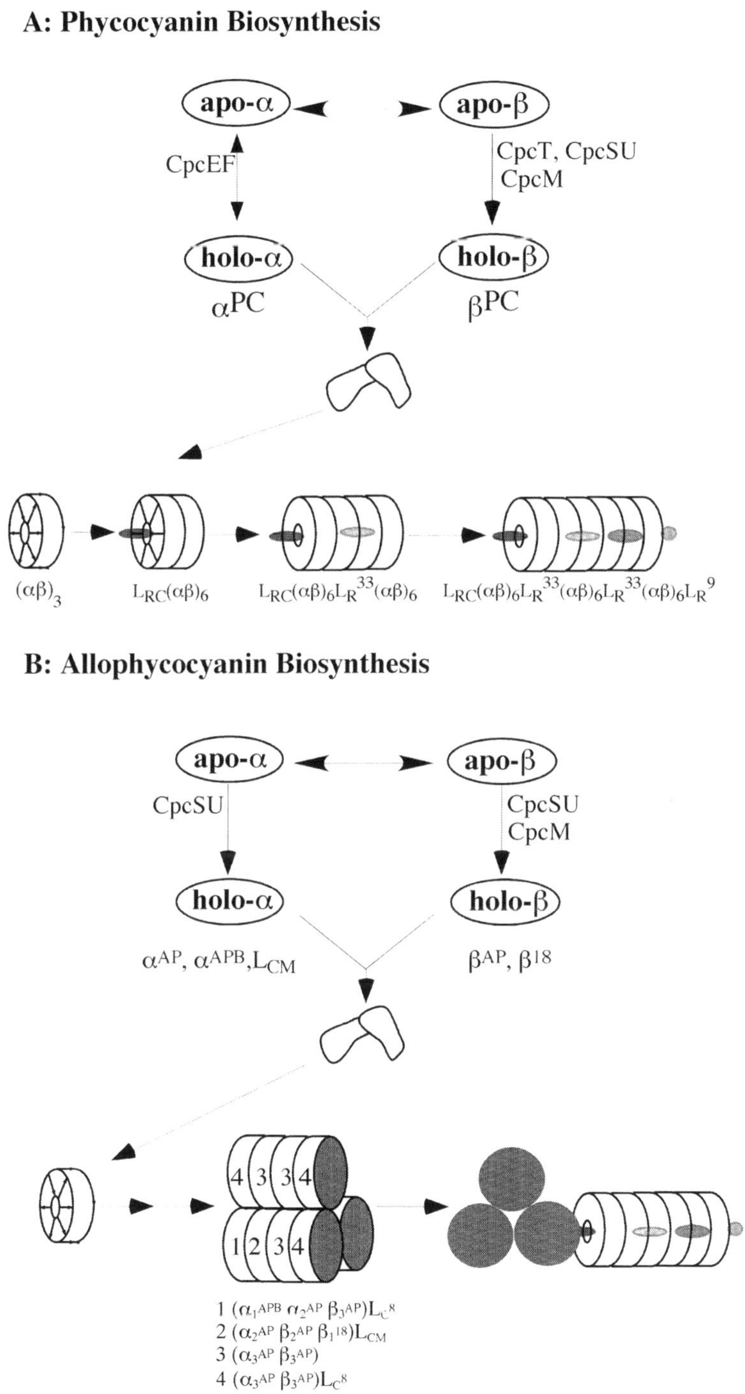

Fig. 12.2 (continued)

the core (see Fig. 12.2). Some cyanobacteria additionally contain PE, or phycoerythrocyanin (PEC; λ_{max} ~590 nm), at the distal position on the rods, which increases the efficiency of green light absorption (Glazer 1989).

In order to harvest light, PBPs must be post-translationally modified by the attachment of at least one of the two different bilins: phycoerythrobilin (PEB) or phycocyanobilin (PCB; see Fig. 12.1). These linear tetrapyrroles are derived from heme and are covalently attached via thioether linkages at specific cysteine residues within the α- and β-subunits (see Fig. 12.1). Phycourobilin (PUB) and phycoviolobilin (PVB) are derived from PEB and PCB, respectively, after attachment and isomerization by a lyase enzyme during the attachment process (Zhao et al. 2000; Blot et al. 2009). In some strains of cyanobacteria, three different bilins can be found attached to PBPs (Bryant et al. 1981; Ong and Glazer 1991), whereas in other strains such as *Synechococcus* sp. PCC 7002 and *Synechocystis* sp. PCC 6803 only PCB is present. Even in those strains that only contain PCB, two different stereoisomers occur on the phycocyanin β-subunit at carbon atom C31 of the bilin (Schirmer et al. 1986). The *R*-isomer is the most commonly found isomer (see Fig. 12.1), but the *S*-isomer is present at some positions (bilins attached to the Cys-153 on PC β-subunits).

A second type of post-translational modification, methylation of the γ-nitrogen of asparagine, occurs for almost all PBP β-subunits of cyanobacteria, red algae, and cryptomonads (Klotz et al. 1986; Klotz and Glazer 1987; Rümbeli et al. 1987; Wilbanks et al. 1989; Ducret et al. 1994; Apt et al. 2001). The highly conserved γ-*N*-methylasparagine residue occurs at the β-72 position (numbering according to CpcB in *Synechococcus* sp. PCC 7002; see Fig. 12.3), and no such modification occurs on the homologous α-subunit position (Klotz et al. 1986; Klotz and Glazer 1987; Swanson and Glazer 1990).

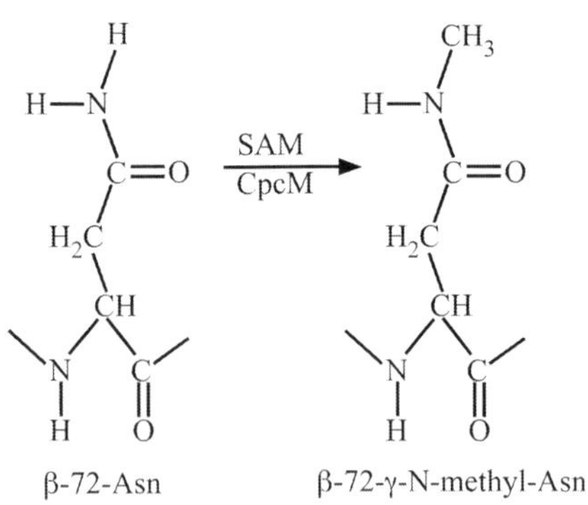

Fig. 12.3 Reaction showing the methylation of the asparagine located on β-subunits at position 72 catalyzed by the methyltransferase CpcM using *S*-adenosylmethionine (SAM) as the donor for the methyl group

Much progress has been made in the last 6 years in identifying and characterizing the enzymes required for the biosynthesis of PBPs, and this chapter reviews the studies that characterized the enzymes required for the biosynthesis of all PBPs in *Synechococcus* sp. PCC 7002.

12.2 Post-translational Modifications of Phycobiliproteins

PBPs have two major types of post-translational modifications: addition of bilin chromophores to specific Cys residues on each subunit and the methylation of an asparagine residue on β-subunits. The chromophorylation process, in most cases, requires enzymes called bilin lyases. It seems that these enzymes are likely required for most PBPs in order to ensure that the appropriate stereochemistry of the bilin linkage to the proper Cys and, probably most importantly, to ensure that the appropriate bilin is attached to the correct Cys of a particular PBP in cases for which more than one bilin is present within the cyanobacterium (e.g., when both PCB and PEB occur in a cell). For example, in in vitro reactions between apo-PBPs and purified bilins, many incorrect products can be produced. Bilins that were more oxidized than the native bilin could be attached to Cys residues and individual Cys residues exhibited virtually no selectivity for one bilin over another (Arciero et al. 1988a, b, c; Fairchild and Glazer 1994a, b). Furthermore, in some cyanobacteria bilin lyases are additionally required to attach and then isomerize one bilin into a different one. For example, PVB is formed by isomerization of PCB, and PUB is formed by isomerization of PEB by specific bilin lyases (Zhao et al. 2000; Storf et al. 2001; Blot et al. 2009). The second type of post-translational modification that occurs on PBPs is N-methylation of an asparagine residue, which is performed by an *S*-adenosylmethionine-dependent methyltranferase that is specific for β-subunits (Swanson and Glazer 1990). This chapter will focus on the enzymes from *Synechococcus* sp. PCC 7002, a marine cyanobacterium that contains a simple PBS comprised of only PC and AP, PBPs that only carry PCB chromophores (Bryant, 1992).

12.2.1 Bilin Attachment

12.2.1.1 CpcE/F-Type Bilin Lyases

It appears that for most PBPs, bilin lyase enzymes are required for the attachment, isomerization, and detachment of the bilin chromophores from the Cys residues of the PBP (Arciero et al. 1988a; Fairchild et al. 1992; Zhou et al. 1992; Fairchild and Glazer 1994b; Dolganov and Grossman 1999; Zhao et al. 2000, 2007a). The first bilin lyase to be characterized was CpcE/CpcF (Zhou et al. 1992). The *cpcE* and *cpcF* genes occur downstream of the *cpcBA* structural genes that encode the β-and α-subunits of PC, respectively, and from the *cpcC* and *cpcD* genes, encoding PC-associated rod linkers (see Table 12.1). When either the *cpcE* or the *cpcF* genes were inactivated in *Synechococcus* sp. PCC 7002, the mutants were yellow-green in color, had increased doubling times, and had low levels of PC (Zhou et al. 1992). About 90% of the PC synthesized in these mutants lacked a PCB chromophore on the α-subunit; however, the β-subunit of PC had the correct bilin incorporated at Cys residues 82 and 153 (Swanson et al. 1992). Recombinant CpcE and CpcF were shown to comprise a heterodimeric lyase (CpcE/CpcF) that specifically attaches

Table 12.1 Characteristics of Phycobiliproteins from *Synechococcus* sp. PCC 7002

PBP	Gene	Attachment sites – bilin[a]	Abs/Fl maxima (nm)[b]	Bilin lyase
α^{PC}	*cpcA*	Cys-84-PCB	625/642	CpcE/CpcF (Zhou et al. 1992; Fairchild et al. 1992)
β^{PC}	*cpcB*	Cys-82-PCB Cys-153-PCB	622/643 597/623	CpcS-I/CpcU (Shen et al. 2008; Saunée et al. 2008) CpcT (Shen et al. 2006)
α^{AP}	*apcA*	Cys-81-PCB	614/634	CpcS-I/CpcU (Saunée et al. 2008)
β^{AP}	*apcB*	Cys-81-PCB	614/634	CpcS-I/CpcU (Saunée et al. 2008)
β^{18}	*apcF*	Cys-81-PCB	616/637	CpcS-I/CpcU (Biswas et al. unpublished)
$\alpha^{AP\text{-}B}$	*apcD*	Cys-81-PCB	672/675	CpcS-I/CpcU (Biswas et al. unpublished)
$L_{CM}{}^{99}$	*apcE*	Cys-186	662/670	None (Biswas et al. unpublished)

[a]Terminal acceptor bilin in PC is underlined (Ong and Glazer 1991)
[b]Absorbance (Abs) and fluorescence (Fl) emission maxima of this chromophore when properly attached to the corresponding monomeric PBP

PCB to Cys-84 of α-PC (CpcA) (Fairchild et al. 1992; Zhou et al. 1992; Fairchild and Glazer 1994b) (see Table 12.1). CpcE/CpcF was also able to catalyze the reverse reaction, transferring the bilin from the holo-α-subunit of PC to an apo-α-subunit of the same or different species (Fairchild et al. 1992) (see Fig. 12.2). This implies that the heterodimeric bilin lyase is able to gain access to and cleave the thioether bond between the protein and the bilin. To date, this is the only bilin lyase that can perform both the bilin addition and removal reactions. CpcE/CpcF can catalyze the addition of PEB to the α-subunit of apo-PC as well, but it exhibited a strong preference for PCB over PEB in both affinity and kinetics (Fairchild and Glazer 1994b).

The PecE/PecF and CpeY/CpeZ bilin lyases are similar in sequence to CpcE/CpcF but are associated with different PBP-encoding operons and are active on different substrates (Jung et al. 1995; Kahn et al. 1997; Zhao et al. 2000, 2002; Storf et al. 2001;). This family of lyase genes has collectively been called the E/F type. These putative PBP lyases contain a short region of high similarity dubbed the E-Z motif (Wilbanks and Glazer 1993a). Morimoto et al. (2002, 2003) noted that a CpcE-like protein in *Synechocystis* sp. PCC 6803 called Slr1098 has HEAT-repeat motifs that occur multiple times within its sequence, and Dolganov and Grossman (1999) noted that some of these regions were important for the function of a CpcE-like protein called NblB. Some of these CpcE-like proteins may function in PBP degradation during nutrient starvation (Dolganov and Grossman 1999) or in assembly or disassembly of PBS (Schluchter and Glazer 1999). All E/F type bilin lyases contain 5–6 HEAT-repeat motifs. HEAT-repeat motifs occur in many proteins from diverse eukaryotic organisms in which they are generally believed to facilitate protein–protein interactions (Andrade et al. 2001; Takano and Gusella 2002).

12.2.1.2 T-Type Bilin Lyases

The *cpcT* gene from *Synechococcus* sp. PCC 7002 is a paralog of *cpeT,* which was first sequenced as a gene of the *cpeCDESTR* operon in *Fremyella diplosiphon* (also known as *Calothrix* sp. PCC 7601) (Cobley et al. 2002). CpcT was shown to be involved in PCB attachment to β-PC (Shen et al. 2004). Compared to wild-type cells, cells of the *cpcT* mutant contained reduced levels of PC, and most of the PC that was produced was missing a PCB chromophore at Cys-153 of β-PC (Shen et al. 2006). Recombinant CpcT could attach PCB specifically to Cys-153 on β-PC, and it is active as a monomer (see Table 12.1) (Shen et al. 2006). Shen et al. hypothesized that the CpcT bilin lyase was required to form the *S*-stereoisomer at this position (see Fig. 12.1); this further suggests that this lyase can bind PCB in the appropriate conformation to allow formation of the *S*-stereoisomer (Shen et al. 2006).

Similar results were obtained with CpcT from *Nostoc* sp. PCC 7120, and for the first time, an order for post-translational modifications was proposed (Zhao et al. 2007b). Interestingly, a phage that contains an ortholog of *cpeT* in its genome has recently been shown to increase phycoerythrin I and II levels during infection of marine *Synechococcus* sp. WH 7803 (Shan et al. 2008). The cryptophyte *Guillardia theta* also encodes a functional CpcT protein in its nucleomorph genome, which can functionally complement a *cpcT* mutant of *Synechocystis* sp. PCC 6803 (Bolte et al. 2008). BLAST analyses identified *cpeT* orthologs and/or paralogs in the sequenced genomes of all cyanobacteria except *Prochlorococcus marinus* MED4, and phylogenetic analyses showed that there are two main groups. One group only contains organisms that synthesize PE, and these proteins are designated CpeT (see 7601 CpeT location in phylogeny shown in Fig. 12.4). The distribution of these sequences strongly suggests that this protein subgroup plays a role in cyanobacterial-type phycoerythrin (C-PE) biosynthesis, probably by attaching PEB at the Cys-153 equivalent position of β-PE. The second group contains at least two sub-groups (upper portion of tree shown in Fig. 12.4) with the CpeT paralog from *Synechococcus* sp. PCC 7002 (denoted 7002 CpcT) along with other CpcT-like proteins from organisms that do not synthesize PE and only synthesize PC and AP (e.g., *Synechocystis* sp. PCC 6803, *Thermosynechococcus elongatus* BP-1, *Synechococcus elongatus* PCC 7942). This distribution suggests that this group of proteins attaches PCB to the Cys-153 equivalent of PCs (Shen et al. 2006).

12.2.1.3 S/U-Type Bilin Lyases

Three genes in *Synechococcus* sp. PCC 7002 showed highest sequence similarity to *cpeS* from *F. diplosiphon* (Cobley et al. 2002). These paralogs were named *cpcS-I*, *cpcU*, and *cpcV* (Shen et al. 2004, 2008b; Saunée et al. 2008), and they are members of the third family of PBP lyases called the S/U type (Scheer and Zhao 2008).

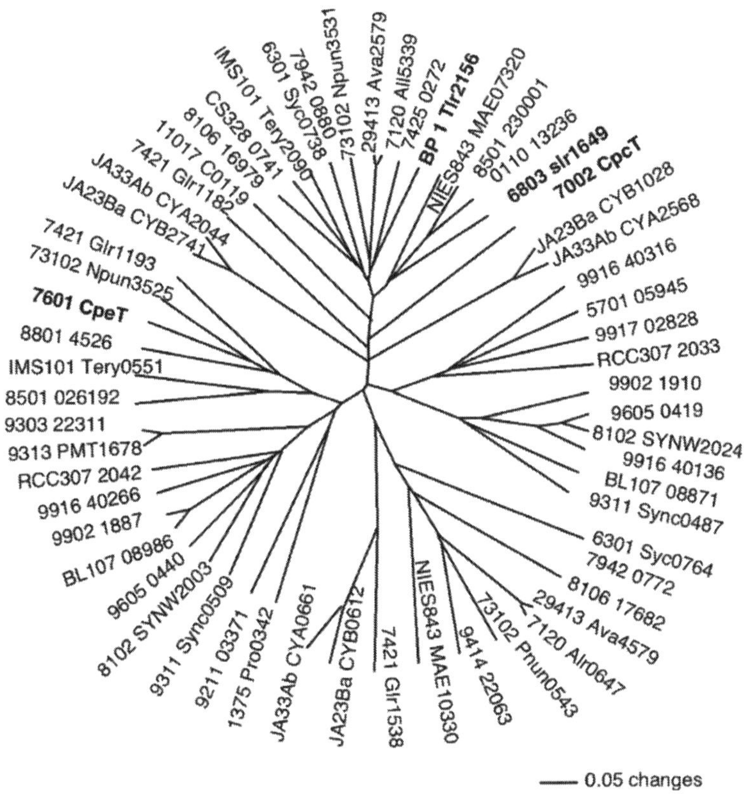

Fig. 12.4 Phylogenetic analysis of CpcT and CpeT proteins from different cyanobacteria. Sequence alignments and comparison were made on CpcT, CpeT, and their homologs from *Synechococcus* 7002 (7002), *Synechocystis* 6803 (6803), *T. elongatus* strain BP-1 (BP 1), *Gloeobacter violaceus* strain PCC 7421 (7421), *Nostoc* sp. strain PCC 7120 (7120), *N. punctiforme* strain PCC 73102 (73102), *Anabaena variabilis* strain ATCC 29413 (29413), *S. elongatus* strain PCC 7942 (7942), *Synechococcus* sp. strain PCC 6301 (6301), *Microcystis aeruginosa* NIES-843 (NIES843), *Cyanothece* sp. strain CCY 0110 (0110), *Cyanothece* sp. PCC 7425 (7425), *Crocosphaera watsonii* strain WH8501 (8501), *Trichodesmium erythraeum* strain IMS101 (IMS101), *Arthropira maxima* CS328 (CS328), *Lyngbya* sp. strain PCC 8106 (8106), *Synechococcus* sp. strain JA-3-3Ab (JA33Ab), *Synechococcus* sp. strain JA-2-3Ba (JA23Ba), *Synechococcus* sp. strain RCC307 (RCC307), *Synechococcus* sp. strain WH5701 (5701), *Synechococcus* sp. strain CC9902 (9902), *Cyanothece* sp. PCC 8801 (8801), *Synechococcus* sp. strain BL107 (BL107), *Synechococcus* sp. strain WH8102 (8102), *Synechococcus* sp. strain CC9605 (9605), *Synechococcus* sp. strain CC9311 (9311), *Synechococcus* sp. strain RS9917 (9917), *Nodularia spumigena* strain CCY9414 (9414), *P. marinus* MIT9211 (9211), *P. marinus* strain MIT9303 (9303), *P. marinus* strain MIT9313 (9313), *Prochlorococcus* sp. strain CC9605 (9605), and *Prochlorococcus* sp. strain CC9902 (9902). The sequence alignment was generated using the ClustalW module within the MacVector program, version 10.6 (MacVector, Inc., Cary, NC). The phylogenetic tree was generated using the phylogenetic analysis program PAUP (Sinauer Associates, Sunderland, MA)

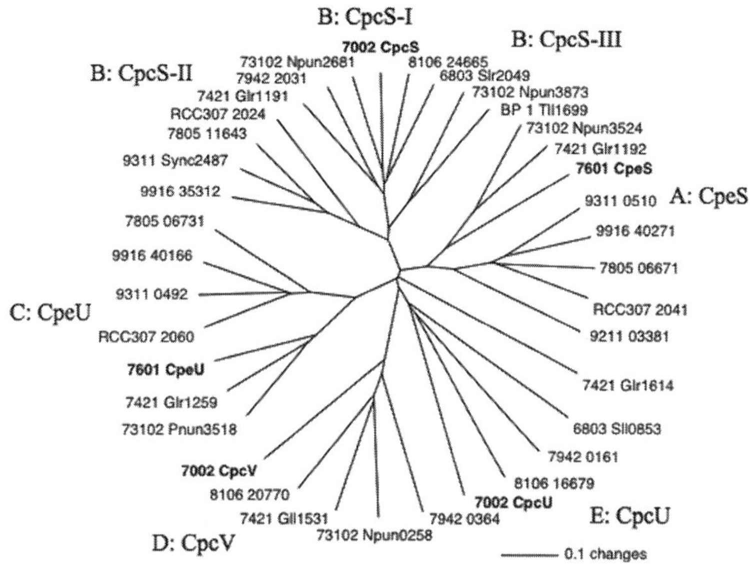

Fig. 12.5 Phylogenetic analysis for proteins paralogous to CpcS, CpcU, CpcV, CpeS, and CpeU. For clarity in comparison and phylogenetic tree generation, representative paralogous proteins were selected from only a few cyanobacteria, including from *Synechococcus* 7002 (7002), *Synechocystis* 6803 (6803), *T. elongatus* strain BP-1 (BP 1), *G. violaceus* strain PCC 7421 (7421), *N. punctiforme* strain PCC 73102 (73102), *S. elongatus* strain PCC 7942 (7942), *Lyngbya* sp. strain PCC 8106 (8106), *Synechococcus* sp. strain RCC307 (RCC307), *Synechococcus* sp. strain WH7805 (7805), *Synechococcus* sp. strain CC9311 (9311), *Synechococcus* sp. strain RS9916 (9916), and *P. marinus* MIT9211 (9211). Amino acid sequences were compared using the ClustalW module within the MacVector program, version 10.6 (MacVector, Inc., Cary, NC). The phylogenetic tree was generated using the phylogenetic analysis program PAUP (Sinauer Associates, Sunderland, MA)

As shown in Fig. 12.5, homologs of the CpcS and CpeS proteins from cyanobacteria can be classified into five groups. Two of these groupings, groups A and C, only include the strains that can synthesize PE. This observation, and the observation that these genes are usually clustered with other genes known to play a role in PE biosynthesis or assembly, suggests that these lyase proteins, denoted CpeS and CpeU, respectively, probably play roles in PEB attachment to PE subunits. All of the other groups contain at least some members that only synthesize proteins with PCB chromophores, and thus at least some of these proteins cannot be involved in PE synthesis. The largest of these groups, group B, can be further divided into three subfamilies, which we have denoted CpcS-I, CpcS-II, and CpcS-III. *Synechococcus* sp. PCC 7002 has a protein within the CpcS-I clade. With only a single exception, *Nostoc punctiforme*, all organisms with a protein in the CpcS-I clade also have a paralogous protein in group E, which we have designated CpcU (Shen et al. 2008b).

Null mutants for *cpcS-I* and *cpcU* from *Synechococcus* sp. PCC 7002 have reduced amounts of PC, and most of the PC produced lacks a chromophore at

Cys-82 of β-PC. In fact, as shown by HPLC/electrospray MS analysis, some of this PC still contained non-covalently bound PCB in the binding pocket (Shen et al. 2008b). The characterization of these mutants also suggested that CpcS-I and CpcU are required for PCB attachment to Cys-81 of the α- and β-subunits of AP (Shen et al. 2008b). Recombinant CpcS-I and CpcU forms a 1:1 heterodimer, which attaches PCB to Cys-82 of β-PC and to Cys-81 on α- and β-subunits of AP (Saunée et al. 2008) (see Table 12.1). Neither CpcS-I nor CpcU alone is able to perform this addition reaction. Recently, a heterologous, multi-plasmid expression system in *Escherichia coli* was used to show that both CpcS-I and CpcU are also required to attach PCB correctly to Cys-81 on the ApcF (β^{18}) and ApcD (α^{APB}) of *Synechococcus* sp. PCC 7002 (Biswas et al. manuscript in preparation; see Table 12.1).

The S/U family of lyases does not appear to perform the transfer or removal of bilins (Zhao et al. 2006; Saunée et al. 2008), and because it can recognize many different PBPs and attach bilins at their Cys-82 equivalent positions, this family has a broader substrate specificity than the E/F type lyases (Scheer and Zhao 2008). The third member of this S/U type lyase family in *Synechococcus* sp. PCC 7002 was named CpcV (group D in Fig. 12.5). Reverse genetics and in vitro biochemistry experiments have not yet uncovered the function of CpcV (Saunée et al. 2008; Shen et al. 2008b), but *Synechocystis* sp. PCC 6803 and many other cyanobacteria do not have orthologs of this gene. Similarly, *Nostoc* sp. PCC 7120 also apparently encodes a paralog of CpcS that lacks bilin lyase activity (Zhao et al. 2006).

In *Nostoc* sp. PCC 7120, the product of open reading frame *alr0617*, which the authors called CpeS but which on the basis of phylogenetic analyses we refer to as CpcS-III (see Fig. 12.5), is another member of this lyase family (Zhao et al. 2006). Zhao et al. (2006, 2007a, b) have shown that CpcS-III alone attaches PCB to Cys-82 on β-PC and β-PEC as well as to several AP subunits in *Nostoc* sp. PCC 7120 (). Therefore, some members of the S/U family are active in the absence of other subunits (CpcS-III), while the other S/U family members are only active as heterodimers (e.g., CpcS-I/CpcU; groups C and D in Fig. 12.5). The equivalent CpcS and CpcU orthologs in *Synechocystis* sp. PCC 6803 are both required for bilin addition to Cys-82 on CpcB and presumably also form a heterodimer (Miller 2007).

The X-ray crystal structure of CpcS-III from *T. elongatus* BP1 (tll1699) was very recently solved and entered into the PDB database (Kuzin et al. 2007). Interestingly, this protein crystallized as a dimer (see Fig. 12.6), and through a collaborative effort, we have successfully demonstrated that Tll1699, or CpcS-III, is a functional bilin lyase specific for Cys-82 on β-PC and α- and β-AP subunits (A. Biswas and W. M. Schluchter unpublished results). The CpcS-III bilin lyase belongs to the lipocalin structural family. All of these proteins adopt a eight-stranded anti-parallel, β-barrel structure with an α-helix; they have various oligomeric states, occurring as monomers, homodimers, heterodimers, or tetramers, and they bind a diverse set of ligands including fatty acids, retinols, carotenoids, pheromones, prostaglandins, and biliverdin (Flower 1996; Bishop 2000; Hieber et al. 2000; Newcomer and Ong 2000; Charron et al. 2005; Grzyb et al. 2006). In fact, the most similar structure to CpcS-III in the protein structure database is the bilin-binding protein of the insect

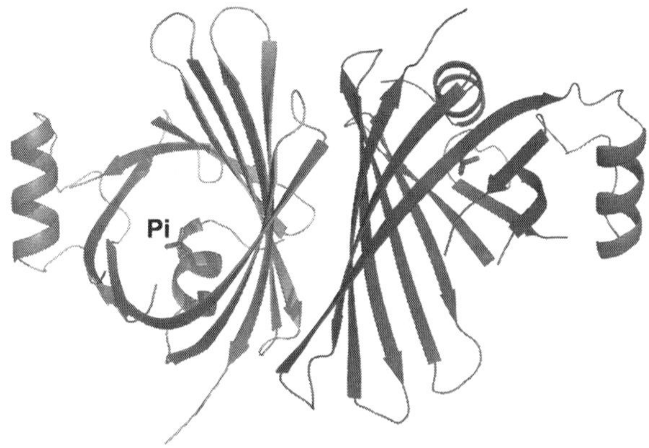

Fig. 12.6 Structure of Tll1699/CpcS-III from *T. elongatus* BP-1 (PDB ID: 3BDR). The structure of the homodimer that crystallized is shown here. This protein has an eight-stranded anti-parallel β-barrel with an α-helix. A phosphate ion co-crystallized with each subunit in the structure (Kuzin et al. 2007)

Pieris brassicae (Huber et al. 1987a, b). This insect protein binds biliverdin IXγ and is responsible for coloration of various stages of some insects, most commonly in Lepidopterans (Sehringer and Kayser 2006). Co-crystals of CpcS-III with biliverdin IXα have recently been obtained (Kuzin et al. unpublished results), and perhaps a structure of the protein with the bound bilin will soon be available. A structure with the bound bilin should allow a better understanding of the conformation adopted by the bilin when bound to the lyase. It appears that the role of the bilin lyase is to bind the appropriate chromophore in a suitable conformation, so that bilin addition to the appropriate Cys residue of the apo-PBP can occur (Schluchter and Glazer 1999; Scheer and Zhao 2008).

12.2.1.4 Autocatalytic Bilin Addition

Targeted mutagenesis results in *Synechococcus* sp. PCC 7002 show that all of the major, highly abundant PBPs like PC and AP require one of the three bilin lyases mentioned above. For example, a triple mutant lacking *cpcS*, *cpcU*, and *cpcT* produces almost no PBP (Shen et al. 2008b); this suggests that, at least in vivo, all of the major PBPs require bilin lyases for chromophorylation. However, one report suggested that the core membrane linker protein ($L_{CM}{}^{99}$), also called ApcE, has autocatalytic bilin addition activity (Zhao et al. 2005). However, the protein had limited solubility, and the bilin addition reactions had to be performed in 4 M urea. ApcE ($L_{CM}{}^{99}$) contains an AP-like domain at its amino-terminus and several repeated linker domains, which bind together the AP core trimers; ApcE is also thought to bind the PBS to the membrane (Capuano et al. 1991; Gindt et al. 1994). The AP-like domain of ApcE from *Synechococcus* sp. PCC 7002 binds PCB at Cys-186, has a red-shifted absorbance at 665 nm, and plays a role in accepting the energy from the chromophores within the core of the PBS and transferring it to the reaction centers (Gindt et al. 1994). We used a construct that fuses glutathione *S*-transferase to amino acids 1–228 of the AP-like domain of ApcE in order to produce soluble protein. This ApcE domain has intrinsic bilin lyase activity for attaching PCB in the appropriate conformation to yield a red-shifted PCB product with an absorbance maximum at 662 nm and fluorescence emission at 672 nm (Biswas et al. 2010; Table 12.1).

12.2.2 Methylation of Asparagine on the β-Subunits of PBPs

PBPs contain a unique, post-translational modification of a conserved Asn present at β-72. This modification has never been observed in other proteins, but it was recently identified in cytotoxic linear peptides in a marine sponge (Hamada et al. 2005). Interestingly, the homologous Asn residue within the α-subunits is never modified, which implied that this modification serves an important functional role. The methylated Asn is located very close to the bilin chromophore present at β-82, the terminal energy acceptor within the trimer, suggesting that this modification might play a role in changing the absorption and energy transfer properties of this particular bilin. Mutants unable to modify their PBPs exhibited less efficient energy transfer from PC to AP (Swanson and Glazer 1990), lower electron transfer rates through photosystem II (PS II) under light regimes that are preferentially absorbed by PBS (Thomas et al. 1993), and changes in bilin ring geometry (Thomas et al. 1994). The conclusion from these studies was that this post-translational modification changes the environment of the β-84 chromophore to minimize the rates of non-radiative energy losses within PBS.

We identified the gene encoding this enzyme and named it *cpcM* (Miller et al. 2008); Shen et al. 2008a). The *cpcM* gene is present in all organisms that contain PBPs, and *cpcM* mutants lacked the modification on their β-subunits and were extremely sensitive to high light intensities (Shen et al. 2008a). We hypothesized that the PBPs lacking this modification might generate more reactive oxygen species. We also showed that recombinant CpcM methylated only β-subunits and not α-subunits, and from these in vitro experiments using various substrates, we inferred that the enzyme probably methylates β-subunits after chromophorylation but before trimer assembly (Miller et al. 2008). Given the remarkable primary, secondary, and tertiary structural similarity between α- and β-subunits, it is unclear whether this enzyme recognizes a specific conserved motif present in β-subunits (Clarke 2002) or whether it recognizes a larger region over the entire β-subunits (Miller et al. 2008).

12.3 Formation of Unnatural PBPS

Previous studies suggested that the co-expression of genes for CpcA, the CpcE/CpcF PBP lyase, and the enzymes for synthesis of PCB [heme oxygenase and PCB synthase (PcyA)] in *E. coli* could produce holo-PC α in reasonable yield (Tooley et al. 2001). The recent identification of enzymes for the synthesis of PEB (Frankenberg et al. 2001; Dammeyer et al. 2008) and phytochromobilin

(PΦB) (Kohchi et al. 2001) led us to perform experiments to explore the production of unnatural recombinant CpcA variants (Alvey et al. manuscript in preparation). When the *cpcA* gene was co-expressed with *cpcE*, *cpcF*, *hox1*, and *pcyA*, holo-PC α was formed as previously described (Tooley et al. 2001). By substituting *pebS* for *pcyA*, a highly fluorescent variant of recombinant PC α harboring PEB could be produced with a very high yield. Similarly, substituting *pcyA* with the HY2 gene of *Arabidopsis thaliana* led to the production of PC α-subunits carrying PΦB. By substituting the *cpcE* and *cpcF* genes with the *pecE* and *pecF* genes from *Nostoc* sp. PCC 7120 (or in some cases just *pecF* alone), it was possible to produce recombinant PC α-subunits harboring PVB (if *pcyA* was also present), PUB (if *pebS* was also present), or ΦVB (phytoviolobilin), a chromophore that has never been observed in natural PBPs. Attempts to get the CpcE/CpcF lyase to attach biliverdin to PC α-subunits by leaving out *pcyA* in *E. coli* together, and an attempt to attach 15,16-dihydrobiliverdin (by substituting *pebA* for *pcyA*), are in progress. By using a system for overexpression of genes in *Synechococcus* sp. PCC 7002 (Xu et al. 2009), we have successfully generated strains which overexpress the *pebS* and *HY2* genes. Both systems led to efficient incorporation of foreign bilins into PBPs, although CpcA had higher levels of PEB than any other PBP; it is unclear whether this reflects a greater stability of this protein (CpcA-PEB) or a greater discrimination by the CpcSU and CpcT lyases. Remarkably, when HY2 is overexpressed in a *pcyA* null background, all PBPs can be produced with PΦB and still assembled into PBS. Because of the high fluorescence quantum yields of PBPs, and the efficiency with which these proteins can be produced recombinantly, it is possible that some of these unnatural PBPs could have important applications in bioimaging, cell sorting, and other biotechnological applications.

12.4 Concluding Remarks

The beautifully colored and highly fluorescent PBPs have captured the imagination of many scientists for over 150 years. They are currently and widely used as fluorescent tags when conjugated to other proteins such as antibodies (Glazer and Stryer 1983, 1984, 1990; Glazer 1994; Sekar and Chandramohan 2008). All of the enzymes required for post-translational modification of all PBPs in a single cyanobacterium, *Synechococcus* sp. PCC 7002, have now been identified (Table 12.1). This organism produces "simple" PBS containing only one type of bilin (PCB), and chromophorylation of all of its PBP subunits requires only two heterodimeric bilin lyases, CpcE/CpcF and CpcS/CpcU, and one monomeric bilin lyase, CpcT (see Fig. 12.2). In addition, the Asn methyltransferase, CpcM, is the only enzyme required to transfer a methyl group from *S*-adenosylmethionine to the γ-nitrogen atom of Asn-72 on all PBP β-subunits (see Figs. 12.2 and 12.3). Although more work is required on this aspect, it appears that these post-translational modifications occur in a specific order (Zhao et al. 2007b; Miller et al. 2008). The role(s) of CpcV (Saunée et al. 2008; Shen et al. 2008b) and several proteins related in sequence to CpcE (Schluchter and

Glazer 1999) in *Synechococcus* sp. PCC 7002 remain to be discovered. Finally, in organisms that produce more than one bilin (e.g., PCB and PEB), the situation is much more complex. Genomes from marine *Synechococcus* spp. strains that contain PCB, PEB, and PUB attached to their PBP contain genes for multiple bilin lyases belonging to these three categories, and most of these genes are clustered in close proximity to the PBP structural gene operons (Wilbanks and Glazer 1993a, b; Six et al. 2007). However, much more work needs to be done to characterize the specificity of the putative bilin lyases in organisms producing "complex PBPs."

Acknowledgments This research was supported by National Science Foundation grants to W. M. S. (MCB-0133441 and MCB-0843664) and to D. A. B. (MCB-0077586 and MCB-0519743). We would like to thank Farhad Forouhar and John Hunt from Columbia University for preparing the image shown in Fig. 12.6.

References

Andrade MA, Petosa C, O'Donoghue SI, Müller CW, Bork P (2001) Comparison of ARM and HEAT protein repeats. J Mol Biol 309:1–18

Apt KE, Metzner S, Grossman AR (2001) The gamma subunits of phycoerythrin from a red alga: position in phycobilisomes and sequence characterization. J Phycol 37:64–70

Arciero DM, Bryant DA, Glazer AN (1988a) *In vitro* attachment of bilins to apophycocyanin I: specific covalent adduct formation at cysteinyl residues involved in phycocyanobilin binding in C-phycocyanin. J Biol Chem 263:18343–18349

Arciero DM, Dallas JL, Glazer AN (1988b) *In vitro* attachment of bilins to apophycocyanin: II: determination of the structures of tryptic bilin peptides derived from the phycocyanobilin adduct. J Biol Chem 263:18350–18357

Arciero DM, Dallas JL, Glazer AN (1988c) *In vitro* attachment of bilins to apophycocyanin: III: properties of the phycoerythrobilin adduct. J Biol Chem 263:18358–18363

Betz M (1997) One century of protein crystallography: the phycobiliproteins. Biol Chem 378: 167–176

Bishop RE (2000) The bacterial lipocalins. Biochim Biophys Acta-Protein Struct Mol Enzymol 1482:73–83

Biswas A, Vasquez YM, Dragomani TM, Kronfel ML, Williams SR, Alvey RM, Bryant DA, Schluchter WM (2010) HYPERLINK "http://www.ncbi.nlm.nih.gov/pubmed/20228104" Biosynthesis of cyanobacterial phycobiliproteins in *Escherichia coli*: chromophorylation efficiency and specificity of all bilin lyases from *Synechococcus* sp. strain PCC 7002. Appl Environ Microbiol 76:2729–2739.

Blot N et al (2009) Phycourobilin in Trichromatic phycocyanin from oceanic cyanobacteria is formed post-translationally by a phycoerythrobilin lyase-isomerase. J Biol Chem 284: 9290–9298

Bolte K, Kawach O, Prechtl J, Gruenheit N, Nyalwidhe J, Maier UG (2008) Complementation of a phycocyanin-bilin lyase from *Synechocystis* sp PCC 6803 with a nucleomorph-encoded open reading frame from the cryptophyte *Guillardia theta*. Bmc Plant Biol 8

Bryant DA (1992) Puzzles of chloroplast ancestry. Curr Biol 5:240–242.

Bryant DA, Cohen-Bazire G, Glazer AN (1981) Characterization of the biliproteins of *Gloeobacter violaceus*: chromophore content of a cyanobacterial phycoerythrin carrying phycourobilin chromophore. Arch Microbiol 129:190–198

Capuano V, Braux AS, Tandeau de Marsac N, Houmard J (1991) The "anchor polypeptide" of cyanobacterial phycobilisomes. Molecular characterization of the *Synechococcus* sp. PCC 6301 *apcE* gene. J Biol Chem 266:7239–7247

Charron JBF, Ouellet F, Pelletier M, Danyluk J, Chauve C, Sarhan F (2005) Identification, expression, and evolutionary analyses of plant lipocalins. Plant Physiol 139:2017–2028

Clarke S (2002) The methylator meets the terminator. Proc Natl Acad Sci USA 99:1104–1106

Cobley JG et al (2002) CpeR is an activator required for expression of the phycoerythrin operon (*cpeBA*) in the cyanobacterium *Fremyella diplosiphon* and is encoded in the phycoerythrin linker- polypeptide operon (*cpeCDESTR*). Mol Microbiol 44:1517–1531

Dammeyer T, Hofmann E, Frankenberg-Dinkel N (2008) Phycoerythrobilin synthase (PebS) of a marine virus – Crystal structures of the biliverdin complex and the substrate-free form. J Biol Chem 283:27547–27554

Dolganov N, Grossman AR (1999) A polypeptide with similarity to phycocyanin α-subunit phycocyanobilin lyase involved in degradation of phycobilisomes. J Bacteriol 181: 610–617

Ducret A, Sidler W, Frank G, Zuber H (1994) The complete amino acid sequence of R-Phycocyanin-I α and β subunits from the red alga *Porphyridium cruentum*: structural and phylogenetic relationships of the phycocyanins within the phycobiliproteins families. Eur J Biochem 221:563–580

Fairchild CD, Glazer AN (1994a) Nonenzymatic bilin addition to the α subunit of an apophycoerythrin. J Biol Chem 269:28988–28996

Fairchild CD, Glazer AN (1994b) Oligomeric structure, enzyme kinetics, and substrate specificity of the phycocyanin alpha subunit phycocyanobilin lyase. J Biol Chem 269:8686–8694

Fairchild CD, Zhao J, Zhou J, Colson SE, Bryant DA, Glazer AN (1992) Phycocyanin α subunit phycocyanobilin lyase. Proc Natl Acad Sci USA 89:7017–7021

Flower DR (1996) The lipocalin protein family: structure and function. Biochem J 318:1–14

Frankenberg N, Mukougawa KK, Lagarias JC (2001) Functional genomic analysis of the HY2 family of ferredoxin-dependent bilin reductases from oxygenic photosynthetic organisms. Plant Cell 13:965–978

Gindt YM, Zhou J, Bryant DA, Sauer K (1994) Spectroscopic studies of phycobilisome subcore preparations lacking key core chromophores: assignment of excited state energies to the L_{cm}, β^{18} and $\alpha^{AP\text{-}B}$ chromophores. Biochim Biophys Acta 1186:153–162

Gingrich JC, Lundell DJ, Glazer AN (1983) Core substructure in cyanobacterial phycobilisomes. J Cell Biochem 22:1–14

Glazer AN (1984) Phycobilisome: a macromolecular complex optimized for light energy transfer. Biochim Biophys Acta 768:29–51

Glazer AN (1985) Light harvesting by phycobilisomes. Annu Rev Biophys Biophys Chem 14:47–77

Glazer AN (1989) Light guides: directional energy transfer in a photosynthetic antenna. J Biol Chem 264:1–4

Glazer AN (1994) Phycobiliproteins – A family of valuable, widely used fluorophores. J. Appl. Phycol. 6:105–112

Glazer AN, Lundell DJ, Yamanaka G, Williams RC (1983) The structure of a "simple" phycobilisome. Ann Microbiol (Inst Pasteur) 134B:159–180

Glazer AN, Stryer L (1983) Fluorescent tandem phycobiliprotein conjugates. Emission wavelength shifting by energy transfer. Biophys J 43:383–386

Glazer AN, Stryer L (1984) Phycofuor probes. Trends Biochem Sci 9:423–427

Glazer AN, Stryer L (1990) Phycobiliprotein avidin and phycobiliprotein biotin conjugates. Methods Enzymol 184:188–194

Grzyb J, Latowski D, Strzalka K (2006) Lipocalins – A family portrait. J Plant Physiol 163: 895–915

Hamada T, Matsunaga S, Yano G, Fusetani. N (2005) Polytheonamides A and B, highly cytotoxic, linear polypeptides with unprecedented structural features, from the marine sponge, *Theonella swinhoei*. J Am Chem Soc 127:110–118

Hieber AD, Bugos RC, Yamamoto HY (2000) Plant lipocalins: violaxanthin de-epoxidase and zeaxanthin epoxidase. Biochim Biophys Acta-Protein Struct Mol Enzymol 1482:84–91

Huber R et al (1987a) Crystallization, crystal-structure analysis and preliminary molecular-model of the bilin binding protein from the insect *Pieris brassicae* J Mol Biol 195:423–434

Huber R et al (1987b) Molecular structure of the bilin binding protein (BBP) from *Pieris brassicae* after refinement at 2.0 A resolution. J Mol Biol 198:499–513

Jung LJ, Chan CF, Glazer AN (1995) Candidate genes for the phycoerythrocyanin α subunit lyase: biochemical analysis of *pecE* and *pecF* interposon mutants. J Biol Chem 270: 12877–12884

Kahn K, Mazel D, Houmard J, Tandeau de Marsac N, Schaefer MR (1997) A role for cpeYZ in cyanobacterial phycoerythrin biosynthesis. J Bacteriol 179:998–1006

Klotz AV, Glazer AN (1987) gamma-N-methylasparagine in phycobiliproteins. Occurrence, location, and biosynthesis. J Biol Chem 262:17350–17355

Klotz AV, Leary JA, Glazer AN (1986) Post-translational methylation of asparaginyl residues. Identification of beta-71 gamma-N-methylasparagine in allophycocyanin. J Biol Chem 261:15891–15894

Kohchi T, Mukougawa K, Frankenberg N, Masuda M, Yakota A, Lagarias JC (2001) The *Arabidopsis* HY2 gene encodes phytochromobilin synthase, a ferredoxin-dependent biliverdin reductase. Plant Cell 13:425–436

Kuzin AP et al (2007) Crystal structure of fatty acid-binding protein-like Ycf58 from *Thermosynecoccus elongatus*. Northeast structural genomics consortium target Ter13. In: pdb ID: 3BDR http://www.pdb.org/pdb/explore.do?structureId=3BDR

Miller CA (2007) Identification and characterization of enzymes involved in post-translational modifications of phycobiliproteins in the cyanobacterium *Synechocystis* sp. PCC 6803. In: Department of Biological Sciences. University of New Orleans, New Orleans, LA p 130

Miller CA et al (2008) Biogenesis of phycobiliproteins.III. CpcM is the asparagine methyltransferase for phycobiliprotein β-subunits in cyanobacteria. J Biol Chem 283 19293–19300

Morimoto K, Nishio K, Nakai M (2002) Identification of a novel prokaryotic HEAT-repeats-containing protein which interacts with a cyanobacterial IscA homolog. FEBS Lett 519: 123–127

Morimoto K, Sato S, Tabata S, Nakai M (2003) A HEAT-repeats containing protein, IaiH, stablizes the iron-sulfur cluster bound to the cyanobacterial IscA homologue, IscA2. J Biochem (Tokyo) 134:211–217

Newcomer ME, Ong DE (2000) Plasma retinol binding protein: structure and function of the prototypic lipocalin. Biochim Biophys Acta-Protein Struct Mol Enzymol 1482:57–64

Ong LJ, Glazer AN (1991) Phycoerythrins of marine unicellular cyanobacteria. I. Bilin types and locations and energy transfer pathways in *Synechococcus* spp. phycoerythrins. J Biol Chem 266:9515–9527

Rümbeli R, Suter F, Wirth M, Sidler W, Zuber H (1987) [gamma]-N-methylasparagine in phycobiliproteins from the cyanobacteria *Mastigocladus laminosus* and Calothrix. FEBS Lett 221:1–2

Saunée NA, Williams SR, Bryant DA, Schluchter WM (2008) Biogenesis of phycobiliproteins. II. CpcS-I and CpcU comprise the heterodimeric bilin lyase that attaches phycocyanobilin to Cys-82 of beta -phycocyanin and Cys-81 of allophycocyanin subunits in *Synechococcus* sp. PCC 7002. J Biol Chem 283:7513–7522

Scheer H, Zhao KH (2008) Biliprotein maturation: the chromophore attachment. Mol Microbiol 68:263–276

Schirmer T, Bode W, Huber R, Sidler W, Zuber H (1985) X-ray crystallographyic structure of the light-harvesting biliprotein C-phycocyanin from the thermophilic cyanobacterium *Mastigocladus laminosus* and its resemblance to globin structures. J Mol Biol 184:257–277

Schirmer T, Huber R, Schneider M, Bode W, Miller M, Hackert ML (1986) Crystal structure analysis and refinement at 2.5 Å of hexameric C-phycocyanin from the cyanobacterium *Agmenellum quadruplicatum*: the molecular model and its implications from light-harvesting. J Mol Biol 188:651–676

Schluchter WM, Glazer AN (eds) (1999) Biosynthesis of phycobiliproteins in cyanobacteria. Plenum, New York

Sehringer B, Kayser H (2006) Butterfly wings, a new site of porphyrin synthesis and cleavage: studies on the expression of the lipocalin bilin-binding protein in *Pieris brassicae*. Insect Biochem Mol Biol 36:482–491

Sekar S, Chandramohan M (2008) Phycobiliproteins as a commodity: trends in applied research, patents and commercialization. J Appl Phycol 20:113–136

Shan J, Jia J, Clokie MR, Mann NH (2008) Infection by the "photosynthetic" phage S-PM2 induces increased synthesis of phycoerythrin in *Synechococcus* sp. WH7803. FEMS Microbiol Lett 283:154–161

Shen GZ, Saunee NA, Gallo E, Begovic Z, Schluchter WM, Bryant DA (2004) Identification of novel phycobiliprotein lyases in cyanobacteria. In: Niederman RA, Blankenship RE, Frank H, Robert B, van Grondelle R (eds) Photosynthesis 2004 Light-Harvesting Systems Workshop, Saint Adele, Quebec, Canada, pp 14–15

Shen G, Saunee NA, Williams SR, Gallo EF, Schluchter WM, Bryant DA (2006) Identification and characterization of a new class of bilin lyase: the *cpcT* gene encodes a bilin lyase responsible for attachment of phycocyanobilin to Cys-153 on the beta subunit of phycocyanin in *Synechococcus* sp. PCC 7002. J Biol Chem 281:17768–17778

Shen G, Leonard HS, Schluchter WM, Bryant DA (2008a) CpcM post-translationally methylates asparagine-71/72 of phycobiliprotein beta subunits in *Synechococcus* sp. PCC 7002 and *Synechocystis* sp. PCC 6803. J Bacteriol 190:4808–4817

Shen G, Schluchter WM, Bryant DA (2008b) Biogenesis of phycobiliproteins. I. *cpcS-I* and *cpcU* mutants of the cyanobacterium *Synechococcus* sp. PCC 7002 define a heterodimeric phycocaynobilin lyase specific for beta -phycocyanin and allophycocyanin subunits. J Biol Chem 28:7503–7512

Six C et al (2007) Diversity and evolution of phycobilisomes in marine Synechococcus spp.: a comparative genomics study. Genome Biol 8:R259

Storf M et al (2001) Chromophore attachment to biliproteins: specificity of PecE/PecF, a lyase-isomerase for the photoactive 3(1)-cys-alpha 84-phycoviolobilin chromophore of phycoerythrocyanin. Biochemistry 40:12444–12456

Swanson R, Glazer A (1990) Phycobiliprotein methylation: effect of the γ-*N*-methylasparagine residue on energy transfer in phycocyanin and the phycobilisome. J Mol Biol 214:787–796

Swanson RV et al (1992) Characterization of phycocyanin produced by *cpcE* and *cpcF* mutants and identification of an intergenic suppressor of the defect in bilin attachment. J Biol Chem 267:16146–16154

Takano H, Gusella J (2002) The predominantly HEAT-like motif structure of huntingtin and its association and coincident nuclear entry with dorsal, an NF-kB/Rel/dorsal family transcription factor. BMC Neurosci 14:15

Thomas BA, Bricker TM, Klotz AV (1993) Post-translational methylation of phycobilisomes and oxygen evolution efficiency in cyanobacteria. Biochim Biophys Acta 1143:104–108

Thomas BA, McMahon LP, Klotz AV (1994) Gamma-N-methylasparagine- A Posttranslational modification that improves energy-transfer efficiency in phycobiliproteins. Biophys J 66: A164–A164

Tooley A, Cai Y, Glazer A (2001) Biosynthesis of a fluorescent cyanobacterial C-Phycocyanin holo-α subunit in a heterologous host. Proc Natl Acad Sci USA 98:10560–10565

Wilbanks SM, Glazer AN (1993a) Rod structure of a phycoerythrin II-containing phycobilisome I: organization and sequence of the gene cluster encoding the major phycobiliprotein rod components in the genome of marine *Synechococcus* sp. WH8020. J Biol Chem 268:1226–1235

Wilbanks SM, Glazer AN (1993b) Rod structure of a phycoerythrin II-containing phycobilisome II: complete sequence and bilin attachment site of a phycoerythrin γ subunit. J Biol Chem 268:1236–1241

Wilbanks SM, Wedemayer G, Glazer AN (1989) Posttranslational modifications of the b-Subunit of a cryptomonad phycoerythrin. J Biol Chem 264:17860–17867

Xu Y, Alvey R, Byrne PO, Graham JE, Shen G, Bryant DA (2010) Expression of genes in cyanobacteria: adaptation of endogenous plasmids as platforms for high-level gene expression in *Synechococcus* sp. PCC 7002. In: Carpentier R (ed) Photosynthesis research protocols. Humana, Totowa, NJ (in press)

Zhao KH et al (2000) Novel activity of a phycobiliprotein lyase: both the attachment of phycocyanobilin and the isomerization to phycobiliviolin are catalyzed by the proteins PecE and PecF encoded by the phycoerythrocyanin operon. FEBS Lett 469:9–13

Zhao KH et al (2002) Characterization of phycoviolobilin phycoerythrocyanin-alpha 84-cystein-lyase-(isomerizing) from *Mastigocladus laminosus*. Eur J Biochem 269:4542–4550

Zhao KH et al (2005) Reconstitution of phycobilisome core-membrane linker, L-CM, by autocatalytic chromophore binding to ApcE. Biochim Biophys Acta-Bioenerg 1706:81–87

Zhao KH et al (2006) Chromophore attachment to phycobiliprotein beta-subunits: phycocyanobilin:cystein-beta84 phycobiliprotein lyase activity of CpeS-like protein from *Anabaena* sp. PCC7120. J Biol Chem 281:8573–8581

Zhao K et al (2007a) Phycobilin:cystein-84 biliprotein lyase, a near-universal lyase for cysteine-84-binding sites in cyanobacterial phycobiliproteins. Proc Natl Acad Sci USA 104:14300–14305

Zhao KH et al (2007b) Lyase activities of CpcS- and CpcT-like proteins from *Nostoc* PCC7120 and sequential reconstitution of binding sites of phycoerythrocyanin and phycocyanin beta-subunits. J Biol Chem 282:34093–34103

Zhou J, Gasparich GE, Stirewalt VL, de Lorimier R, Bryant DA (1992) The *cpcE* and *cpcF* genes of *Synechococcus* sp. PCC 7002: construction and phenotypic characterization of interposon mutants. J Biol Chem 267:16138–16145

Chapter 13
The Tetrapyrrole Biosynthetic Pathway and Its Regulation in *Rhodobacter capsulatus*

Sébastien Zappa, Keran Li, and Carl E. Bauer

Abstract The purple anoxygenic photosynthetic bacterium *Rhodobacter capsulatus* is capable of growing in aerobic or anaerobic conditions, in the dark or using light, etc. Achieving versatile metabolic adaptations from respiration to photosynthesis requires the use of tetrapyrroles such as heme and bacteriochlorophyll, in order to carry oxygen, to transfer electrons, and to harvest light energy. A third tetrapyrrole, cobalamin (vitamin B_{12}), is synthesized and used as a cofactor for many enzymes. Heme, bacteriochlorophyll, and vitamin B_{12} constitute three major end products of the tetrapyrrole biosynthetic pathway in purple bacteria. Their respective synthesis involves a plethora of enzymes, several that have been characterized and several that are uncharacterized, as described in this review. To respond to changes in metabolic requirements, the pathway undergoes complex regulation to direct the flow of tetrapyrrole intermediates into a specific branch(s) at the expense of other branches of the pathway. Transcriptional regulation of the tetrapyrrole synthesizing enzymes by redox conditions and pathway intermediates is reviewed. In addition, we discuss the involvement of several transcription factors (RegA, CrtJ, FnrL, AerR, HbrL, Irr) as well as the role of riboswitches. Finally, the interdependence of the tetrapyrrole branches on each other synthesis is discussed.

13.1 Introduction

Cyclic tetrapyrroles encompass porphyrins, such as heme, chlorophyll and bacteriochlorophyll (Bchl), and porphynoids, which are more reduced. The porphynoid class consists of the corrinoids (cobalamin), siroheme, heme d_1, and coenzyme F_{430} (Frankenberg et al. 2003). Many species are capable of synthesizing numerous tetrapyrrole end products. One of the best studied organisms is *Rhodobacter capsulatus*, an anoxygenic photosynthetic bacterium of the α-proteobacteria subfamily. It is capable of growing under a variety of different environmental conditions such

C.E. Bauer (✉)
Biology Department, Indiana University, Bloomington, IN 47405, USA
e-mail: bauer@indiana.edu

P.C. Hallenbeck (ed.), *Recent Advances in Phototrophic Prokaryotes*, Advances in Experimental Medicine and Biology 675, DOI 10.1007/978-1-4419-1528-3_13,

as dark aerobic, dark and light anaerobic. This versatile metabolism requires the capability of synthesizing the tetrapyrroles heme and Bchl that are needed for respiratory or photosynthetic growth. Heme is needed for electron transfer during both respiration and photosynthesis while Bchl is the major pigment involved in collecting and converting light energy into biochemical energy during photosynthesis. In addition to synthesizing heme and Bchl, *R. capsulatus* also synthesizes cobalamin [vitamin B_{12} (vitB_{12})]. VitB_{12} and its derivatives are the most complex tetrapyrrole synthesized in nature and are involved in various cellular functions such as DNA repair and methionine synthesis (Martens et al. 2002). The synthesis of siroheme in *R. capsulatus* strain has been reported as well (Olmo-Mira et al. 2006).

The four tetrapyrrole end products synthesized by *R. capsulatus* are derived from different branches that split off from a common tetrapyrrole biosynthetic pathway "core" (Fig. 13.1). A common trunk of the pathway used by all three branches involves a multistep conversion of δ-aminolevulinic acid (δ-ALA) into uroporphyrinogen III (uro'gen III). At uro'gen III, the "corrin branch" splits off leading to the synthesis of cobalamin. Uro'gen III is also the potential substrate for siroheme synthesis. The central core of the pathway continues to synthesize intermediates up to protoporphyrin IX (proto IX) that are common to both the heme and Bchl branches. At proto IX, the heme and Bchl branches differentiate either through the insertion of Fe to form heme or the insertion of Mg to form Mg–proto IX which is the first committed step in the Mg branch that leads to the production of Bchl.

The biosynthesis of heme, vitB_{12}, and Bchl is tightly controlled in order to achieve an adequate Bchl/heme/vitB_{12} ratio to meet the metabolic requirements of the cell. Heme and cobalamin levels appear to be maintained at rather stable levels while the amount of Bchl swings dramatically depending on the presence or absence of environmental oxygen. Moreover, a number of the intermediates in each of the three branches are potentially toxic as they can generate reactive singlet oxygen as a byproduct of light absorption. Tuning the relative amount of Bchl, heme, and vitB_{12} as a function of the metabolic requirements, while coping with the potential toxicity of intermediate products, requires this pathway to be highly regulated. Early work on the regulation of pigment synthesis in purple bacteria was initiated in the 1950s by Cohen-Bazire (Cohen-Bazire et al. 1957). More recently it has been demonstrated that the tetrapyrrole biosynthetic pathway in purple bacteria is highly complex and its regulation extremely sophisticated. This review is centered on the model species *R. capsulatus* and attempts to identify the actors involved in this pathway, both synthesizing enzymes and regulators.

13.2 Topology of Genes Involved in Synthesis of Tetrapyrroles

Examination of the complete genome of *R. capsulatus* has enabled the identification of the genes involved in each branch of tetrapyrrole synthesis (cf. annotations available at http://www.ergo-light.com and http://onco.img.cas.cz/rhodo/results/index.html). While genes involved in the vitB_{12} and Bchl synthesis

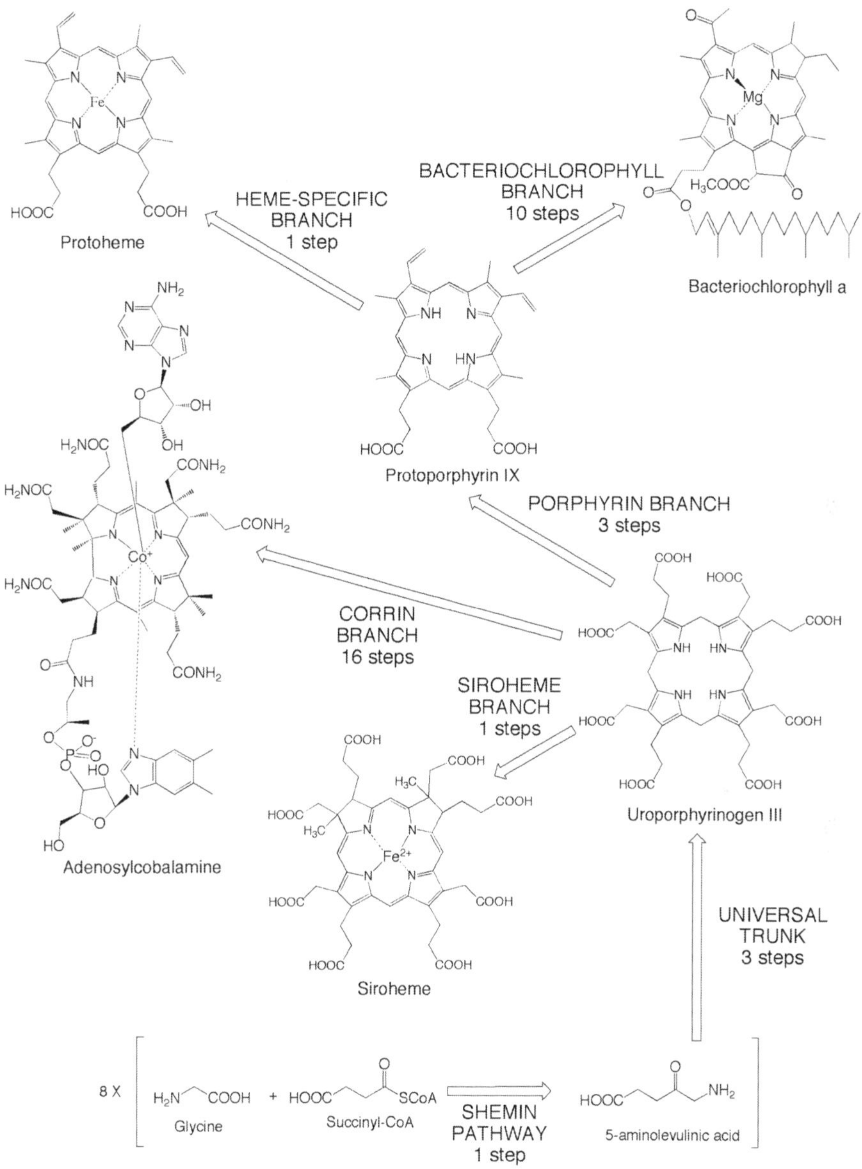

Fig. 13.1 Overview of the tetrapyrrole biosynthetic pathway with its substrates, major intermediates, and products. The number of enzyme reactions for each branch is indicated

are found in large operons, genes encoding the enzymes involved in heme synthesis are scattered over the genome (Young et al. 1989; McGoldrick et al. 2002; Warren et al. 2002; Smart et al. 2004). This feature is not unusual, though a few organisms show an operon organization for some of their *hem* genes (Avissar and Moberg

1995). The one exception in *R. capsulatus* is the *hemC* and *hemE* genes that are divergently transcribed and separated by 111 bp (Smart et al. 2004).

13.3 The Universal Trunk: From δ-ALA to Urog'en III

13.3.1 Feeding the Universal Trunk: The Synthesis of δ-ALA

The biosynthetic pathway to produce uro'gen III from δ-ALA is highly conserved in nature and is therefore considered universal. However, there are two different pathways to synthesize δ-ALA, which is the precursor of all known tetrapyrroles. One is the C-5 pathway where two enzymatic reactions transform glutamyl-tRNA^{Glu} into δ-ALA. The other is the C-4 or Shemin pathway that involves the condensation of glycine and succinyl-CoA to produce δ-ALA, CO_2, and CoA (Avissar and Moberg 1995; Frankenberg et al. 2003). As a typical representative of the α-proteobacteria subfamily, *R. capsulatus* synthesizes δ-ALA using the C-4 pathway. This reaction is performed by δ-ALA synthase (EC 2.3.1.37, ALAS), encoded by the *hemA* gene, previously identified and cloned in 1988 (Biel et al. 1988). *R. capsulatus* ALAS is a piridoxal-5′-phosphate-dependent homodimer, consisting of two 44 kDa subunits. It shows 49% identity and 70% similarity with the human erythroid-specific ALAS. The structure, solved in 2005 by Astner et al., provided insight on mutations that generate dysfunctional forms of the human eALAS (Astner et al. 2005). Moreover, despite a very low primary sequence identity of 18%, ALAS shows remarkable three-dimensional homology with the glutamate-1-semialdehyde-2,1-aminomutase (GSAM) of *Thermosynechococcus elongatus*. The GSAM is the last enzyme of the C-5 pathway and, as such, synthesizes δ-ALA. It is likely that the C-4 pathway originated from the evolution of GSAM into ALAS, followed by the loss of the C-5 pathway in common ancestors of α-proteobacteria and of early mitochondria (Schulze et al. 2006). ORF encoding a putative GSAM, *hemL*, was found in the genome of *R. capsulatus*. In prokaryotes, the occurrence of two uncoupled δ-ALA biosynthetic pathways was shown in *Streptomyces nodosus* subsp. *asukaensis*, where the C-5 pathway produces δ-ALA for heme synthesis, while the C-4 pathway generates δ-ALA targeted to secondary metabolism (Petricek et al. 2006). In *R. capsulatus* there is no evidence for an alternative δ-ALA synthesis pathway as the first enzyme of the C-5 pathway, glutamyl-tRNA reductase, is not found in the genome. Furthermore, the annotation of this ORF as a *hemL* gene has to be confirmed experimentally as GSAM-encoding genes share a great deal of similarity with genes coding for other types of aminotranferases (Panek and O'Brian 2002). Interestingly, when a δ-ALA-requiring mutant containing a single insertion was characterized, it was found that a low level (10%) of the ALAS activity of the parental strain remained (Wright et al. 1987). Finally, the synthesis of δ-ALA appears as one of the crucial regulation points of the tetrapyrrole biosynthetic pathway, the transcription of *hemA* being up-regulated under low oxygen tension (Wright et al. 1991; Smart et al. 2004) and down-regulated in the presence of excess of exogenous heme (Smart and Bauer 2006).

13.3.2 Entering the Universal Trunk: Synthesis of Porphobilinogen

The first common step in tetrapyrrole synthesis is achieved by porphobilinogen synthase (EC 4.2.1.24, PBGS), or δ-ALA dehydratase, that converts two δ-ALA molecules into a monopyrrole porphobilinogen. In *R. capsulatus*, *hemB* encodes a 35.8 kDa protein which presents unusual characteristics among the PBGS studied so far. Indeed, most PBGS are known to be homooctameric metalloenzymes presenting a high variability regarding their cations requirement in terms of catalysis (Zn^{2+}-, Mg^{2+}-, or K^{+}- binding site at the active site) and activity regulation, most of them containing an extra allosteric Mg^{2+}-binding site. The biochemical characterization of *R. capsulatus* revealed that its enzyme is a homohexamer clearly independent of cations (Zn^{2+}, Mg^{2+}, K^{+}) or chelators (EDTA, 1,10-phenathroline). *R. capsulatus* PBGS seems to switch between a highly active homohexameric form and a less active homodimeric form (Nandi and Shemin 1973; Bollivar et al. 2004). Regarding metal dependence, this feature was partially predicted from the primary sequence. Indeed, among 130 PBGS sequences representing the entire tree of life, only the ones from the *Rhodobacter* genus, namely *R. capsulatus* and *R. sphaeroides*, lack the cysteine-rich motif involved in Zn binding at the active site and the conserved arginine and glutamate residues that coordinate the allosteric Mg. It is postulated that PBGS containing Zn at the active site evolved to other configurations (metal free, Mg- and/or K-containing active site) during the set up of the photosynthetic process and the early synthesis of Bchl. Indeed, Zn is a competitor of Mg in the metallation step of proto IX (Jaffe 2003; Frère et al. 2005). If the PBGSs of the *Rhodobacter* genus fit well in this model regarding their active site, the evolutionary pressure(s) that allowed them to totally lose metal content, including the Mg in the allosteric site, remain(s) unknown. Moreover, despite porphobilinogen accumulation under low oxygen tension in *R. capsulatus*, the activity of PBGS is unregulated by oxygen, proto IX, and hemin in *R. capsulatus*. In addition, no or only a slight increase in *hemB* transcription was observed upon decreasing the oxygen tension (Biel et al. 2002; Smart et al. 2004). However, the conversion of δ-ALA into porphobilinogen seems to be a major control point in the tetrapyrrole biosynthesis pathway (Biel 1992). This may be achieved by controlling the level of porphobilinogen through the availability of the substrate of PBGS, i.e., δ-ALA. Indeed, the vast majority of the δ-ALA synthesized is transformed into aminohydroxyvalerate by an oxygen-dependent δ-ALA dehydrogenase (Biel et al. 2002). Lastly, although not occurring in a manner as dramatic as for *hemA*, *hemC*, or *hemE*, it was shown that the transcription of *hemB* is repressed in the presence of exogenous heme (Smart and Bauer 2006).

13.3.3 Reaching the First Crossroad: Synthesis of the Tetrapyrrole Ring, Uro'gen III

The transformation of monopyrrole porphobilinogen into the cyclic tetrapyrrole uro'gen III is catalyzed by the enzymes porphobilinogen deaminase (EC 4.2.1.24, PBGD) and uro'gen III synthase (4.2.1.75, UROS) encoded by *hemC* and *hemD*, respectively. During the first step, PBGD polymerizes four molecules of porphobilinogen into a linear tetrapyrrole, pre-uro'gen or 1-hydroxymethylbilane, and four NH_3. The main product, pre-uro'gen, is very unstable and, if exposed to solvent, will be converted chemically into a toxic cyclic porphyrin, uro'gen I. To avoid toxicity pre-uro'gen is passed to UROS that presumably interacts with PBGD to limit exposure of the cell to free pre-uro'gen (Shoolingin-Jordan 1995). UROS catalyzes the formation of uro'gen III, by circularizing the linear tetrapyrrole while inverting ring D. An asymmetric ring D is a signature of biologically functional porphyrins.

PBGD was purified from *R. capsulatus* and the N-terminal sequence was determined in order to identify the corresponding gene, *hemC*. The latter was also expressed heterologously in *Escherichia coli*, confirming that this ORF is sufficient to produce an active form of PBGD (Biel et al. 2002). No biochemical characterization of the *R. capsulatus* enzyme has been performed so far, but analysis of the sequence indicates that *hemC* corresponds to a 34.1 kDa protein with 47% homology with PBGDs from *E. coli* and *Pseudomonas aeruginosa* and 43% with the *Bacillus subtilis* homolog. In addition, previous characterizations of PBGDs highlighted the use of a unique cofactor, dipyrromethane which consists of two porphobilinogen molecules. This cofactor acts as a primer during the polymerization of a hexapyrrole, which finally releases the tetrapyrrolic reaction product and the intact cofactor. *R. capsulatus* PBGD contains the conserved cysteine residue that is involved in cofactor binding. More work has been done regarding the genetics. Indeed, characterization of a *hemC* mutant confirmed that it was incapable of synthesizing either siroheme or vitB_{12} and grew only in media supplemented with cysteine and methionine, in addition to hemin (Biel et al. 2002). As well, *hemC* expression was shown to be regulated by oxygen, with a fivefold increase of the transcription level when changing from aerobic to anaerobic growth conditions (Smart et al. 2004). Finally, the transcription of *hemC* was repressed two-thirds upon the addition of exogenous hemin (Smart and Bauer 2006).

UROS is only very poorly characterized unclear as it has never been purified from *R. capsulatus*. A high variability in the primary sequences of UROSs has made it difficult to identify the *hemD* gene in the *R. capsulatus* genome. In addition, the lack of genetic organization such as a *hemCD* operon or a *hem* gene cluster, as in *E. coli* or *B. subtilis*, renders the task even harder (Avissar and Moberg 1995). Recently, an ORF in the *R. capsulatus* genome was annotated as encoding a UROS but because *hemD* genes have been miss-annotated in many other organisms, experimental work has to be undertaken to confirm this annotation (Panek and O'Brian 2002).

13.4 The Porphyrin Branch: From Uro'gen III to Heme

13.4.1 The Synthesis of Coproporphyrinogen III and Its Puzzling Secretion

The first step after the uro'gen III crossroad consists in the synthesis of coproporphyrinogen III (copro'gen III). This activity is performed by uro'gen III decarboxylase (EC 4.1.1.37, UROD) which catalyzes the sequential decarboxylation of the four acetate side chains into methyl groups. The reaction is ordered, starting with the ring D and then decarboxylating rings A, B, and finally ring C. Numerous UROD-encoding genes from various biological sources have been identified showing a rather high level of conservation and presumably a common mechanism of action (Frankenberg et al. 2003; Heinemann et al. 2008). The *R. capsulatus* gene encoding UROD, *hemE*, was identified in 1995 (Ineichen and Biel 1995). The deduced protein sequence shows more than 34% identity with previously characterized UROD and contains a signature sequence of this class of enzyme (PXWXMRQAGR) at the amino-terminal. No biochemical work has been performed with *R. capsulatus* UROD but a bit is known regarding the genetics and its regulation. An interesting feature of *hemE* is its linkage to *hemC*, while the other *hem* genes are scattered on the genome. These two genes are divergently transcribed and separated by an intergenic region of 111 bp. Furthermore, the transcription of *hemE* is strongly dependent on redox conditions, as seen using a *hemE::lacZ* fusion. Expression of *hemE* increases by 112- and 52-fold under semi-aerobic and anaerobic conditions, respectively (Smart et al. 2004). As with *hemC*, the transcription level of *hemE* is decreased by two-thirds in the presence of exogenous heme (Smart and Bauer 2006).

An interesting phenomenon about this step of the pathway is that the cells excrete copious amounts of the product of the reaction, copro'gen III under certain growth conditions. Cooper (1956, 1963) highlighted this feature in *R. capsulatus*, using cells grown under iron starvation or with excess exogenous methionine. More recently, various *R. capsulatus* mutants, either unable to synthesize Mg–protoporphyrin monomethyl ester or lacking *c*-type cytochromes, were studied and showed a similar phenotype. Moreover, it was discovered that coproporphyrin is excreted as a complex with a 66 kDa protein (Biel and Biel 1990; Biel 1991) that turned out to be the major outer membrane porin (Bollivar and Bauer 1992).

13.4.2 Reaching the Second Crossroad: The Synthesis of Protoporphyrin IX

The antepenultimate and penultimate steps of heme synthesis are achieved by copro'gen III oxidase (EC 1.3.99.22, CPO) and proto'gen IX oxidase (EC 1.3.3.4, PPO), respectively. These steps lead to the formation of protoporphyrin IX which is a precursor for the synthesis of heme and Bchl. Oxygen-dependent and oxygen-independent CPOs conventionally named HemF and HemN (or HemZ), respectively, are known to catalyze this reaction. HemF is prevalent in eukaryotes and in a few bacterial groups (primarily cyanobacteria and proteobacteria) while HemN is widely distributed among prokaryotes. These two forms of CPOs share no obvious primary sequence homology indicating that they arose from distinct evolutionary processes. Both enzymes convert copro'gen III into proto'gen IX by decarboxylating the propionate side chains of rings A and B consecutively to yield vinyl groups (Panek and O'Brian 2002; Frankenberg et al. 2003; Heinemann et al. 2008). Careful examination of the genome of *R. capsulatus* revealed no HemF-encoding gene but three ORFs that may code for putative HemNs. This is in contrast to the genome of the phylogenetically related species *R. sphaeroides* that contains one *hemF* and three *hemN* genes (locus tags in img.jgi.doe.gov: RSP_0682, RSP_0699, RSP_1224, RSP_0317).

HemN contains a [4Fe–4S] cluster for catalysis and requires *S*-adenosyl-L-methionine (SAM) as a cofactor or co-substrate. As such, this enzyme belongs to a family of "radical SAM" enzymes that carry the signature sequence CXXXCXXC, where the cysteine residues are involved in the coordination of the Fe atoms (Layer et al. 2005). This motif is present in the deduced protein sequences of the three putative CPOs. Occurrence of multiple *hemN* genes in one organism has been reported and several bacterial oxygen-independent CPOs have been shown to be expressed and active in the presence of oxygen (Rompf et al. 1998; Schobert and Jahn 2002). The expression of *hemN1*, designated *hemZ* in the cited articles, was studied as a function of aeration, which showed only a moderate influence. Indeed, a roughly twofold variation was recorded, with a maximum transcription level under semi-aerobic conditions (Smart et al. 2004). In addition, the presence of exogenous heme under these growth conditions decreased the expression of the gene by twofold factor, bringing it back to its basal level of transcription (Smart and Bauer 2006). Finally, in silico analysis of α-proteobacterial genomes revealed an "iron-rhodo-box" in the promoter region of the *hemN2* gene. Thus, *hemN2* is predicted to be part of the iron regulon (Rodionov et al. 2006).

The next step in the pathway is the aromatization of proto'gen IX by removal of six electrons to yield proto IX. This is performed by proto'gen IX oxidase (PPO) that exists as two forms, oxygen-dependent (HemY) and oxygen-independent (HemG). Most of our knowledge about PPOs comes from studies of eukaryotic enzymes that are oxygen dependent and inhibited by diphenyl ether herbicides in plants (Dailey 2002; Heinemann et al. 2008). In prokaryotes, the situation is very unclear as the PPO-encoding gene is unidentifiable in many genomes and, so far, *hemG* seems to be limited to six genera within the γ-proteobacteria group (Panek and O'Brian 2002; Frankenberg et al. 2003). As is the case with many other heme-synthesizing prokaryotes, a PPO-encoding gene has not been identified in the genome of either *R. capsulatus* or *R. sphaeroides* (cf. img.jgi.doe.gov). Even if the oxidation of proto'gen IX into proto IX can occur chemically, an enzyme-mediated catalysis is most probably required. Indeed, on the one hand, *hemY* and *hemG* mutants have been shown to be heme deficient in *B. subtilis* and *E. coli*, respectively (Panek and O'Brian 2002). On the other hand, this oxidation has to occur under anaerobic conditions as well, when the need for Bchl peaks. So there must be either an unknown PPO-encoding gene or an already known gene carrying out such an activity. The ERGO annotation of the *R. capsulatus* genome recently indicated an ORF encoding a "NAD (FAD) utilizing dehydrogenase with a possible PPO activity." This has yet to be experimentally confirmed.

13.4.3 Delivering the Final Product: The Synthesis of Heme

The synthesis of heme from proto IX is performed by the enzyme ferrochelatase (EC 4.99.1.1, FC) that chelates ferrous iron and inserts it into the center of the tetrapyrrole ring. The core of the enzyme is highly conserved between bacteria, plants, and mammals, although bacterial FCs do not contain [2Fe–2S] clusters (Frankenberg et al. 2003; Heinemann et al. 2008). Interestingly, FC and the anaerobic cobalt chelatase CbiK are both class 2 metal ion chelatases that exhibit a similar three-dimensional structure (Schubert et al. 2002). Recently, Masoumi et al. confirmed in vivo what Koch et al. predicted in silico that PPO and FC form a stable PPO–FC complex (Koch et al. 2004; Masoumi et al. 2008). Such a configuration may ensure the channeling of the potentially toxic proto IX to FC. Finally, it has been suggested that this complex might include the CPO to form a tripartite association CPO–PPO–FC (Dailey 2002; Koch et al. 2004).

In *R. capsulatus,* the FC-encoding gene, *hemH*, was described and cloned in the 1990s with the deduced protein sequence exhibiting 41% (59%) and 21% (44%) identity (similarity) with *E. coli* and *B. subtilis* FCs (Kanazireva and Biel 1996). Moreover, homologous expression of a second copy of *hemH* causes a dramatic reduction of the tetrapyrrole pool, with 2.3-fold and 5-fold decreases of the porphyrin and the Bchl concentrations, respectively. In addition, the ALAS activity is also decreased 2.3-fold under these conditions (Kanazireva and Biel 1995). This phenotype is a signature of negative feedback by heme operating on the entire pathway, as has been studied more recently (Smart and Bauer 2006). The transcription of *hemH* was analyzed as well, highlighting a unique responsive behavior to oxygen among the other *R. capsulatus hem* genes. Indeed, while the expression level is rather constant between aerobic and anaerobic conditions, it is threefold lower under semi-aerobic conditions (Smart et al. 2004). Finally, in semi-aerobic conditions, *hemH* transcription was found to be independent of the exogenously added heme (Smart and Bauer 2006).

13.4.4 Regulation of hem Gene Expression

13.4.4.1 RegA

The RegA–RegB system is a global regulator in *R. capsulatus* that has been shown to regulate synthesis of numerous cellular processes as a function of redox conditions (reviewed in Elsen et al. 2004; Wu and Bauer 2008). As shown in Fig. 13.2, studies of the transcription of various *hem* genes in a *regA* mutant strain highlighted that *hem* genes are part of the RegA–RegB regulon (Smart et al. 2004). Indeed, the most significant effect was observed on *hemE* and *hemH* where transcription was activated 5- to 10-fold by RegA. A strong activating role of RegA was also observed with *hemA* but only in semi-aerobic and anaerobic conditions. Moderate influences were reported on *hemC* and *hemZ*. Finally, the transcription of *hemB* does not seem to be regulated by RegA (Smart et al. 2004). Overall, the dramatic transcriptional activation of *hemE* and *hemH* highlights the crucial intervention of the RegA–RegB system at key steps of the tetrapyrrole synthetic pathway. Activating the transcription of these genes ensures a constant stock of mRNA for a quick synthesis of UROD and FC when the flow of tetrapyrrole synthesis needs to be directed toward the porphyrin branch rather than the corrin branch and when heme needs to be produced at the expense of Bchl. Finally, the RegA-induced increase

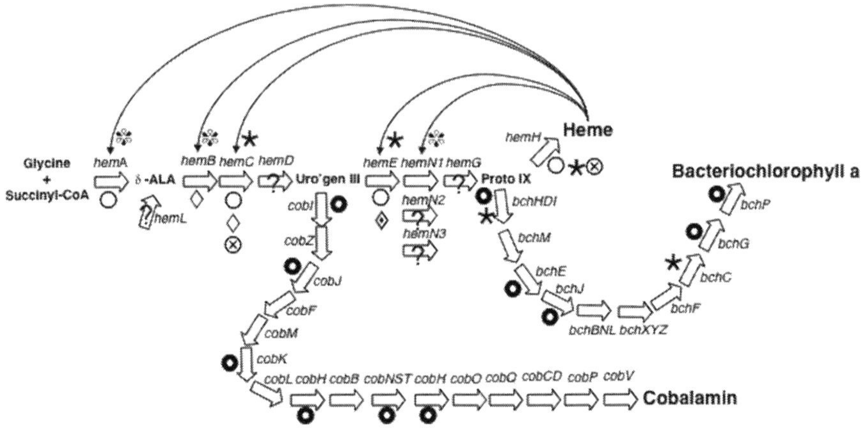

Fig. 13.2 Overview of the transcriptional regulation of the tetrapyrrole biosynthesis pathway. Transcription factors involved are RegA (○, based on gene expression studies; based on both gene expression and DNA-binding studies), FnrL (◇, repressor; , activator), AerR (⊗), CrtJ (★), HbrL ( ). *Arrows* indicate a feedback control by heme operated through CrtJ or HbrL. Genes identified from bioinformatics analysis only, with no experimental studies, are specified with a question mark

in the transcription of *hemA* under semi-aerobic and anaerobic conditions underlines a function of the RegA–RegB system, namely stimulation of the production of the general tetrapyrrole precursor δ-ALA in coordination with the synthesis of the pigment-binding proteins of the photosystem.

13.4.4.2 CrtJ

Present in the photosynthesis gene cluster is a redox-responding transcription factor called CrtJ or PpsR depending upon the species. Studies have demonstrated that CrtJ/PpsR is an aerobic repressor of heme (*hem*), Bchl (*bch*), and carotenoid (*crt*) biosynthesis genes in *R. capsulatus* (Ponnampalam and Bauer 1997; Elsen et al. 1998; Smart et al. 2004). CrtJ also aerobically represses synthesis of light-harvesting II peptides (*puc*) that bind Bchl and carotenoids in these species (Ponnampalam et al. 1995). CrtJ cooperatively binds to two copies of the palindromic sequence TGT-N_{12}-ACA that is present in many *hem* promoter regions (Ponnampalam and Bauer 1997; Elsen et al. 1998; Ponnampalam et al. 1998; Smart et al. 2004).

The characterization of a *crtJ* mutant strain has revealed its involvement in the transcriptional control of several *hem* genes, summarized in Fig. 13.2 (Smart et al. 2004). The strongest regulation is the activation of *hemH* expression, whatever the redox condition is. CrtJ is also required to induce the translation of *hemZ* under semi-aerobic conditions. In the same redox conditions, CrtJ exhibits a dual role of repressor of *hemC* and activator of *hemE* at the same time. This is puzzling as these genes share a common promoter region, containing a CrtJ-binding site (Smart et al. 2004). Elucidating how, in constant redox conditions, the interaction with this promoter induces two opposite events will be a challenge.

13.4.4.3 FnrL

The redox regulator FnrL appeared to have only moderate effect on transcription of *hem* genes (Fig. 13.2), as either an activator or a repressor, depending on the gene and/or on redox conditions (Smart et al. 2004). This makes it hard to clearly define its mode of action and function in the regulation of tetrapyrrole synthesis. On the one hand, FnrL significantly represses *hemA* in aerobic conditions and *hemB* and *hemC* in semi-aerobic and anaerobic conditions. On the other hand, it dramatically activates the transcription of *hemE* in semi-aerobic conditions. FnrL seems to be mostly involved in the early steps of the tetrapyrrole synthesis, the universal trunk, and the first step of the porphyrin branch. It was shown in other photosynthetic bacteria, such as *R. sphaeroides* and *Rubrivivax gelatinosus*, to have a crucial role in the transcription of *hemZ*, which did not appear to occur in *R. capsulatus* (Smart et al. 2004; Ouchane et al. 2007).

13.4.4.4 AerR

The aerobic repressor AerR (Dong et al. 2002) has a limited although important role in the transcriptional regulation of the tetrapyrrole synthesis, consisting in the activation of *hemE* and *hemH* whatever the redox conditions and the repression of *hemC* in semi-aerobic conditions are (Fig. 13.2). Overall, AerR seems to back up CrtJ in its target genes, with the exception of *hemZ* (Smart et al. 2004).

13.4.4.5 HbrL

The heme-binding regulatory LysR-type (HbrL) protein was identified in 2006 from the screening of a cosmid library for suppressors defective in heme synthesis using *hemB* as a reporter (Smart and Bauer 2006). From these experiments, an 804 bp ORF was isolated and cloned and the deduced amino acid sequence showed a LysR-type transcriptional regulator type (LTTR) signature: A helix-turn-helix DNA-binding domain at the N-terminal followed by an LTTR substrate-binding motif or coinducer-binding motif. In addition to being ubiquitous in prokaryotes, the LTTRs represent the largest family of DNA-binding protein. They regulate functions as diverse as metabolism, cell division, quorum sensing, oxidative stress, virulence, nitrogen fixation, etc. (Maddocks and Oyston 2008). As shown in Fig. 13.2, HbrL appears to be involved in the heme-mediated control of the expression of *hemA*, *hemB*, and *hemN1* (*hemZ*). Whereas it acts as a strong activator of *hemA* and *hemN1* in the absence of exogenous heme, it is a moderate repressor of *hemB*. Recombinant HbrL was obtained by heterologous expression in *E. coli* and purification. It exhibited heme-binding properties either in vivo by supplementing the expression medium with δ-ALA or in vitro by adding heme to the crude lysate containing the apoprotein. Finally, recombinant HbrL was found to interact with the promoter region of the genes it regulates and the DNA-binding properties were affected by the presence or absence of heme (Smart and Bauer 2006). Interestingly, out of seven strains from the laboratory freezer, only four displayed an intact HbrL-encoding gene (unpublished observation).

13.4.4.6 Irr

The iron response regulator (Irr) is a transcription factor related to the ferric-uptake regulator (Fur) family of metalloregulators. It has been extensively studied in *Bradyrhizobium japonicum* where it is a global regulator of iron homeostasis that acts in particular as a repressor of heme synthesis in iron-limited conditions. It is constitutively expressed and its activity mostly controlled by post-translational modification. Indeed, its degradation is triggered by binding to both redox states of heme: The ferric form binds to an N-terminal heme regulatory motif (HRM), while the ferrous form interacts with a histidine-rich domain (Small et al. 2009). In the genome of *R. capsulatus*, an ORF was annotated as a Fur transcription factor (rcc02670 in http://onco.img.cas.cz/rhodo and RRC01140 in http://www.ergo-light.com/). Analysis of the sequence revealed that it belongs to the Irr subgroup rather than to the Fur one stricto sensu, as seen from divergence in key residues involved in metal coordination in the Fur group (Rudolph et al. 2006). Compared to the well-characterized *B. japonicum* Irr, it does not carry any HRM, which indicates a presumably different mechanism of action. On the other hand, the DNA-binding helix is very well conserved. It is noteworthy that, among the α-proteobacteria subfamily, *R. capsulatus* is the only representative that carries Irr as the sole regulator involved in the iron-responsive network. The usual master regulators Fur and RirA are indeed missing (Rodionov et al. 2006). This fact indicates a potential major role of Irr in *R. capsulatus* regarding the management of iron metabolism. Preliminary studies have shown that an *irr*-deleted strain of *R. capsulatus* presents abnormal levels of heme, approximately 25% less and 25% more than the wild-type strain in aerobic and photosynthetic conditions, respectively (unpublished data).

13.5 The Porphynoid Branch

13.5.1 The Siroheme Sub-branch: The Other Fe Branch

Siroheme, the prosthetic group of various reductases, was reported to occur in *R. capsulatus* strain E1F1. A 17 kb region contains genes encoding an assimilatory nitrate reduction system, which consists of (i) putative regulatory genes *nsrR* and *nasTS*; (ii) an ABC-type nitrate transporter coded by *nasFED*; (iii) genes coding for the apo-nitrate and apo-nitrite reductases, *nasA* and *nasB*; (iv) a gene encoding a siroheme synthase, *cysG*, responsible for synthesis of the nitrite reductase cofactor (Pino et al. 2006). Although the siroheme synthase was not biochemically studied, the nitrite reductase NasB has been heterologously expressed in *E. coli* and purified recombinant protein exhibited a siroheme spectral signature (Olmo-Mira et al. 2006). CysG (EC 2.1.1.107) was characterized in *Salmonella enterica* and in *E. coli* where it was shown to be a bimodular homodimer that catalyzes the four reactions that converts uro'gen III into siroheme (Stroupe et al. 2003).

13.5.2 The Corrin Sub-branch: From Uro'gen III to VitB$_{12}$

13.5.2.1 An Unusual Pathway

Cobalamin synthesis requires more than 30 enzymatic steps with genes that encode these enzymes taking up approximately 1% of a typical bacterial genome (Roth et al. 1993; Martens et al. 2002). There are two distinct cobalamin biosynthesis pathways that exist in different microorganisms: the well-studied aerobic pathway, represented by *Pseudomonas denitrificans*, and the partially resolved anaerobic pathway, represented by *Salmonella typhimurium, Bacillus megaterium,* and *Propionibacterium shermanii* (Scott and Roessner 2002). The two pathways diverge at precorrin-2, a compound synthesized by dimethylation of uro'gen III (Raux et al. 1999). These two pathways primarily differ in their requirement for molecular oxygen and the timing of cobalt insertion. In the aerobic pathway, molecular oxygen is incorporated before the ring contraction, where the carbon bridge (C20) between the rings A and D is excised, and cobalt chelation occurs nine steps after the synthesis of precorrin-2. In contrast, in the anaerobic pathway, cobalt is inserted directly into precorrin-2. Ring contraction is likely facilitated by the different oxidation states of the cobalt ion (Martens et al. 2002; Warren et al. 2002). The existence of distinct cobalamin synthesis pathways is also reflected at the genetic level with organisms containing genes unique to one pathway or the other. Interestingly, *R. capsulatus* carries genetic hallmarks for an aerobic pathway although cobalamin synthesis occurs under both aerobic and anaerobic conditions (McGoldrick et al. 2002). A *cobG* gene, encoding a mono-oxygenase, is missing in *R. capsulatus.* This mono-oxygenase catalyzes the contraction of the corrin ring, which is a characteristic reaction in the aerobic pathway. Instead, an isofunctional enzyme, CobZ, was identified (McGoldrick et al. 2005), which may mediate the ring contraction process under both aerobic and anaerobic conditions. Details of enzymes at additional steps of this pathway are beyond the scope of this review but can be obtained from a recent review of this complex branch in the pathway by Warren (Warren and Deery 2009).

13.5.2.2 *cob* Gene Regulation

Riboswitch: Regulation at the Post-transcriptional Level

The regulation of cobalamin biosynthesis is mostly reported to occur post-transcriptionally via an RNA structure called a riboswitch. Riboswitches are metabolite-binding regions located within the 5′-UTR of messenger RNAs that function as regulatory elements for target genes (Mandal et al. 2003). Riboswitches are widespread in prokaryotes and interact with various target metabolites. In *B. subtilis*, seven different metabolites, including guanine, adenine, lysine, thiamine pyrophosphate, FMN, SAM, and adenosylcobalamin, regulate at least 68 genes via a riboswitch mechanism (Mandal et al. 2003). In cobalamin biosynthesis, adenosylcobalamin is known to be an effector regulating translation of several cobalamin genes (Nahvi et al. 2002; Rodionov et al. 2003). A B_{12}-binding element exists within the 5′-UTR of *btuB* mRNA in *E. coli* and within the 5′-UTR of cobalamin biosynthesis operon in *S. typhimurium* that binds adenosylcobalamin (Nahvi et al. 2004). At elevated adenosylcobalamin concentrations, the translation of cobalamin transport protein (BtuB) and *cob* operon is repressed by adenosylcobalamin binding to the B_{12} box (Nahvi et al. 2004). Such interactions induce a conformational change in the mRNA secondary structure at the B_{12} box that prevents ribosome binding (Nou and Kadner 2000). Rodionov et al. (2003) employed the combination of a comparative approach of gene regulation, positional clustering, and phylogenetic profiling for identifying cobalamin-regulated biosynthesis/transport genes. They identified a conserved RNA structure called the B_{12} element that is widely distributed in eubacteria. For example, 13 B_{12} elements are present in UTR regions of the *R. capsulatus* genome with potential target genes, including *cobW*, *cbiMNQO*, *btuFCD*, *btuD*, and *btuB*, as well as upstream of several non-cobalamin biosynthesis/transport genes (Rodionov et al. 2003).

Transcriptional Regulation

Examples of cobalamin genes regulated at the transcriptional level are not abundant. In *S. typhimurium*, the global regulators Crp/cAMP and ArcA/ArcB are responsible for indirect redox and carbon controls of cobalamin biosynthesis by controlling synthesis of the positive regulatory protein, PocR. PocR subsequently uses propanediol as an inducer of *cob* gene expression (Rondon and Escalante-Semerena 1992, 1996). Other global regulators directly regulating *cob* genes are not known at present.

Recently, we observed that the global RegA–RegB signal transduction cascade not only controls expression of the Bchl and heme branches but also controls expression of enzymes at two steps in the cobalamin branch, the *cobK* gene and the *cobWNHIJ* operon (Li 2009). Thus, expression of cobalamin enzymes appears to be regulated by feedback control via a riboswitch mechanism as well as in response to redox via the RegA–RegB signal transduction cascade.

13.6 The Bchl Branch: From Proto IX to Bchl

13.6.1 The Bchl Biosynthesis Gene Cluster

Analysis of genes involved in the Mg branch of the tetrapyrrole biosynthesis pathway was primarily advanced by the work of Marrs and coworkers who used a generalized transducing agent (GTA) from *R. capsulatus* to map the location of Bchl biosynthesis genes (Marrs 1974; Yen and Marrs 1976). These studies established that carotenoid and Bchl biosynthesis genes were present in the *R. capsulatus* chromosome in a tight linkage order. The transduction mapping studies were followed by the isolation of an R′ plasmid that was shown by marker rescue and transposition mapping analyses to contain all of the essential genetic information needed to synthesize Bchl (Marrs 1981; Biel and Marrs 1983; Taylor et al. 1983; Zsebo and Hearst 1984). Structural and functional information regarding these loci was refined by complete sequence analysis of the R′ by Hearst and coworkers (Alberti et al. 1995) and by construction of defined interposon mutations in each of the sequenced open reading frames (Giuliano et al. 1988; Young et al. 1989; Yang and Bauer 1990; Bollivar et al. 1994a, b). Similar clustering of Bchl biosynthesis genes occurs in other purple bacterial species (Igarashi et al. 2001; Jaubert et al. 2004).

Several of the enzymes encoded by one or more of the Bchl biosynthesis genes have been expressed in *E. coli* and shown to exhibit enzymatic activity at specific steps of the Mg branch. The first committed step of the Mg branch involves insertion of Mg into proto IX by the enzyme Mg chelatase (EC 6.6.1.1) to form Mg–proto IX. Gibson et al. (1995) were the first to successfully heterologously express subunits of Mg chelatase encoded by the *bchH*, *bchI*, and *bchD* genes in *E. coli* to definitively establish the subunit composition of this enzyme. Activity is dependent on Mg, ATP, and proto IX. The multimeric Mg chelatase is a typical class 1 metal chelatase (Schubert et al. 2002) where BchH binds proto IX (Gibson et al. 1995) while the structures of BchI and BchD show that they both have ATPase domains of the AA+ class (Fodje et al. 2001).

The next step of the pathway involves SAM-dependent methylation of the carboxyl group of Mg–proto IX by the enzyme SAM Mg–proto IX-*O*-methyltransferase (EC 2.1.1.11) to form Mg–proto IX monomethyl ester (MPE). Confirmation that the *R. capsulatus bchM* encodes this enzyme occurred when this gene was heterologously expressed in *E. coli* with cell lysate extracts exhibiting activity (Bollivar et al. 1994a, b).

Oxidative cyclization to form the fifth ring of Bchl is catalyzed by the enzyme MPE oxidative cyclase (EC 1.14.13.81) that is coded by the *bchE* gene to form the product protochlorophyllide (PChlide). This enzyme has not been well characterized biochemically owing to the difficulty in observing activity in vitro. However, analysis of the *R. capsulatus* BchE peptide sequence indicates that it may be a SAM enzyme that also uses vitB_{12} as a cofactor (Gough et al. 2000). Indeed depletion of vitB_{12} in vivo leads to accumulation of MPE in this branch of the pathway (Gough et al. 2000).

Reduction of the D ring of PChlide to form chlorophyllide (Chlide) occurs by the enzyme NADPH–PChlide oxidoreductase (EC 1.3.1.33). In nature there are two unrelated enzymes that can catalyze this reduction. One is a light-dependent version that is present in cyanobacteria, alga, gymnosperms, and angiosperms, while the other is light-independent and present in anoxygenic bacteria, cyanobacteria, alga, and gymnosperms (Fujita and Bauer 2003; Heyes and Hunter 2005). The dark operative form is present in *R. capsulatus*, encoded by *bchL*, *bchN*, and *bchB*. Interestingly, it has a high degree of primary sequence similarity to nitrogenase

(Fujita and Bauer 2003). Biochemical characterization of the *R. capsulatus* enzyme indicates that it contains iron sulfur centers and requires ATP, ferredoxin, and a reducing agent for catalysis (Fujita and Bauer 2003; Nomata et al. 2005). Reduction of ring B of Chlide by the enzyme Chlide a reductase (EC 1.3.1.75), encoded by *bchX*, *bchY*, and *bchZ*, also uses an enzyme that is very similar to dark PChlide reductase (Nomata et al. 2006).

There are several latter steps of the pathway that have not been biochemically characterized. The one exception is the esterification of the propionate on ring IV by the enzyme Bchl *a* synthase, encoded by *bchG*. This enzyme is membrane bound and difficult to isolate. However, it was heterologously expressed in *E. coli* with membrane fractions shown to exhibit this activity (Oster et al. 1997).

13.6.2 Regulation of Bchl Biosynthesis Gene Expression

13.6.2.1 CrtJ/PpsR

Studies have demonstrated that CrtJ/PpsR is an aerobic repressor of Bchl genes, *bch*, in *R. capsulatus* (Ponnampalam and Bauer 1997; Elsen et al. 1998). As discussed below, anaerobic induction of Bchl, carotenoid, and light-harvesting genes also requires phosphorylated RegA. So together, CrtJ and RegA regulate synthesis of Bchl biosynthesis genes by coordinating aerobic repression and anaerobic activation, respectively (Fig. 13.2).

CrtJ cooperatively binds to two copies of the palindromic sequence TGT-N_{12}-ACA that is present in all of the characterized *bch* promoters (Alberti et al. 1995; Ponnampalam and Bauer 1997; Elsen et al. 1998; Ponnampalam et al. 1998). The palindromic sequence is found either 8 bp apart, or at sites that are distantly separated. For example, the *R. capsulatus bchC* promoter region has a CrtJ recognition palindrome that spans the –35 promoter region and a second CrtJ palindrome located 8 bp away that spans the –10 promoter region (Ponnampalam et al. 1998). Binding to these two palindromes is cooperative so if the 8 bp space between the two palindromes in the *bchC* promoter region is altered by the addition or deletion of just a few base pairs, then CrtJ is unable to bind to either palindrome effectively (Ponnampalam et al. 1998).

CrtJ also cooperatively binds to two palindromes at other promoters but these palindromes are separated by more than 100–150 bp (Elsen et al. 1998). An example of this type of binding occurs in the intergenic region between *crtA* and *crtI* that contains two promoters, one that is responsible for driving expression of the *crtA–bchI–bchD* operon and a second divergent promoter >100 bp away that is responsible for expression of the *crtI–crtB* operon (Elsen et al. 1998). The promoter for the *crtA–bchI—bchD* transcript has a single CrtJ-binding site that spans the –10 promoter sequences while the *crtI–crtB* promoter also has a single CrtJ recognition sequence that spans the –35 recognition sequence. Cooperative binding of CrtJ to these two palindromes coordinately represses expression of both the *crtA–bchI–bchD* and *crtI–crtB* operons (Elsen et al. 1998). This affects synthesis of both Bchl and carotenoids since BchI and BchD are subunits of Mg chelatase (Bollivar et al. 1994a, b) and *crtI* and *crtB* code for phytoene dehydrogenase and phytoene synthase that are enzymes for the first two committed steps of carotenoid biosynthesis, respectively (Armstrong et al. 1990). Presumably, binding to distant sites involves looping of the DNA so that tetrameric CrtJ can bind cooperatively to both of the recognition palindromes (Elsen et al. 1998).

13.6.2.2 RegA–RegB

The Mg branch of the tetrapyrrole biosynthetic pathway is also anaerobically activated by the global two-component system RegA–RegB (Willett et al. 2007). Specifically, expression of the large *bchEJG-orf428-bchP-idi* operon that encodes numerous enzymes in Bchl biosynthesis, as well as an enzyme involved in carotenoid biosynthesis, requires phosphorylated RegA for maximal expression (Bollivar et al. 1994a, b; Alberti et al. 1995; Hahn et al. 1996; Suzuki et al. 1997; Bollivar 2006; Willett et al. 2007). In addition, phosphorylated RegA is also required for expression of the *crtA–bchIDO* operon that codes for early enzymes involved in the Bchl branch as well as early enzymes in the carotenoid biosynthesis pathway (Alberti et al. 1995).

13.7 Co-dependent Syntheses of Tetrapyrrole End Products

The trifurcated tetrapyrrole biosynthetic pathways not only share common early intermediates but also form an intricate network where the end products cobalamin and heme are involved in each others and Bchl biosyntheses. For example, in the Bchl branch, Gough et al. (2000) identified a cyclase encoded by *bchE* that catalyzes the conversion of MPE to PChlide. The sequence similarity between this MPE-cyclase and a cobalamin-dependent P-methylase from *Streptomyces hygroscopicus* indicates an involvement of cobalamin in MPE-cyclase activity. Although no in vitro cyclase assays have been reported, MPE-cyclase activity was demonstrated to be cobalamin dependent in vivo as cobalamin depletion or a *bchE* knockout results in an accumulation of MPE.

In addition to the dependence on cobalamin for Bchl biosynthesis, the synthesis of cobalamin is dependent on the presence of heme. Specifically, one of the most complicated steps in the *R. capsulatus* cobalamin branch is the corrin ring contraction that is catalyzed by the cofactor-rich enzyme, CobZ. CobZ contains heme as a cofactor as well as flavin and two Fe–S centers (McGoldrick et al. 2005). The dependence on heme as a cofactor of CobZ thus makes cobalamin synthesis heme dependent.

Finally, synthesis of heme is also cobalamin dependent as heme synthesis requires *S*-adenosylmethionine as a methyl group donor and its synthesis involves a cobalamin-dependent enzyme (Drennan et al. 1994; Layer et al. 2003; McGoldrick et al. 2005). The interdependent relationship among all three tetrapyrrole compounds as discussed in McGoldrick et al. (2005) suggests that codependence of these different tetrapyrrole branches on different tetrapyrrole end products is a form of regulation used to coordinate the levels of these different tetrapyrrole derivatives.

13.8 Conclusion

The tetrapyrrole biosynthetic pathway of *R. capsulatus* is quite complex involving several branches that are responsible for synthesis of heme, vitB$_{12}$, and Bchl *a*. The flow of tetrapyrroles into each branch in the trifurcated pathway is highly regulated with an intricate network of transcriptional and post-transcriptional events controlling the synthesis of these various end products. This includes the control of enzyme activity by interaction with tetrapyrrole end products and transcriptional control either in response to redox poise or in response to an interaction with a tetrapyrrole. There is also a riboswitch control of translation via interaction of mRNA with a tetrapyrrole. Many players in this complex regulatory web have been identified, although it will not be surprising to find that there are additional players involved in controlling this complex pathway.

Among many challenges going forward will be to determine how multiple transcription factors coordinately regulate the synthesis of individual enzymes in the pathway. For example, the *hemC* promoter appears to be controlled by at least four different activators/repressors. Where these factors bind to the *hemC* promoter and how binding of individual transcription factors affects binding of other transcription factors remain to be determined. The study of this complex pathway has been an ongoing for over five decades and, owing to its complexity, will likely remain an area of focus for some time.

References

Alberti M, Burke DH, Hearst JE (1995) Anoxygenic photosynthetic bacteria. Kluwer Academic Publishers, Dordrecht

Armstrong GA, Schmidt A, Sandmann G et al. (1990) Genetic and biochemical characterization of carotenoid biosynthesis mutants of *Rhodobacter capsulatus*. J Biol Chem 265:8329–8338

Astner I, Schulze JO, van den Heuvel J et al (2005) Crystal structure of 5-aminolevulinate synthase, the first enzyme of heme biosynthesis, and its link to XLSA in humans. EMBO J 24:3166–3177

Avissar YJ, Moberg PA (1995) The common origins of the pigments of life – early steps of chlorophyll biosynthesis. Photosynth Res 44:221–242

Biel AJ, Marrs BL (1983) Transcriptional regulation of several genes for bacteriochlorophyll biosynthesis in *Rhodopseudomonas capsulata* in response to oxygen. J Bacteriol 156:686–694

Biel AJ (1991) Characterization of a coproporphyrin-protein complex from *Rhodobacter capsulatus*. FEMS Microbiol Lett 65:43–47

Biel AJ (1992) Oxygen-regulated steps in the *Rhodobacter capsulatus* tetrapyrrole biosynthetic pathway. J Bacteriol 174:5272–5274

Biel AJ, Canada K, Huang D et al (2002) Oxygen-mediated regulation of porphobilinogen formation in *Rhodobacter capsulatus*. J Bacteriol 184:1685–1692

Biel SW, Wright MS, Biel AJ (1988) Cloning of the *Rhodobacter capsulatus hemA* gene. J Bacteriol 170:4382–4384

Biel SW, Biel AJ (1990) Isolation of a *Rhodobacter capsulatus* mutant that lacks *c*-type cytochromes and excretes porphyrins. J Bacteriol 172:1321–1326

Bollivar DW, Bauer CE (1992) Association of tetrapyrrole intermediates in the bacteriochlorophyll a biosynthetic pathway with the major outer-membrane porin protein of *Rhodobacter capsulatus*. Biochem J 282:471–476

Bollivar DW, Jiang ZY, Bauer CE et al (1994a) Heterologous expression of the *bchM* gene product from *Rhodobacter capsulatus* and demonstration that it encodes S-adenosyl-L-methionine:Mg-protoporphyrin IX methyltransferase. J Bacteriol 176:5290–5296

Bollivar DW, Suzuki JY, Beatty JT et al (1994b) Directed mutational analysis of bacteriochlorophyll a biosynthesis in *Rhodobacter capsulatus*. J Mol Biol 237:622–640

Bollivar DW, Clauson C, Lighthall R et al (2004) *Rhodobacter capsulatus* porphobilinogen synthase, a high activity metal ion independent hexamer. BMC Biochem. doi: 10. 1186/1471-2091-5-17

Bollivar DW (2006) Recent advances in chlorophyll biosynthesis. Photosynth Res 90:173–194

Cohen-Bazire G, Sistrom WR, Stanier RY (1957) Kinetic studies of pigment synthesis by non-sulfur purple bacteria. J Cell Physiol 49:25–68

Cooper R (1956) The production of coproporphyrin precursors by a *Rhodopseudomonas* sp. Biochem J 63:25

Cooper R (1963) The biosynthesis of coproporphyrinogen, magnesium protoporphyrin monomethyl ester and bacteriochlorophyll by *Rhodopseudomonas capsulata*. Biochem J 89:100–108

Dailey HA (2002) Terminal steps of haem biosynthesis. Biochem Soc Trans 30:590–595

Dong C, Elsen S, Swem LR et al (2002) AerR, a second aerobic repressor of photosynthesis gene expression in *Rhodobacter capsulatus*. J Bacteriol 184:2805–2814

Drennan CL, Huang S, Drummond JT et al (1994) How a protein binds B_{12}: A 3.0 A X-ray structure of B12-binding domains of methionine synthase. Science 266:1669–1674

Elsen S, Ponnampalam SN, Bauer CE (1998) CrtJ bound to distant binding sites interacts cooperatively to aerobically repress photopigment biosynthesis and light harvesting II gene expression in *Rhodobacter capsulatus*. J Biol Chem 273:30762–30769

Elsen S, Swem LR, Swem DL et al (2004) RegB/RegA, a highly conserved redox-responding global two-component regulatory system. Microbiol Mol Biol Rev 68:263–279

Fodje MN, Hansson A, Hansson M et al (2001) Interplay between an AAA module and an integrin I domain may regulate the function of magnesium chelatase. J Mol Biol 311:111–122

Frankenberg N, Moser J, Jahn D (2003) Bacterial heme biosynthesis and its biotechnological application. Appl Microbiol Biotechnol 63:115–127

Frère F, Reents H, Schubert WD et al (2005) Tracking the evolution of porphobilinogen synthase metal dependence in vitro. J Mol Biol 345:1059–1070

Fujita Y, Bauer CE (2003) The light-independent protochlorophyllide reductase: a nitrogenase-like enzyme catalyzing a key reaction for greening in the dark. In: Smith K, Kadish K, Guilard R (eds) The porphyrin handbook. Academic, Amsterdam, the Netherlands

Gibson LC, Willows RD, Kannangara CG et al (1995) Magnesium-protoporphyrin chelatase of *Rhodobacter sphaeroides*: reconstitution of activity by combining the products of the *bchH*, *-I*, and *-D* genes expressed in *Escherichia coli*. Proc Natl Acad Sci USA 92:1941–1944

Giuliano G, Pollock D, Stapp H et al (1988) A genetic-physical map of the *Rhodobacter capsulatus* carotenoid biosynthesis gene cluster. Mol Gen Genet MGG 213:78–83

Gough SP, Petersen BO, Duus JO (2000) Anaerobic chlorophyll isocyclic ring formation in *Rhodobacter capsulatus* requires a cobalamin cofactor. Proc Natl Acad Sci USA 97: 6908–6913

Hahn FM, Baker JA, Poulter CD (1996) Open reading frame 176 in the photosynthesis gene cluster of *Rhodobacter capsulatus* encodes *idi*, a gene for isopentenyl diphosphate isomerase. J Bacteriol 178:619–624

Heinemann IU, Jahn M, Jahn D (2008) The biochemistry of heme biosynthesis. Arch Biochem Biophys 474:238–251

Heyes DJ, Hunter CN (2005) Making light work of enzyme catalysis: protochlorophyllide oxidoreductase. Trends Biochem Sci 30:642–649

Igarashi N, Harada J, Nagashima S et al (2001) Horizontal transfer of the photosynthesis gene cluster and operon rearrangement in purple bacteria. J Mol Evol 52:333–341

Ineichen G, Biel AJ (1995) Nucleotide sequence of the *Rhodobacter capsulatus hemE* gene. Plant Physiol 108:423

Jaffe EK (2003) An unusual phylogenetic variation in the metal ion binding sites of porphobilinogen synthase. Chem Biol 10:25–34

Jaubert M, Zappa S, Fardoux J et al (2004) Light and redox control of photosynthesis gene expression in *Bradyrhizobium*: dual roles of two PpsR. J Biol Chem 279:44407–44416

Kanazireva E, Biel AJ (1995) Cloning and overexpression of the *Rhodobacter capsulatus hemH* gene. J Bacteriol 177:6693–6694

Kanazireva E, Biel AJ (1996) Nucleotide sequence of the *Rhodobacter capsulatus hemH* gene. Gene 170:149–150

Koch M, Breithaupt C, Kiefersauer R et al (2004) Crystal structure of protoporphyrinogen IX oxidase: a key enzyme in haem and chlorophyll biosynthesis. EMBO J 23:1720–1728

Layer G, Moser J, Heinz DW et al (2003) Crystal structure of coproporphyrinogen III oxidase reveals cofactor geometry of Radical SAM enzymes. EMBO J 22:6214–6224

Layer G, Kervio E, Morlock G et al (2005) Structural and functional comparison of HemN to other radical SAM enzymes. Biol Chem 386:971–980

Li K (2009) Regulation of and by cobalamin in *Rhodobacter capsulatus*. PhD thesis, Indiana University, BLoomington, Indiana

Maddocks SE, Oyston PCF (2008) Structure and function of the LysR-type transcriptional regulator (LTTR) family proteins. Microbiology 154:3609–3623

Mandal M, Boese B, Barrick JE et al (2003) Riboswitches control fundamental biochemical pathways in *Bacillus subtilis* and other bacteria. Cell 113:577–586

Marrs B (1974) Genetic recombination in *Rhodopseudomonas capsulata*. Proc Natl Acad Sci USA 71:971–973

Marrs B (1981) Mobilization of the genes for photosynthesis from *Rhodopseudomonas capsulata* by a promiscuous plasmid. J Bacteriol 146:1003–1012

Martens JH, Barg H, Warren MJ et al (2002) Microbial production of vitamin B_{12}. Appl Microbiol Biotechnol 58:275–285

Masoumi A, Heinemann IU, Rohde M et al (2008) Complex formation between protoporphyrinogen IX oxidase and ferrochelatase during haem biosynthesis in *Thermosynechococcus elongatus*. Microbiology 154:3707–3714

McGoldrick H, Deery E, Warren M et al (2002) Cobalamin (vitamin B(12)) biosynthesis in *Rhodobacter capsulatus*. Biochem Soc Trans 30:646–648

McGoldrick HM, Roessner CA, Raux E et al (2005) Identification and characterization of a novel vitamin B_{12} (cobalamin) biosynthetic enzyme (CobZ) from *Rhodobacter capsulatus*, containing flavin, heme, and Fe-S cofactors. J Biol Chem 280:1086–1094

Nahvi A, Sudarsan N, Ebert MS et al (2002) Genetic control by a metabolite binding mRNA. Chem Biol 9:1043–1049

Nahvi A, Barrick JE, Breaker RR (2004) Coenzyme B_{12} riboswitches are widespread genetic control elements in prokaryotes. Nucleic Acids Res 32:143–150

Nandi DL, Shemin D (1973) δ-Aminolevulinic acid dehydratase of *Rhodopseudomonas capsulata*. Arch Biochem Biophys 158:305–311

Nomata J, Swem LR, Bauer CE et al (2005) Overexpression and characterization of dark-operative protochlorophyllide reductase from *Rhodobacter capsulatus*. Biochim Biophys Acta 1708:229–237

Nomata J, Mizoguchi T, Tamiaki H et al (2006) A second nitrogenase-like enzyme for bacteriochlorophyll biosynthesis: reconstitution of chlorophyllide a reductase with purified X-protein (BchX) and YZ-protein (BchY-BchZ) from *Rhodobacter capsulatus*. J Biol Chem 281: 15021–15028

Nou X, Kadner RJ (2000) Adenosylcobalamin inhibits ribosome binding to *btuB* RNA. Proc Natl Acad Sci USA 97:7190–7195

Olmo-Mira MF, Cabello P, Pino C et al (2006) Expression and characterization of the assimilatory NADH-nitrite reductase from the phototrophic bacterium *Rhodobacter capsulatus* E1F1. Arch Microbiol 186:339–344

Oster U, Bauer CE, Rudiger W (1997) Characterization of chlorophyll a and bacteriochlorophyll a synthases by heterologous expression in *Escherichia coli*. J Biol Chem 272:9671–9676

Ouchane S, Picaud M, Therizols P et al (2007) Global regulation of photosynthesis and respiration by FnrL: the first two targets in the tetrapyrrole pathway. J Biol Chem 282:7690–7699

Panek H, O'Brian MR (2002) A whole genome view of prokaryotic haem biosynthesis. Microbiology 148:2273–2282

Petricek M, Petrickova K, Havlicek L et al (2006) Occurrence of two 5-aminolevulinate biosynthetic pathways in *Streptomyces nodosus* subsp. *asukaensis* is linked with the production of asukamycin. J Bacteriol 188:5113–5123

Pino C, Olmo-Mira F, Cabello P et al (2006) The assimilatory nitrate reduction system of the phototrophic bacterium *Rhodobacter capsulatus* E1F1. Biochem Soc Trans 34:127–129

Ponnampalam SN, Buggy JJ, Bauer CE (1995) Characterization of an aerobic repressor that coordinately regulates bacteriochlorophyll, carotenoid, and light harvesting-II expression in *Rhodobacter capsulatus*. J Bacteriol 177:2990–2997

Ponnampalam SN, Bauer CE (1997) DNA binding characteristics of CrtJ. A redox-responding repressor of bacteriochlorophyll, carotenoid, and light harvesting-II gene expression in *Rhodobacter capsulatus*. J Biol Chem 272:18391–18396

Ponnampalam SN, Elsen S, Bauer CE (1998) Aerobic repression of the *Rhodobacter capsulatus bchC* promoter involves cooperative interactions between CrtJ bound to neighboring palindromes. J Biol Chem 273:30757–30761

Raux E, Schubert HL, Roper JM et al (1999) Vitamin B_{12}: insights into biosynthesis's mount improbable. Bioorg Chem 27:100–118

Rodionov DA, Vitreschak AG, Mironov AA et al (2003) Comparative genomics of the vitamin B_{12} metabolism and regulation in prokaryotes. J Biol Chem 278:41148–41159

Rodionov DA, Gelfand MS, Todd JD et al (2006) Computational reconstruction of iron- and manganese-responsive transcriptional networks in alpha-proteobacteria. PLoS Comput Biol. doi: 10.1371/journal.pcbi.0020163

Rompf A, Hungerer C, Hoffmann T et al (1998) Regulation of *Pseudomonas aeruginosa hemF* and *hemN* by the dual action of the redox response regulators Anr and Dnr. Mol Microbiol 29:985–997

Rondon MR, Escalante-Semerena JC (1992) The *poc* locus is required for 1,2-propanediol-dependent transcription of the cobalamin biosynthetic (*cob*) and propanediol utilization (*pdu*) genes of *Salmonella typhimurium*. J Bacteriol 174:2267–2272

Rondon MR, Escalante-Semerena JC (1996) In vitro analysis of the interactions between the PocR regulatory protein and the promoter region of the cobalamin biosynthetic (*cob*) operon of *Salmonella typhimurium* LT2. J Bacteriol 178:2196–2203

Roth JR, Lawrence JG, Rubenfield M et al (1993) Characterization of the cobalamin (vitamin B_{12}) biosynthetic genes of *Salmonella typhimurium*. J Bacteriol 175:3303–3316

Rudolph G, Hennecke H, Fischer HM (2006) Beyond the Fur paradigm: iron-controlled gene expression in rhizobia. FEMS Microbiol Rev 30:631–648

Schobert M, Jahn D (2002) Regulation of heme biosynthesis in non-phototrophic bacteria. J Mol Microbiol Biotechnol 4:287–294

Schubert HL, Raux E, Matthews MA et al (2002) Structural diversity in metal ion chelation and the structure of uroporphyrinogen III synthase. Biochem Soc Trans 30:595–600

Schulze JO, Schubert WD, Moser J et al (2006) Evolutionary relationship between initial enzymes of tetrapyrrole biosynthesis. J Mol Biol 358:1212–1220

Scott AI, Roessner CA (2002) Biosynthesis of cobalamin (vitamin B(12)). Biochem Soc Trans 30:613–620

Shoolingin-Jordan PM (1995) Porphobilinogen deaminase and uroporphyrinogen III synthase: structure, molecular biology, and mechanism. J Bioenerg Biomembr 27:181–195

Small SK, Puri S, O'Brian MR (2009) Heme-dependent metalloregulation by the iron response regulator (Irr) protein in *Rhizobium* and other Alpha-proteobacteria. Biometals 22:89–97

Smart JL, Willett JW, Bauer CE (2004) Regulation of hem gene expression in *Rhodobacter capsulatus* by redox and photosystem regulators RegA, CrtJ, FnrL, and AerR. J Mol Biol 342:1171–1186

Smart JL, Bauer CE (2006) Tetrapyrrole biosynthesis in *Rhodobacter capsulatus* is transcriptionally regulated by the heme-binding regulatory protein, HbrL. J Bacteriol 188:1567–1576

Stroupe ME, Leech HK, Daniels DS et al (2003) CysG structure reveals tetrapyrrole-binding features and novel regulation of siroheme biosynthesis. Nat Struct Biol 10:1064–1073

Suzuki JY, Bollivar DW, Bauer CE (1997) Genetic analysis of chlorophyll biosynthesis. Annu Rev Genet 31:61–89

Taylor DP, Cohen SN, Clark WG et al (1983) Alignment of genetic and restriction maps of the photosynthesis region of the *Rhodopseudomonas capsulata* chromosome by a conjugation-mediated marker rescue technique. J Bacteriol 154:580–590

Warren MJ, Raux E, Schubert HL et al (2002) The biosynthesis of adenosylcobalamin (vitamin B_{12}). Nat Prod Rep 19:390–412

Warren MJ, Deery E. 2009. Vitamin B_{12} (cobalamin) biosynthesis in the purple bacteria. In: Hunter CN, Daldal F, Thurnauer MC, Beatty JT (eds) Advances in Photosynthesis and Respiration, vol 28. Springer, Dordrecht

Willett J, Smart JL, Bauer CE (2007) RegA control of bacteriochlorophyll and carotenoid synthesis in *Rhodobacter capsulatus*. J Bacteriol 189:7765–7773

Wright MS, Cardin RD, Biel AJ (1987) Isolation and characterization of an aminolevulinate-requiring *Rhodobacter capsulatus* mutant. J Bacteriol 169:961–966

Wright MS, Eckert JJ, Biel SW et al (1991) Use of a lacZ fusion to study transcriptional regulation of the *Rhodobacter capsulatus hemA* gene. FEMS Microbiol Lett 62:339–342

Wu J, Bauer CE (2008) RegB/RegA, A global redox-responding two-component regulatory system. In: Utsumi R (ed) Advances in experimental medicine and biology, vol 631. Landes Bioscience Eurekah, Georgetown

Yang ZM, Bauer CE (1990) *Rhodobacter capsulatus* genes involved in early steps of the bacteriochlorophyll biosynthetic pathway. J Bacteriol 172:5001–5010

Yen HC, Marrs B (1976) Map of genes for carotenoid and bacteriochlorophyll biosynthesis in *Rhodopseudomonas capsulata*. J Bacteriol 126:619–629

Young DA, Bauer CE, Williams JC et al (1989) Genetic evidence for superoperonal organization of genes for photosynthetic pigments and pigment-binding proteins in *Rhodobacter capsulatus*. Mol Gen Genet 218:1–12

Zsebo KM, Hearst JE (1984) Genetic-physical mapping of a photosynthetic gene cluster from *R. capsulata*. Cell 37:937–947

Part IV
Environmental Sensing and Signal Transduction

Chapter 14
The Gene Transfer Agent of *Rhodobacter capsulatus*

Molly M. Leung, Sarah M. Florizone, Terumi A. Taylor, Andrew S. Lang, and J. Thomas Beatty

Abstract When *Rhodobacter capsulatus* cultures enter the stationary phase of growth, particles of the gene transfer agent (RcGTA) are released from cells. The morphology of RcGTA resembles that of a small, tailed bacteriophage, with a protein capsid surrounding a ~4 kb linear, double-stranded fragment of DNA. However, the DNA present consists of random segments of the *R. capsulatus* genome, which may be transferred to another strain of *R. capsulatus*. The recipient in RcGTA-mediated gene transduction may acquire new alleles and thus express a new phenotype. The genes encoding the structural proteins of the RcGTA are clustered on the *R. capsulatus* chromosome, whereas genes that encode proteins that regulate the production of RcGTA are scattered around the chromosome. These regulatory proteins include a homoserine lactone synthase (GtaI) that produces a quorum-sensing signal, a two-component sensor-kinase protein (CckA), and a two-component response regulator protein (CtrA). We review the proposed evolutionary origin of RcGTA, as well as environmental and cellular factors involved in the induction of this unusual process of genetic exchange.

14.1 Horizontal Gene Transfer and Gene Transfer Agents

Horizontal gene transfer is one mechanism that allows bacteria to adapt to a changing environment and plays an important role in the evolution of prokaryotic genomes. In a recent study of 116 prokaryotic genomes, it was estimated that about 14% of the DNA was a result of horizontal gene transfer, where the proportion of horizontally transferred genes per genomes ranged from 0.05 to 25.2% (van Passel et al. 2005). Currently known mechanisms of horizontal gene transfer in bacteria include conjugation, transformation, and transduction. Transduction has been found to be important in the spread of virulence (Cheetham and Katz 1995) and antibiotic resistance (Davies 1994). A GTA has been defined as "a virus-like particle

J.T. Beatty (✉)
Department of Microbiology and Immunology, University of British Columbia, Vancouver, BC, Canada, V6T 1Z3
e-mail: jbeatty@interchange.ubc.ca

P.C. Hallenbeck (ed.), *Recent Advances in Phototrophic Prokaryotes*, Advances in Experimental Medicine and Biology 675, DOI 10.1007/978-1-4419-1528-3_14,

that only carries random pieces of the genome of the producing cell in a process similar to generalized transduction" (Lang and Beatty 2007). It is estimated that there are up to one billion viral particles per milliliter of marine water (Suttle 2005) and it is possible that a significant portion of these viral particles are gene transfer agents (GTAs). The first GTA was found in *Rhodobacter capsulatus* (Marrs 1974), and to avoid confusion, hereafter gene transfer agents in general will be called "GTAs," and the *R. capsulatus* gene transfer agent will be called "RcGTA." Other GTAs have been identified, including Dd1 in *Desulfovibrio desulfuricans* (Rapp and Wall 1987), VSH-1 in *Brachyspira hyodysenteriae* (Humphrey et al. 1997), VTA in *Methanococcus voltae* (Eiserling et al. 1999), and an RcGTA-like GTA in *Ruegeria* (formerly *Silicibacter*) *pomeroyi* (Biers et al. 2008). A summary of the properties of these GTAs is found in Table 14.1.

Table 14.1 Summary of GTA properties

Species	Gene transfer agent	Head diameter, tail length (nm)	DNA packaged (kb)	Size of coding region (kb)	References
R. capsulatus	RcGTA	30, 50	4.5	14.1	Yen et al. (1979)
B. hyodysenteriae	VSH-1	45, 64	7.5	16.3	Humphrey et al. (1997)
D. desulfuricans	Dd1	43, 7	13.6	n/a	Rapp and Wall (1987)
M. voltae	VTA	40, 61	4.4	n/a	Eiserling et al. (1999)
R. pomeroyi	GTA	~50–70, 0	n/a	14.7	Biers et al. (2008)

n/a, not known

14.2 Properties of RcGTA

Electron microscopy of the RcGTA particle revealed that it looks very similar to a tailed phage (Fig. 14.1) (Yen et al. 1979), and it was found that RcGTA transfers DNA in a mechanism similar to transduction. Thus, unlike conjugation RcGTA-mediated genetic exchange does not require direct contact between the donor cell and the recipient cell, and unlike transformation the extracellular DNA that is transferred is protected from DNase degradation (Marrs 1974). However, RcGTA differs from typically lytic tailed bacteriophages because it causes no observable lysis in *R. capsulatus*. Also, RcGTA particles contain ~4 kb of double-stranded DNA (Yen et al. 1979), which is insufficient to transfer the ~14-kb RcGTA gene cluster (Lang and Beatty 2000). Rather than specifically transferring its own genes, RcGTA randomly packages genomic DNA (Yen et al. 1979).

14.3 RcGTA Gene Clusters

Figure 14.2 gives a representation of the RcGTA gene cluster and surrounding genes. The RcGTA gene cluster consists of 15 ORFs. and about half of these ORFs have sequence similarity to phage elements such as a terminase, a prohead protease,

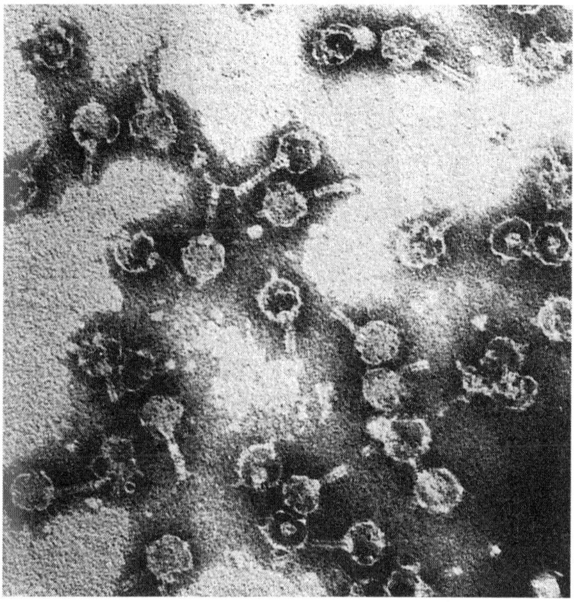

Fig. 14.1 Electron micrograph of RcGTA particles. Electron microscopy reveals that the RcGTA particle resembles a phage particle with a head of ~30 nm diameter and a tail of ~50 nm length (Yen et al. 1979)

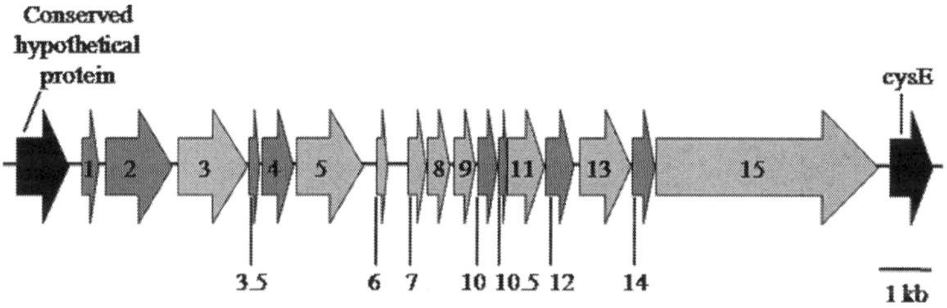

Fig. 14.2 Representation of the RcGTA gene cluster. Each ORF in the RcGTA gene cluster has been numbered and shaded in *gray*. Genes surrounding the RcGTA gene cluster are in *black*. In *light gray* are ORF-encoding gene products that have been identified by mass spectrometry of purified RcGTA particles (Chen et al. 2009). The following ORFs have sequence similarity to phage components: 2, terminase; 3, portal protein; 4, prohead protease; 5, capsid protein; 7, head–tail adaptor; 9, major tail protein; 11, tail tape measure; and 15, host specificity protein

a capsid protein, and a major tail protein (Lang and Beatty 2007). Recently, many of the proteins predicted to be present in RcGTA particles were found using a proteomics approach (Chen et al. 2009).

Although only *R. capsulatus* (Solioz et al. 1975) and *R. pomeroyi* (Biers et al. 2008) have been found to produce functional RcGTA-like particles, RcGTA-like gene homologs are widespread in α-proteobacteria (Lang et al. 2002; Lang and Beatty 2007; Biers et al. 2008; Paul 2008). The occurrence of RcGTA gene homologs and their organization relative to the RcGTA gene cluster can be separated into four major categories: (1) a complete RcGTA-like gene cluster containing genes 1–15 is present in a single location; (2) a partial and/or rearranged RcGTA-like gene cluster is present in a single location; (3) several RcGTA gene homologs are scattered in various locations in the genome; and (4) no detectable RcGTA gene homologs are present (Lang et al. 2002; Lang and Beatty 2007; Biers et al. 2008). When these categories are superimposed upon a 16S rDNA phylogenetic tree of selected representatives from the major orders of α-proteobacteria for which complete genome sequences have been determined, a correlation between the pattern of RcGTA-like gene distribution and the phylogenetic tree can be seen (Fig. 14.3). This suggests that RcGTA-like gene clusters descended along with the 16S rRNA genes from a common ancestor, with a high degree of conservation in the Rhodobacterales order and partial or complete loss in most other groups (Lang and Beatty 2007).

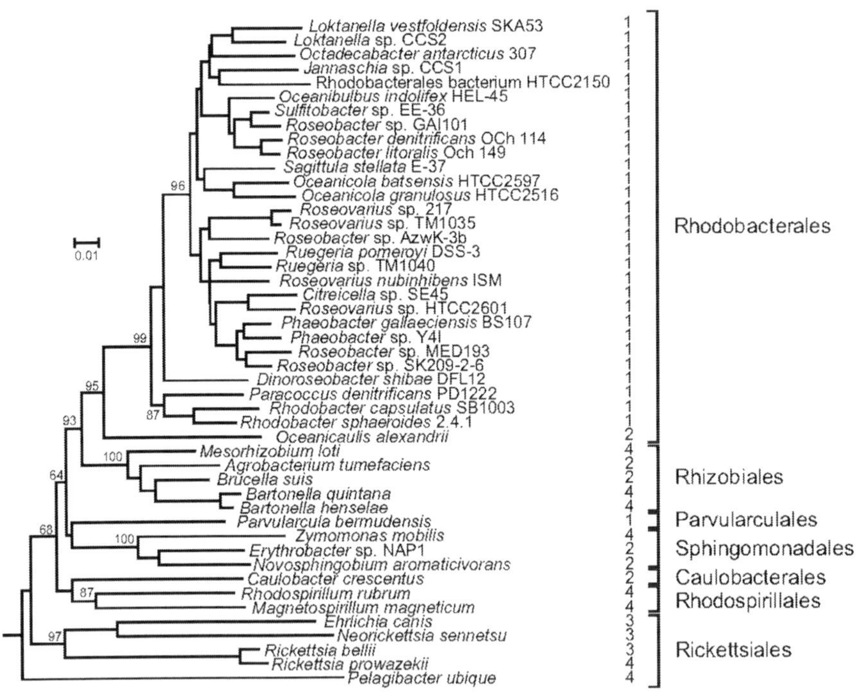

Fig. 14.3 A 16S rDNA phylogenetic tree of selected genome-sequenced α-proteobacteria, with categories of RcGTA-like gene cluster indicated by the numbers on the *right*: (1) a complete RcGTA-like gene cluster containing genes 1–15 is present in a single location; (2) a partial and/or a rearranged RcGTA-like gene cluster is present in a single location; (3) several RcGTA gene homologs are scattered in various locations in the genome; and (4) no detectable RcGTA gene homologs are present. Bootstrap values for the neighbor-joining tree are shown for the major lineages next to branches (percentages based on 10,000 replicates). The *scale bar* indicates the number of base substitutions per site

14.4 Regulation of the RcGTA Gene Cluster Transcription by Cellular Systems

14.4.1 CtrA

14.4.1.1 CtrA in *C. crescentus*

CtrA is a response regulator first discovered in *C. crescentus*, with homologs found in many bacteria including *Sinorhizobium meliloti* (Barnett et al. 2001), *Rhodopseudomonas palustris, Brucella abortus* (Bellefontaine et al. 2002), and *R. capsulatus* (Lang and Beatty 2000). In *C. crescentus*, CtrA acts as a master regulator of the cell cycle and the levels of this protein increase as a cell develops though its cell cycle (Laub et al. 2000, 2002). CtrA is involved in regulating processes such as polar morphogenesis, DNA replication initiation, DNA methylation, and cell division. Microarray analysis of *C. crescentus* has shown that CtrA regulates the expression of 144 genes, of which at least 95 are regulated directly (Laub et al. 2000, 2002). Fourteen of these genes are regulatory genes (including CtrA itself), 10 of which encode two-component system proteins. Analysis of the regulatory region of these 144 genes has revealed two alternative consensus sequences: one gapped (TTAA-N7-TTAAC) and one ungapped (TTAACCAT) (Laub et al. 2000, 2002). Footprinting experiments on several of these upstream sequences have shown that CtrA does in fact bind this sequence (Reisenauer et al. 1999). However, the binding of CtrA to DNA is still not completely understood, because CtrA can regulate genes that do not have this consensus sequence, and in other cases CtrA does not regulate genes that have this consensus sequence (Laub et al. 2002).

It is known that the phosphorylated form of CtrA (CtrA~P) binds DNA and this phosphorylation is tightly regulated by a phosphorelay system involving ChpT, CckA, and DivK. CtrA~P activity is controlled by proteolysis, which is thought to be triggered by dephosphorylation of CpdR; unphosphorylated CpdR stimulates CtrA proteolysis (Bowers et al. 2008). The phosphorylation state of DivK is regulated by DivJ (located at the stalk pole) and PleC (located at the swarmer pole). DivK is phosphorylated by DivJ and dephosphorylated by PleC, resulting in the accumulation of CtrA in the swarmer cell and the lack of CtrA in the stalk cells (Fig. 14.4).

CtrA in *C. crescentus* regulates its own transcription, and the *ctrA* gene has two promoters (P1 and P2), each containing one CtrA-binding site. A footprinting experiment showed that CtrA-binding sites are located in the –10 region of P1 (which contains an ungapped consensus sequence) and the –35 region of P2 (which contains a gapped consensus sequence) (Domian et al. 1999). The positioning of these binding sites in conjunction with mutational studies indicates that CtrA acts as a negative regulator of P1 and positive regulator of P2. P1 is active earlier than P2 in the cell cycle. When CtrA~P is at a low concentration in the cell, it will not bind P1, leaving it active, nor will it bind P2, leaving it inactive. But as CtrA~P accumulates in the cell with transcription driven by P1 (the weaker promoter), CtrA~P binds to

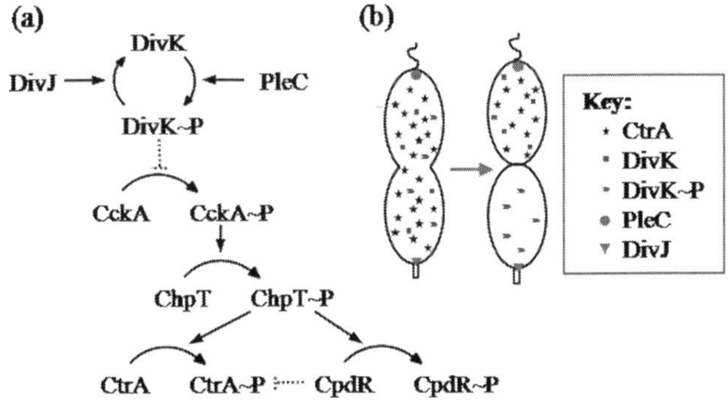

Fig. 14.4 A phosphorelay system regulates CtrA phosphorylation state and proteolysis that results in asymmetric cell division in *C. crescentus*. (**a**) *Curved arrows* represent phosphorylation and dephosphorylation events. *Straight arrows* represent the regulation of phosphoryl transfer events. *Dotted bars* represent inhibition. DivK~P inhibits phosphorylation of CckA by an unknown mechanism and unphosphorylated CpdR inhibits CtrA~P activity by stimulating CtrA proteolysis. (**b**) Prior to cell division (*left*), PleC is located at the swarmer pole and DivJ is located at the stalk pole but DivK, DivK~P, and CtrA are freely diffusible. After cytoplasmic compartmentalization (*right*), DivK~P accumulates at the stalk cell resulting in the proteolysis of CtrA, and DivK accumulates in the swarmer cell resulting in the accumulation of CtrA

and inactivates P1 and activates P2 (the stronger promoter), thus causing an increase in CtrA production (Domian et al. 1999).

14.4.1.2 CtrA in *R. capsulatus*

CtrA is essential for the viability of *C. crescentus* (Quon et al. 1996) and *S. meliloti* (Barnett et al. 2001) but non-essential for *R. capsulatus* (Lang and Beatty 2000). This allows for much more freedom (e.g., disruption of the *ctrA* gene) in studying the role of CtrA in *R. capsulatus*. Like the *C. crescentus* CtrA, the *R. capsulatus* CtrA regulates flagellar genes (Lang and Beatty 2002). Alignment of the *C. crescentus* and *R. capsulatus* CtrA protein sequence shows that *C. crescentus* and *R. capsulatus* CtrA are 71% identical at the amino acid sequence level, with the helix-turn-helix motif being 100% identical in sequence (Lang and Beatty 2000). Although the similarity in CtrA sequences between *C. crescentus* and *R. capsulatus* suggests that they function in a similar manner, there must also be differences that make CtrA essential in *C. crescentus* and non-essential in *R. capsulatus*. We speculate that there are two different types of CtrA systems: (1) in bacteria such as *C. crescentus* with asymmetrical cell division CtrA is essential; and (2) in bacteria such as *R. capsulatus* with symmetrical cell division CtrA is non-essential.

It has been found that CtrA regulates RcGTA expression (Lang and Beatty 2000), but the *C. crescentus* CtrA cannot replace the function of *R. capsulatus* CtrA in regulating RcGTA production (Lang and Beatty 2001). Recent experiments have revealed that, like the GTA gene cluster, *ctrA* expression is also affected by growth and nutrient limitation (unpublished; see below).

14.4.1.3 Regulation of RcGTA by CtrA in *R. capsulatus*

It is thought that the RcGTA gene cluster is transcribed as an operon and that the primary transcript is degraded into RNA of various sizes encoding RcGTA proteins. Indeed, northern blots using orfg2 and orfg4 probes showed heterogeneity in the size of complementary RNA molecules. Also, both northern blots and expression from an *orfg2::lacZ* plasmid-borne gene fusion indicated that transcription of the RcGTA gene cluster is *ctrA* dependent (Lang and Beatty 2000). *Rhodobacter capsulatus ctrA* and *cckA* mutants have decreased RcGTA transduction efficiencies of less than 0.1% and about 1% respectively (unpublished). Northern blots of the *ctrA* mutant showed decreased amounts of RcGTA mRNA, revealing that CtrA is needed for induction of transcription of the RcGTA gene cluster (Lang and Beatty 2000). Furthermore, unpublished western blots showed that *ctrA* mutants lacked detectable RcGTA capsid protein, which is consistent with the northern blot and RcGTA transduction results. Western blots of the *cckA* mutant showed an intracellular accumulation of the capsid protein but little or no extracellular capsid protein, whereas the parental strain had RcGTA capsid protein present both in the intracellular and the extracellular portion of the culture (unpublished). These results indicate that CtrA is involved in regulating RcGTA gene cluster transcription, but CckA is involved in regulating RcGTA release into the extracellular environment.

14.4.2 Quorum Sensing and Regulation of RcGTA

14.4.2.1 Introduction to Quorum-Sensing Systems

Quorum sensing is a mechanism used by bacterial cells to detect surrounding population densities. A basic quorum-sensing system consists of a signal and a signal receptor. At low population densities, bacterial cells produce low amounts of quorum-sensing signal, but as cell population density increases, production of the signaling molecule also increases. When the signal accumulates to a threshold concentration, it binds to a quorum-sensing receptor. This leads to a change in transcription of a variety of genes, depending on the species. For example, quorum sensing regulates bioluminescence in *Vibrio fischeri* (Engebrecht et al. 1983), antibiotic biosynthesis in *Streptomyces* spp. (Waters and Bassler 2005), conjugation in *Agrobacterium tumefaciens* (Piper et al. 1993), competence and sporulation in *Bacillus subtilis* (Waters and Bassler 2005), the production of virulence factors in *Staphylococcus aureus* (Waters and Bassler 2005), and biofilm formation in *Pseudomonas aeruginosa* (Davies et al. 1998).

There are a variety of quorum-sensing systems and many different types of quorum-signaling molecules, including oligopeptides, *Pseudomonas* quinolone signals (PQS), autoinducer-2 (AI-2), and acyl-homoserine lactones (acyl-HSLs) (Camilli and Bassler 2006). In gram-positive bacteria, the oligopeptide signal is detected by a membrane-bound receptor of a two-component system (Waters and Bassler 2005). Binding of the signal activates a phosphorylation cascade which leads to a change in transcription of target genes. The highly hydrophobic quinolone signals were recently discovered in *P. aeruginosa* and found to be transported between cells in endogenously produced membrane vesicles. Also recently discovered was AI-2, which is made by LuxS (the AI-2 synthase) in a wide variety of bacteria and has been suggested as a means of interspecies communication (Waters and Bassler 2005; Camilli and Bassler 2006). The focus of this section will be on the gram-negative type of quorum-sensing system involving an acyl-HSL signal (Waters and Bassler 2005). In this system LuxI (or a homolog), the homoserine lactone synthase, produces an acyl-HSL signal and LuxR (or a homolog) is the cognate response regulator that detects this signal (Fig. 14.5). As in the well-studied *V. fischeri* system, upon binding acyl-HSL, the LuxR protein binds DNA to modify transcription of target genes. The LuxI/LuxR quorum-sensing system is autoregulatory because production of acyl-HSL results in the inhibition of *luxR* transcription and the activation of *luxI* transcription. The major function of the LuxI/LuxR quorum-sensing system appears to be the activation of transcription of the genes involved in bioluminescence (*luxCDABE*), which are in the same operon as *luxI* (Fig. 14.5). It is not unusual to find more than one LuxI/LuxR type of quorum-sensing system in a single species. For example, *P. aeruginosa* has both the LasI/LasR and RhlI/RhlR quorum-sensing systems (Waters and Bassler 2005), and *Erwinia carotovora* has the CarI/CarR and ExpI/ExpR quorum-sensing systems, all of which are homologous to the LuxI/LuxR system (Welch et al. 2000; von Bodman et al. 2003). Interestingly, CarI and ExpI produce the same HSL autoinducer (*N*-(3-oxohexanoyl)-HSL), which may provide a mechanism for coordinating the

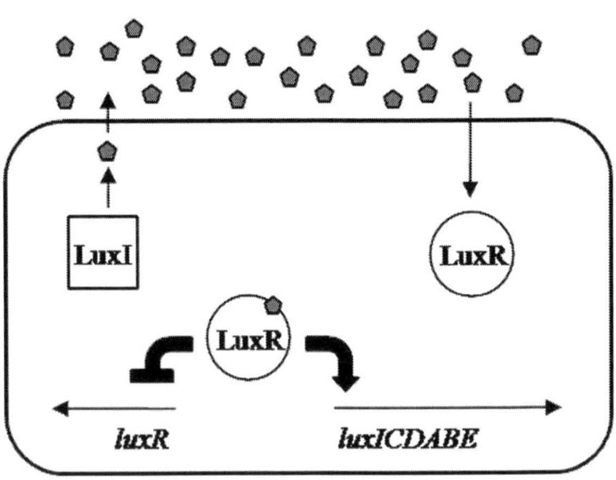

Fig. 14.5 The LuxI/LuxR quorum-sensing system in *V. fischeri*. In *V. fischeri*, the quorum-sensing system consists of LuxI (*square*), which is responsible for synthesizing the HSL autoinducer *N*-(3-oxohexanoyl)-homoserine lactone (*pentagon*), and LuxR (*circle*), which is responsible for regulating target genes upon binding the HSL autoinducer. As the cell population density increases, the HSL autoinducer accumulates and when a threshold concentration of HSL autoinducer is reached, the LuxR protein binds to it. The HSL–LuxR complex then binds the *luxCDABE* promoter to activate transcription and binds the *luxR* promoter to inhibit transcription. This results in an exponential increase in both autoinducer synthesis and light production

two systems. One variation in the LuxI/LuxR-type quorum-sensing system is found in the *Erwinia chrysanthemi* CarI/CarR system. Gel mobility shift experiments showed that in vitro, CarR bound target DNA in the absence of *N*-(3-oxohexanoyl)-HSL and released DNA in the presence of the HSL signal (Welch et al. 2000).

14.4.2.2 The GtaI/GtaR Quorum-Sensing System in *R. capsulatus*

Although a preliminary genome sequence of *R. capsulatus* contained a *luxI* homolog (now called *gtaI*), acyl-HSL in *R. capsulatus* could not be detected by bioassay (Schaefer et al. 2002). However, Schaefer et al. (2002) developed an assay that identified two long-chain acyl-HSLs (C16-HSL and C14-HSL) involved in quorum sensing in *R. capsulatus*. Further study of the *R. capsulatus* genome revealed three ORFs encoding signal receptors with significant sequence similarity to known LuxR homologs (Schaefer et al. 2002). These three ORFs, RCC03806, RCC04617, and RCC02401, have 37, 31, and 36% similarity (respectively) to CerR, the *Rhodobacter sphaeroides* LuxR-type protein (Schaefer et al. 2002). RCC03806, the ORF with the highest percentage similarity to CerR, is found immediately upstream of the *gtaI* gene (the *gtaI* start codon and the RCC03806 stop codon are separated by 49 bp). With such close proximity, it is likely that the protein encoded by RCC03806 is the cognate signal receptor (which we name GtaR) of the HSL produced by GtaI and that *gtaI* and RCC03806 are co-transcribed as an operon.

14.4.2.3 Regulation of RcGTA by the Quorum-Sensing System in *R. capsulatus*

The presence of ORFs encoding the GtaI/GtaR homologs of LuxI/LuxR in *R. capsulatus*, in combination with the knowledge that RcGTA expression is maximal in the stationary phase (Solioz et al. 1975), led to the discovery of the relationship between GTA production and quorum sensing in *R. capsulatus*. Using a fusion plasmid in which the GTA promoter was fused to *lacZ*, Schaefer et al. (2002) found that a *gtaI* mutant had a sevenfold reduction in β-galactosidase-specific activity compared to a wild-type strain. Furthermore, the addition of C16-HSL to cultures of this *gtaI* mutant strain restored the expression of the fusion gene to wild-type levels and similar results were obtained in GTA transduction assays (Schaefer et al. 2002).

14.4.3 Effects of Growth Phase and Nutrient Limitation on GTA Expression

It was discovered in the 1970s that the frequency of GTA-mediated gene transduction is maximal in the stationary phase of cultures (Solioz et al. 1975). Our cloning of the RcGTA structural genes led to northern blot and *lacZ* gene fusion experiments which showed that this stationary phase regulation is due to induction of RcGTA structural gene transcription (Lang and Beatty 2000). Also, western blots have revealed that RcGTA appears to accumulate abruptly in wild-type cells in the early stationary phase, whereas extracellular RcGTA accumulates at later time points (unpublished). However, it is not known what aspect of stationary phase cessation of culture growth leads to induction of RcGTA gene transcription. In fact, it is not known why cultures enter the stationary phase in the standard media used for growth of *R. capsulatus*.

Variants of the RCV minimal medium, a defined medium containing phosphate as the P source, malic acid as the C source, and ammonium as the N source (Beatty and Gest 1981), have been used to titrate growth-limiting amounts of nutrients. In this approach, reducing the concentration of one nutrient such that cultures enter the stationary phase at a lower culture density compared to RCV medium, we can separate cell concentration effects (as in quorum sensing) from nutrient depletion effects. *Rhodobacter capsulatus* cultures have been grown in P-, C- and N-deficient variants of the RCV medium and western blots done using the RcGTA capsid protein antibody. Preliminary results showed that an N-deficient culture lacked detectable RcGTA capsid protein compared to the replete RCV culture; in a P-limited culture, intracellular amounts of the capsid protein were elevated, and RcGTA was released in high quantity from cells; in a C-limited culture, elevated levels of the capsid were present in cells, but there was no increase in RcGTA released.

These preliminary experiments indicate interesting signal transduction pathways that regulate RcGTA production and which resemble the phenotypes of *ctrA* and *cckA* mutants: N limitation, like *ctrA* mutation, results in greatly reduced GTA gene expression, whereas P limitation results in increased expression; although C limitation resulted in the accumulation of the capsid protein in cells, there was little or no extracellular RcGTA, similar to the *cckA* mutant. We speculate that N or P limitation differentially affects the phosphorylation of CtrA and that C limitation affects CckA activity.

14.5 Concluding Remarks

It has become clear that a complicated network of systems regulates the production of extracellular RcGTA. The pathways involving CtrA, CckA, and the GtaI/GtaR quorum-sensing system are intricate, involving factors such as growth phase, cell culture density, and nutrient availability. This complexity compounds the difficulty in elucidating the mechanisms of RcGTA regulation. Many questions remain unanswered and will require further study. These questions include the following: How is the RcGTA particle released from the cell? Is there a non-lytic mechanism or does a sub-population lyse to release RcGTA? What is the method of coordination between the CtrA, the CckA, and the GtaI/GtaR system in regulating RcGTA expression? Are there other systems involved in regulating RcGTA expression? What percentage of virus-like particles in the environment are GTA-like particles, and can they mediate interspecies horizontal gene transfer? The *R. capsulatus* RcGTA provides a model for the study of GTAs in general and the role of RcGTA-like particles involved in genetic exchange. It is unknown how many bacterial species are capable of genetic exchange mediated by GTA particles, but with an estimation of up to 10^{30} viral particles in aquatic environments, the possibility that a portion of these are GTAs appears likely.

References

Barnett MJ, Hung DY, Reisenauer A et al (2001) A homolog of the CtrA cell cycle regulator is present and essential in *Sinorhizobium meliloti*. J Bacteriol 183:3204–3210

Beatty JT, Gest H (1981) Generation of succinyl-coenzyme A in photosynthetic bacteria. Arch Microbiol 129:335–340

Bellefontaine AF, Pierreux CE, Mertens P et al (2002) Plasticity of a transcriptional regulation network among alpha-proteobacteria is supported by the identification of CtrA targets in *Brucella abortus*. Mol Microbiol 43:945–960

Biers EJ, Wang K, Pennington C et al (2008) Occurrence and expression of gene transfer agent genes in marine bacterioplankton. Appl Environ Microbiol 74:2933–2939

Bowers LM, Shapland EB, Ryan KR (2008) Who's in charge here? Regulating cell cycle regulators. Curr Opin Microbiol 11:547–552

Camilli A, Bassler BL (2006) Bacterial small-molecule signaling pathways. Science 311:1113–1116

Cheetham BF, Katz ME (1995) A role for bacteriophages in the evolution and transfer of bacterial virulence determinants. Mol Microbiol 18:201–208

Chen F, Spano A, Goodman BE et al (2009) Proteomic analysis and identification of the structural and regulatory proteins of the *Rhodobacter capsulatus* gene transfer agent. J Proteome Res 8:967–973

Davies DG, Parsek MR, Pearson JP et al (1998) The involvement of cell-to-cell signals in the development of a bacterial biofilm. Science 280:295–298

Davies J (1994) Inactivation of antibiotics and the dissemination of resistance genes. Science 264:375–382

Domian IJ, Reisenauer A, Shapiro L (1999) Feedback control of a master bacterial cell-cycle regulator. Proc Natl Acad Sci USA 43:6648–6653

Eiserling F, Pushkin A, Gingery M et al (1999) Bacteriophage-like particles associated with the gene transfer agent of *Methanococcus voltae* PS. J Gen Virol 80:3305–3308

Engebrecht J, Nealson K, Silverman M (1983) Bacterial bioluminescence: isolation and genetic analysis of functions from *Vibrio fischeri*. Cell 32:773–781

Humphrey SB, Stanton TB, Jensen NS et al (1997) Purification and characterization of VSH-1, a generalized transducing bacteriophage of *Serpulina hyodysenteriae*. J Bacteriol 179:323–329

Lang AS, Beatty JT (2000) Genetic analysis of a bacterial genetic exchange element: The gene transfer agent of *Rhodobacter capsulatus*. Proc Natl Acad Sci USA 97:859–864

Lang AS, Beatty JT (2001) The gene transfer agent of *Rhodobacter capsulatus* and "constitutive transduction" in prokaryotes. Arch Microbiol 175:241–249

Lang AS, Beatty JT (2002) A bacterial signal transduction system controls genetic exchange and motility. J Bacteriol 184:913–918

Lang AS, Beatty JT (2007) Importance of widespread gene transfer agent genes in α-proteobacteria. Trends Microbiol 15:54–62

Lang AS, Taylor TA, Beatty JT (2002) Evolutionary implications of phylogenetic analyses of the gene transfer agent (GTA) of *Rhodobacter capsulatus*. J Mol Evol 55:534–543

Laub MT, Chen SL, Shapiro L et al (2002) Genes directly controlled by CtrA, a master regulator of the *Caulobacter* cell cycle. Proc Natl Acad Sci USA 99:4632–4637

Laub MT, McAdams HH, Feldblyum T et al (2000) Global analysis of the genetic network controlling a bacterial cell cycle. Science 290:2144–2148

Marrs BL (1974) Genetic recombination in *R. capsulata*. Proc Natl Acad Sci USA 71:971–973

Paul JH (2008) Prophages in marine bacteria: dangerous molecular time bombs or the key to survival in the seas? ISME J 2:579–589

Piper KR, Beck von Bodman S, Farrand SK (1993) Conjugation factor of *Agrobacterium tumefaciens* regulates Ti plasmid transfer by autoinduction. Nature 362:448–450

Quon KC, Marczynski GT, Shapiro L (1996) Cell cycle control by an essential bacterial two-component signal transduction protein. Cell 84:83–93

Rapp BJ, Wall JD (1987) Genetic transfer in *Desulfovibrio desulfuricans*. Proc Natl Acad Sci USA 84:9128–9130

Reisenauer A, Quon K, Shapiro L (1999) The CtrA response regulator mediates temporal control of gene expression during the *Caulobacter* cell cycle. J Bacteriol 181:2430–2439

Schaefer AL, Taylor TA, Beatty JT et al (2002) Long-chain acyl-homoserine lactone quorum sensing regulation of *Rhodobacter capsulatus* gene transfer agent production. J Bacteriol 184:6515–6521

Solioz M, Yen H-C, Marrs BL (1975) Release and uptake of gene transfer agent by *Rhodopseudomonas capsulata*. J Bacteriol 123:651–657

Suttle C (2005) Viruses in the sea. Nature 437:356–361

van Passel M, Thygesen H, Luyf A et al (2005) An acquisition account of genomic islands based on genome signature comparisons. BMC Genomics 6:163–173

von Bodman SB, Ball JK, Faini MA et al (2003) The quorum sensing negative regulators EsaR and ExpR (Ecc), homologues within the LuxR family, retain the ability to function as activators of transcription. J Bacteriol 185:7001–7007

Waters CM, Bassler BL (2005) Quorum sensing: cell-to-cell communication in bacteria. Annu Rev Cell Dev Biol 21:319–346

Welch M, Todd DE, Whitehead NA et al (2000) *N-*.Aacyl homoserine lactone binding to the CarR receptor determines quorum-sensing specificity in *Erwinia*. EMBO J 19:631–641

Yen HC, Hu NT, Marrs BL (1979) Characterization of the gene transfer agent made by an overproducer mutant of *Rhodopseudomonas capsulata*. J Mol Biol 131:157–168

Chapter 15
Integrative Control of Carbon, Nitrogen, Hydrogen, and Sulfur Metabolism: The Central Role of the Calvin–Benson–Bassham Cycle

Rick Laguna, Gauri S. Joshi, Andrew W. Dangel, Amanda K. Luther, and F. Robert Tabita

Abstract Nonsulfur purple (NSP) photosynthetic bacteria use the Calvin–Benson–Bassham (CBB) reductive pentose phosphate pathway for the reduction of CO_2 via ribulose 1,5-bisphosphate (RuBP) carboxylase–oxygenase (RubisCO), as a means to build cell mass during chemoautotrophic or photoautotrophic conditions. In addition, the CBB pathway plays an important role in maintaining redox balance during photoheterotrophic growth conditions. In this communication we describe protein–protein interactions between two transcriptional regulators CbbR and RegA and the possible role of the CbbX protein in regulating the CBB pathway in *Rhodobacter sphaeroides*. In *Rhodopseudomonas palustris*, the CbbR and the CbbRRS system (a three-protein, two-component regulatory system) regulate the CBB pathway. Moreover, derepression of the nitrogenase complex, and the production of hydrogen gas, appears to be a common mechanism to balance the redox potential in RubisCO-compromised strains of NSP photosynthetic bacteria.

15.1 Introduction

Nonsulfur purple (NSP) photosynthetic bacteria are capable of growing in the presence or the absence of light and oxygen. These organisms build cell mass by utilizing either organic or inorganic carbon, while obtaining their energy either from photochemical reactions or via the oxidation of organic compounds (Anderson and Fuller 1967a, b; Madigan and Gest 1979, Madigan et al. 1979). The great diversity of metabolic processes that are utilized under different environmental growth conditions obviously requires a vast amount of cellular regulation (Dubbs and Tabita 2004). One major biosynthetic route used by the NSP photosynthetic bacteria is the Calvin–Benson–Bassham (CBB) reductive pentose phosphate pathway for the

F.R. Tabita (✉)
Department of Microbiology, The Ohio State University, 484 West 12th Avenue, Columbus, OH 43210-1292, USA
e-mail: tabita.1@osu.edu

P.C. Hallenbeck (ed.), *Recent Advances in Phototrophic Prokaryotes*, Advances in Experimental Medicine and Biology 675, DOI 10.1007/978-1-4419-1528-3_15,

reduction of carbon dioxide. This pathway is also used as a key means to balance the redox potential of the cell during photoheterotrophic growth when organic carbon is used as the electron donor such that CO_2 is used as the preferred electron acceptor (Tichi and Tabita 2000; Wang et al. 1993). The key enzyme of the CBB pathway is ribulose 1,5-bisphosphate (RuBP) carboxylase–oxygenase (RubisCO) (Anderson and Fuller 1969; Benson and Calvin 1947; Calvin and Benson 1948). In this communication we describe how the CBB pathway is regulated in NSP bacteria, using *Rhodobacter sphaeroides* and *Rhodopseudomonas palustris* as model systems. We also describe how these organisms are capable of photoheterotrophic growth in the absence of a functional CBB pathway, resulting in the production of copious quantities of hydrogen gas caused by the recruitment of novel regulatory events.

15.2 CBB Regulation in *R. sphaeroides*

CbbR and RegA (PrrA) are transcriptional regulators of the cbb_I and cbb_{II} operons in *R. sphaeroides* (Dubbs and Tabita 1998, 2003; Dubbs et al. 2000; Gibson and Tabita 1993). CbbR is a LysR-type transcriptional regulator (LTTR) and was shown to positively regulate the expression of both *cbb* operons (Gibson and Tabita 1993). RegA is part of a two-component signal transduction system that involves the membrane-bound histidine kinase RegB (PrrB) (Elsen et al. 2004). RegB catalyzes the phosphorylation of the response regulator RegA to activate its regulatory function (Comolli et al. 2002; Inoue et al. 1995). The DNA-binding sites for CbbR and RegA are found in the regulatory regions of the cbb_I and cbb_{II} operons (Dubbs et al. 2000; Dubbs and Tabita 2003). Within the regulatory region of the cbb_I operon, there is a CbbR-binding site and a RegA-binding site that overlap each other, which may be indicative of possible protein–protein interactions. To examine this, gel mobility shift and chemical cross-linking experiments were conducted and a protein–protein interaction between CbbR and RegA was established (Dangel and Tabita 2009). It was further found that RegA does not need to be bound to DNA for this protein–protein interaction to occur with CbbR (Dangel and Tabita 2009). In contrast, CbbR must be bound to DNA for interactions with RegA to occur (Dangel and Tabita 2009), possibly ensuring that RegA only binds transcriptional regulators such as CbbR at the appropriate promoter site. The presence of RegA enhances the ability of CbbR to bind the cbb_I promoter (Dangel and Tabita 2009). It is thus possible that RegA lowers the activation energy required for CbbR to bind DNA, thereby increasing the DNA-binding affinity of CbbR. Further examination of the RegA-binding sites revealed that there is communication or cooperation between RegA-binding site 1/2 and site 3 that allows RegA/DNA complex formation at a lower concentration of RegA (Dangel and Tabita 2009). An intermediate loop structure could be formed to facilitate the communication between RegA site 1/2 and site 3. It is possible that complex formation between CbbR and RegA~P on the cbb_I promoter facilitates recruitment of RNA polymerase and expression of the cbb_I operon

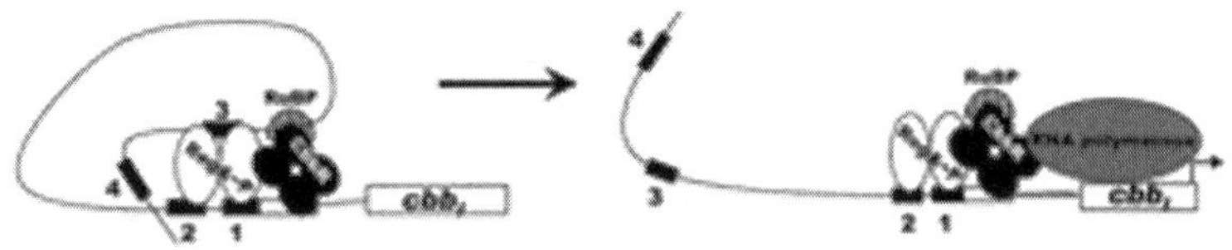

Fig. 15.1 Model illustrating protein complex formation between CbbR and RegA~P (phosphorylated RegA) on the *cbb*$_I$ promoter, subsequently allowing recruitment of RNA polymerase and expression of the *cbb*$_I$ operon. RuBP denotes ribulose 1,5-bisphosphate, a co-inducer of CbbR from *R. sphaeroides*. *Black boxes* represent RegA-binding sites 1, 2, 3, and 4 and the *white box* represents the CbbR-binding site

(Fig. 15.1) (Dangel and Tabita 2009). Current experiments are underway to examine the specific residue interactions between CbbR and RegA.

Previous studies suggest that there may be additional novel components involved in CBB regulation in *R. sphaeroides* (Dubbs and Tabita 2003; Gibson and Tabita 1997). The *cbbXYZ* genes are located immediately downstream of the *cbb*$_I$ operon in *R. sphaeroides*. Previous studies demonstrated that these genes form an operon and are partially subjected to the same regulation as the *cbb*$_I$ operon (Gibson and Tabita 1997). With the exception of CbbZ (which functions as a phosphoglycolate phosphatase), the physiological role of proteins encoded by genes of this operon has yet to be determined in *R. sphaeroides*. However, upon deletion of each of the *cbbX, cbbY,* and *cbbZ* genes, the *cbbX* deletion strain alone displayed a long lag period during photoautotrophic (PA) growth conditions as compared to the wild type (Gibson and Tabita 1997). Recently it was proposed that the CbbX protein of the unicellular red alga *Cyanidioschyzon merolae* functions as a transcriptional regulator of the *cbbLS* genes, which encode a form I RubisCO (Fujita et al. 2008). Current experiments are underway to gain a greater understanding of the function of CbbX in *R. sphaeroides*.

15.3 CBB Regulation in *R. palustris*

Within the NSP photosynthetic bacteria, *R. palustris* exhibits exemplary metabolic versatility and unique organization of its *cbb* operons (VerBerkmoes et al. 2006). The unique feature of the *R. palustris cbb*$_I$ operon is the presence of genes that encode a two-component regulatory system (referred to as the CbbRRS system) juxtaposed between the master transcriptional regulator CbbR and genes encoding form I RubisCO (*cbbLS*) (Fig. 15.2). The CbbRRS system is an atypical two-component system comprising a hybrid sensor kinase (CbbSR) and two response regulator proteins (CbbRR1 and CbbRR2), with no apparent DNA-binding domains on any of these proteins (Romagnoli and Tabita 2006). It was demonstrated that CbbSR undergoes autophosphorylation and transfers phosphate to both CbbRR1 and CbbRR2 (Romagnoli and Tabita 2006, 2007). CbbR is absolutely required for the expression of form I RubisCO in *R. palustris* (Joshi et al. 2009; Romagnoli and Tabita

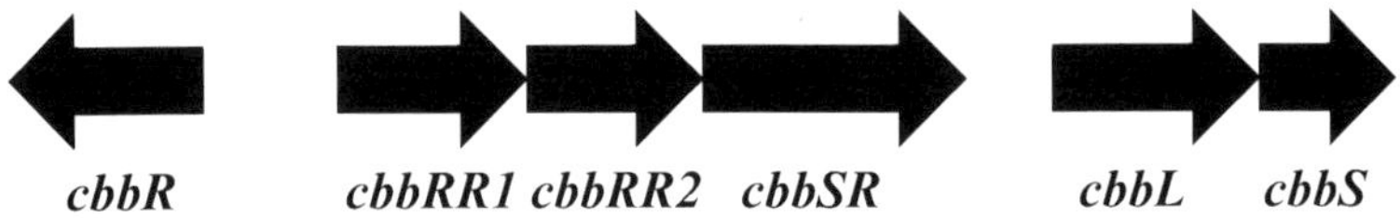

Fig. 15.2 Genetic organization of the cbb_I operon in *R. palustris*

2006). It has been shown that the CbbRRS system modulates the expression of form I RubisCO during photoautotrophic growth possibly in response to a redox signal. However, the CbbRRS system does not appear to play a regulatory role under photoheterotrophic (benzoate) growth conditions, suggesting differential utilization of this system in *R. palustris* (Joshi et al. 2009). The observations stemming from the physiological studies with photoautotrophically grown cells led to the hypothesis that the CbbRRS proteins (especially the response regulators CbbRR1 and CbbRR2) influence the interactions of CbbR at the cbb_I promoter and thereby influence form I RubisCO expression (Romagnoli and Tabita 2006). Current experiments indicate that protein–protein interactions between CbbR and CbbRRS do take place and are quite significant (Joshi and Tabita, manuscripts in preparation).

15.4 Derepression of the Nitrogenase Complex and Hydrogen Production in RubisCO-Compromised NSP Photosynthetic Bacteria

During aerobic chemoautotrophic and anaerobic photoautotrophic growth conditions, NSP photosynthetic bacteria use the CBB cycle for the reduction of CO_2 into fixed carbon. In contrast, during photoheterotrophic growth, CO_2 is primarily used as an electron acceptor via the CBB cycle in order to maintain redox poise (Falcone and Tabita 1991; Hallenbeck et al. 1990; Wang et al. 1993). Over the years we have shown that NSP photosynthetic bacteria possess an array of metabolic and regulatory capabilities that allow for the utilization of alternative redox sinks when the primary electron sink, CO_2, is nullified via the inactivation or the deletion of the RubisCO genes (Falcone and Tabita 1993; Joshi and Tabita 1996; Tichi and Tabita 2000). For example, in many instances the derepression of the nitrogenase (*nifHDK*) complex occurred under normal repressive conditions (the presence of ammonium). Such gain-of-function adaptive mutant strains have been obtained in our laboratory from *Rhodobacter capsulatus, R. sphaeroides, Rhodospirillum rubrum,* and *R. palustris* (Joshi and Tabita 1996; Tichi and Tabita 2000, 2001; Romagnoli and Tabita 2009). Such strains balance their redox potential via nitrogenase-catalyzed reduction of protons to hydrogen gas (Fig. 15.3). The nitrogenase-derepressed mutant strains produce copious quantities of hydrogen gas by virtue of using the nitrogenase enzyme complex exclusively as a hydrogenase. Current experiments are underway to examine the molecular mechanisms that allow the *cbb* and *nif* systems to compete for reducing equivalents and to examine a RubisCO-compromised mutant strain of *R. sphaeroides* that appears to use the reduction of sulfate as an

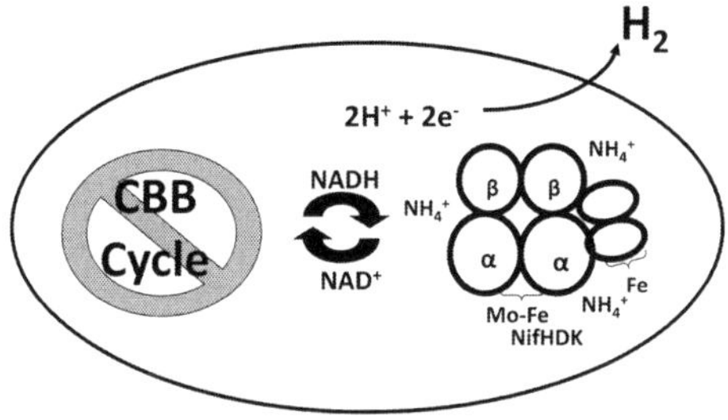

Fig. 15.3 Derepression of the nitrogenase complex (NifHDK) occurs under normal repressive conditions (presence of NH_4^+) due to a nonfunctional CBB cycle. The reduction of protons to hydrogen gas, by the nitrogenase complex, serves to maintain cellular redox poise during photoheterotrophic growth conditions

alternative redox sink. Moreover, previous studies involving reconstitution of the CBB pathway indicate that control of the CBB and other redox balancing pathways is linked, with the CBB route exerting the prime regulatory signal in NSP bacteria (Joshi and Tabita 1996; Tichi and Tabita 2000, 2001).

15.5 Conclusion

Examination of the CBB pathway in NSP photosynthetic bacteria reveals that there are many layers of regulation. In *R. sphaeroides* and *R. Palustris*, overall regulation of the CBB pathway differs, although each organism, along with *R. rubrum* and *R. capsulatus*, appears to emphasize the derepression of the nitrogenase complex in the absence of a functional CBB pathway.

Acknowledgments This work was supported by grants DE-FG02-07ER64489 and DE-FG02-DE-FG02-08ER15976 from the Office of Biological and Environmental Research (Genomics: GTL Program) and the Chemical Sciences, Geosciences, and Biosciences Division of the Office of Basic Energy Sciences, respectively, of the US Department of Energy.

References

Anderson L, Fuller RC (1967a) Photosynthesis in *Rhodospirillum rubrum* II. Photoheterotrophic carbon dioxide fixation. Plant Physiol 42:491–496

Anderson L, Fuller RC (1967b) Photosynthesis in *Rhodospirillum rubrum* I. Autotrophic carbon dioxide fixation. Plant Physiol 42:487–490

Anderson LE, Fuller RC (1969) Photosynthesis in *Rhodospirillum rubrum*. IV. Isolation and characterization of ribulose 1,5-diphosphate carboxylase. J Biol Chem 244:3105–3109

Benson A, Calvin M (1947) The dark reductions of photosynthesis. Science 105:648–649

Calvin M, Benson AA (1948) The path of carbon in photosynthesis. Science 107:476–480

Comolli JC, Carl AJ, Hall C et al (2002) Transcriptional activation of the *Rhodobacter sphaeroides* cytochrome c(2) gene P2 promoter by the response regulator PrrA. J Bacteriol 184:390–399

Dangel AW, Tabita FR (2009) Protein–protein interactions between CbbR and RegA (PrrA), transcriptional regulators of the *cbb* operons of *Rhodobacter sphaeroides*. Mol Microbiol 71:717–729

Dubbs JM, Tabita FR (1998) Two functionally distinct regions upstream of the *cbb*$_I$ operon of *Rhodobacter sphaeroides* regulate gene expression. J Bacteriol 180:4903–4911
Dubbs JM, Tabita FR (2003) Interactions of the *cbb*$_{II}$ promoter-operator region with CbbR and RegA (PrrA) regulators indicate distinct mechanisms to control expression of the two *cbb* operons of *Rhodobacter sphaeroides*. J Biol Chem 278:16443–16450
Dubbs JM, Tabita FR (2004) Regulators of nonsulfur purple phototrophic bacteria and the interactive control of CO_2 assimilation, nitrogen fixation, hydrogen metabolism and energy generation. FEMS Microbiol Rev 28:353–376
Dubbs JM, Bird TH, Bauer CE et al (2000) Interaction of CbbR and RegA* transcription regulators with the *Rhodobacter sphaeroides cbb*$_I$ promoter-operator region. J Biol Chem 275: 19224–19230
Elsen S, Swem LR, Swem DL et al (2004) RegB/RegA, a highly conserved redox responding global two-component regulatory system. Microbiol Mol Biol Rev 68:263–279
Falcone DL, Tabita FR (1991) Expression of endogenous and foreign ribulose 1,5 bisphosphate carboxylase–oxygenase (RubisCO) genes in a RubisCO deletion mutant of *Rhodobacter sphaeroides*. J Bacteriol 173:2099–2108
Falcone DL, Tabita FR (1993) Complementation analysis and regulation of CO_2 fixation gene expression in a ribulose 1,5-bisphosphate carboxylase–oxygenase deletion strain of *Rhodospirillum rubrum*. J Bacteriol 175:5066–5077
Fujita K, Tanaka K, Sadaie Y et al (2008) Functional analysis of the plastid and nuclear encoded CbbX proteins of *Cyanidioschyzon merolae*. Genes Genet Syst 83:135–142
Gibson JL, Tabita FR (1997) Analysis of the *cbbXYZ* operon in *Rhodobacter sphaeroides*. J Bacteriol 179:663–669
Gibson JL, Tabita FR (1993) Nucleotide sequence and functional analysis of *cbbR*, a positive regulator of the Calvin cycle operons of *Rhodobacter sphaeroides*. J Bacteriol 175:5778–5784
Hallenbeck PL, Lerchen R, Hessler P et al (1990) Roles of CfxA, CfxB, and external electron acceptors in regulation of ribulose 1,5-bisphosphate carboxylase/oxygenase expression in *Rhodobacter sphaeroides*. J Bacteriol 172:1736–1748
Inoue K, Kouadio JL, Mosley CS et al (1995) Isolation and *in vitro* phosphorylation of sensory transduction components controlling anaerobic induction of light harvesting and reaction center gene expression in *Rhodobacter capsulatus*. Biochemistry 34:391–396
Joshi GS, Romagnoli S, Verberkmoes NC et al (2009) Differential accumulation of form I RubisCO in *Rhodopseudomonas palustris* CGA010 under photoheterotrophic growth conditions with reduced carbon sources. J Bacteriol 191:4243–4250
Joshi HM, Tabita FR (1996) A global two component signal transduction system that integrates the control of photosynthesis, carbon dioxide assimilation, and nitrogen fixation. Proc Natl Acad Sci USA 93:14515–14520
Madigan MT, Gest H (1979) Growth of the photosynthetic bacterium *Rhodopseudomonas capsulata* chemoautotrophically in darkness with H_2 as the energy source. J Bacteriol 137:524–530
Madigan MT, Wall JD, Gest H (1979) Dark anaerobic dinitrogen fixation by a photosynthetic microorganism. Science 204:1429–1430
Romagnoli S, Tabita FR (2006) A novel three-protein two-component system provides a regulatory twist on an established circuit to modulate expression of the *cbb*$_I$ region of *Rhodopseudomonas palustris* CGA010. J Bacteriol 188:2780–2791
Romagnoli S, Tabita FR (2007) Phosphotransfer reactions of the CbbRRS three-protein two-component system from *Rhodopseudomonas palustris* CGA010 appear to be controlled by an internal molecular switch on the sensor kinase. J Bacteriol 189:325–335
Romagnoli S, Tabita FR (2009) Carbon dioxide metabolism and its regulation in nonsulfur purple photosynthetic bacteria. In: Hunter CN, Daldal F, Thurnauer MC, Beatty JT (eds) The purple phototrophic bacteria. Advances in photosynthesis and respiration, vol. 28. Springer, Dordrecht, pp 563–576
Tichi MA, Tabita FR (2000) Maintenance and control of redox poise in *Rhodobacter capsulatus* strains deficient in the Calvin–Benson–Bassham pathway. Arch Microbiol 174:322–333

Tichi MA, Tabita FR (2001) Interactive control of *Rhodobacter capsulatus* redox balancing systems during phototrophic metabolism. J Bacteriol 183:6344–6354

VerBerkmoes NC, Shah MB, Lankford PK et al (2006) Determination and comparison of the baseline proteomes of the versatile microbe *Rhodopseudomonas palustris* under its major metabolic states. J Proteome Res 5:287–298

Wang X, Falcone DL, Tabita FR (1993) Reductive pentose phosphate-independent CO_2 fixation in *Rhodobacter sphaeroides* and evidence that ribulose bisphosphate carboxylase/oxygenase activity serves to maintain the redox balance of the cell. J Bacteriol 175:3372–3379

Part V
Applied Aspects

Chapter 16
Better Living Through *Cyanothece* – Unicellular Diazotrophic Cyanobacteria with Highly Versatile Metabolic Systems

Louis A. Sherman, Hongtao Min, Jörg Toepel, and Himadri B. Pakrasi

Abstract *Cyanothece* sp. ATCC 51142 is a unicellular, diazotrophic cyanobacterium with a versatile metabolism and very pronounced diurnal rhythms. Since nitrogen fixation is exquisitely sensitive to oxygen, *Cyanothece*utilizes temporal regulation to accommodate these incompatible processes in a single cell. When grown under 12 h light–dark (LD) periods, it performs photosynthesis during the day and N_2 fixation and respiration at night. Genome sequences of *Cyanothece* sp. ATCC 51142 and that of five other *Cyanothece* species have been completed and have produced some surprises. Analysis at both the transcriptomic and the proteomic levels in *Cyanothece* sp. ATCC 51142 has demonstrated the relationship of the metabolic synchrony with gene expression and has given us insights into diurnal and circadian regulation throughout a daily cycle. We are particularly interested in the regulation of metabolic processes, such as H_2 evolution, and the way in which these organisms respond to environmental cues, such as light, the lack of combined nitrogen, and changing O_2 levels. *Cyanothece* strains produce copious amounts of H_2 under different types of physiological conditions. Nitrogenase produces far more H_2 than the hydrogenase, in part because the nitrogenase levels are extremely high under N_2-fixing conditions. With *Cyanothece* 51142 cultures grown in NO_3-free media, either photoautotrophically or mixotrophically with glycerol, we have obtained H_2 production rates over 150 μmol/mg Chl/h.

16.1 Introduction

Unicellular, diazotrophic cyanobacteria are interesting and versatile organisms. They perform oxygenic photosynthesis, but they can also fix atmospheric N_2. Nitrogenase, the enzyme that is responsible for this N_2 fixation into ammonia is

---

L.A. Sherman (✉)
Department of Biological Sciences, Purdue University, West Lafayette, IN 47907, USA
e-mail: lsherman@purdue.edu

P.C. Hallenbeck (ed.), *Recent Advances in Phototrophic Prokaryotes*, Advances in Experimental Medicine and Biology 675, DOI 10.1007/978-1-4419-1528-3_16,

normally rapidly and irreversibly inactivated upon exposure to molecular oxygen. Thus, one interesting question about such organisms is how they regulate nitrogenase activity in the presence of poisonous oxygen. In addition, what other properties do they have that might make them valuable for other kinds of experimentation? We have been interested in studying these organisms and in understanding how they regulate various metabolic properties. Most recently, we have also been interested in the ability of these organisms to use light energy to generate alternative biofuels such as H_2.

We have primarily studied the unicellular diazotrophic cyanobacterium *Cyanothece* sp. ATCC 51142, hereafter *Cyanothece* 51142 (Reddy et al. 1993; Schneegurt et al. 1994), which performs photosynthesis during the light and fixes nitrogen during the dark (Schneegurt et al. 1997a, b, 1998). Our early studies with *Cyanothece* 51142 showed a strong correlation between activity and transcript level for a subset of genes related to photosynthesis (Meunier et al. 1998) and N_2 fixation (Colón-López et al. 1997) during 12 h light–dark (LD) cycles. In addition, differential gene expression was reported for the main photosynthetic genes and the nitrogenase genes in the light and dark, respectively (Colón-López et al. 1997, 1999; Colón-López and Sherman 1998). During a diurnal period, *Cyanothece* 51142 cells actively accumulate and degrade different storage inclusion bodies for the products of photosynthesis and N_2 fixation. This ability to utilize metabolic compartmentalization and energy storage makes *Cyanothece* an ideal system for bioenergy research and for studies on how a unicellular organism balances multiple, often incompatible, processes in the same cell. The genome sequences will help provide a significant basis for future insights into this metabolic "balancing act".

Recently, whole genome microarray experiments were carried out to determine the diurnal gene expression under LD conditions (12 h L/12 h D) (Stöckel et al. 2008) and under continuous light (LL) (Toepel et al. 2008). Both groups found ~30% of the ~5000 genes on the microarray exhibited diurnal oscillations under 12 h LD conditions and Toepel et al. (2008) demonstrated that ~10% of the genes demonstrated circadian behavior during growth in free-running (LL) conditions. Toepel et al. (2008) also demonstrated that nitrogenase transcript abundance and nitrogenase activity were correlated in *Cyanothece* 51142 under LL and that N_2 fixation followed a ~24 h rhythm under these conditions, albeit with reduced rates. Such results indicate a LD-independent expression pattern for nitrogenase genes, consistent with the circadian behavior for nitrogenase-related genes suggested by Sherman et al. (1998).

Photosynthetic activity depends on incident light, although genes encoding photosynthetic proteins can display a diurnal or a circadian-dependent expression pattern (Stöckel et al. 2008; Michael et al. 2008; Toepel et al. 2009). In *Cyanothece* 51142, maximum photosynthetic rates in a LD cycle occurred after 6–8 h light incubation and photosynthetic capacity decreased strongly during the N_2 fixation period (Meunier et al. 1998; Toepel et al. 2008). Toepel et al. (2008) showed that photosynthetic rates were lower during LL growth and demonstrated no circadian-related pattern for photosynthesis genes. In the case of growth under continuous light, the glycogen content stayed at high levels and did not decrease until the cells were again placed in darkness (Toepel et al. 2008). Furthermore, these results indicated that nitrogenase transcription and activity was metabolically or energetically regulated via glycogen breakdown and suggested that photosynthesis is light activated, but probably regulated by the internal carbohydrate level.

The work with *Cyanothece* 51142 has provided some answers as to the regulation of the disparate metabolic processes, but also has led to some surprises. These include the discovery of a small linear chromosome that coexists along with the larger, circular chromosome. In order to better understand the genomic basis of the genus *Cyanothece*, an additional six *Cyanothece* strains have been sequenced by the Department of Energy Joint Genome Initiative. We have begun to compare and contrast the different genomes and have found a *Cyanothece* gene core within this genus using five of the strains so far sequenced. Interestingly, *Cyanothece* 7425 demonstrates significant differences in many genomic properties. We will discuss some of the important attributes of gene regulation that we have determined in *Cyanothece* 51142 as well as discuss key features in regard to H_2 production in these strains.

16.2 Results

16.2.1 Genome Sequencing of Cyanothece 51142

The genome of *Cyanothece* 51142 was sequenced at the Washington University Genome Sequencing Center (St. Louis, MO), and the finished assembly was independently confirmed using an optical restriction map generated by OpGen, Inc. (Welsh et al. 2008). The 5.5 Mb *Cyanothece* 51142 genome consists of a 4.93 Mb circular chromosome, four plasmids ranging in size from 10 to 40 kb, and notably a 430 kb linear chromosome. The finding of a linear element in the *Cyanothece* genome was unanticipated, but was confirmed by two independent genome assembly approaches (Welsh et al. 2008). This also represented the first report of a linear element in the genome of a photosynthetic bacterium, although linear genomic elements have been identified in other bacterial genera, such as *Borrelia* (Ferdows and Barbour 1989), *Streptomyces* (Kinashi et al. 1992), and *Agrobacterium* (Allardet-Servent et al. 1993).

The gene content of the linear chromosome was examined to investigate its possible origin and importance to the organism. Several genes are found on both the circular and the linear chromosomes, including a *coxABC* operon and a cluster containing genes related to glycolysis and fermentation (*ppk*, *pyk*, *pgi*, *eno*, *ackA*, *glgP*). This cluster on the linear chromosome contains the only gene that encodes an L-lactate dehydrogenase, required for the terminal step in lactate fermentation, and suggests that the linear chromosome may play a role in fermentation. Additional genes unique to the linear chromosome are ones encoding the integrase–recombinase protein XerC, an *xseA/xseB* operon, and a *hicA/hicB* operon. Most of the remaining genes on the linear chromosome are either hypothetical, unknown, or of uncertain function. The mechanisms of maintenance and replication of the linear chromosome remain unknown, and sequence analysis did not indicate the presence of any feature, such as inverted repeats or stem–loop structures, known to be related to these functions in previously characterized genera (Volff and Altenbuchner 2000).

The predicted proteome of *Cyanothece* consists of 5269 open reading frames: 34% of known function, 29% of uncertain function, and 37% of unknown/hypothetical (Welsh et al. 2008). The annotation of genes of unknown function was greatly aided by data from the high-throughput proteomic analysis. Proteomic data were used, in conjunction with early draft genomic sequence data, to build an accurate mass and time (AMT) library (Lipton et al. 2002) for use in quantitative proteomic experiments. The combined analysis of proteome and genome data is an important approach that resulted in the inclusion or reclassification of 510 genes and lent an additional level of validation to the genome annotation. The importance of the proteomic data to metabolic objectives will be discussed in later sections.

16.2.2 Nitrogenase Genes

In *Cyanothece*, most of the genes involved in nitrogen fixation are located in a single contiguous cluster containing 28 genes separated by no more than 3 kb, with conserved synteny to those found in most other sequenced nitrogen-fixing cyanobacteria (Fig. 16.1). The cluster is more distantly conserved in *Trichodesmium erythraeum* sp. IMS 101, where 20 genes are present in a single large cluster. In the heterocyst-forming *Anabaena* and *Nostoc* strains, one or more inserts ranging in size from 9 to 24 kb break the cluster into several smaller clusters, with several genes duplicated or missing between clusters. The conserved synteny of the genes within the *nif* clusters of the *Cyanothece* and *Anabaena* families, together with the proteome-wide phylogenetic tree, supports a single acquisition event of the *nif* cluster in a common ancestor. The *Synechococcus* sp. JA-2-3B′a(2-13) and *Synechococcus* sp. JA-3-3Ab strains contain a single contiguous cluster of 20 genes, but the order and orientation of the genes are extensively reorganized relative to that seen in the other strains. *Cyanothece* 51142 contains the largest contiguous cluster of nitrogen fixation-related genes yet observed in cyanobacteria. If nitrogen fixation

Fig. 16.1 Clusters of N_2 fixation-related genes. Shown are genes with conserved synteny between *Cyanothece* 51142 and other nitrogen-fixing cyanobacteria. *Black arrows* represent genes assigned to functional categories and *white arrows* correspond to hypothetical genes and genes of unknown function. A possible inversion event in *Synechococcus*sp. JA-3–3Ab is highlighted in brackets. GenBank accession numbers for the sequences used are as follows: *Cyanothece*sp. ATCC 51142, CP000806; spheroid body of *Rhopalodia gibba*, AY728387; *Crocosphaera watsonii*WH 8501,

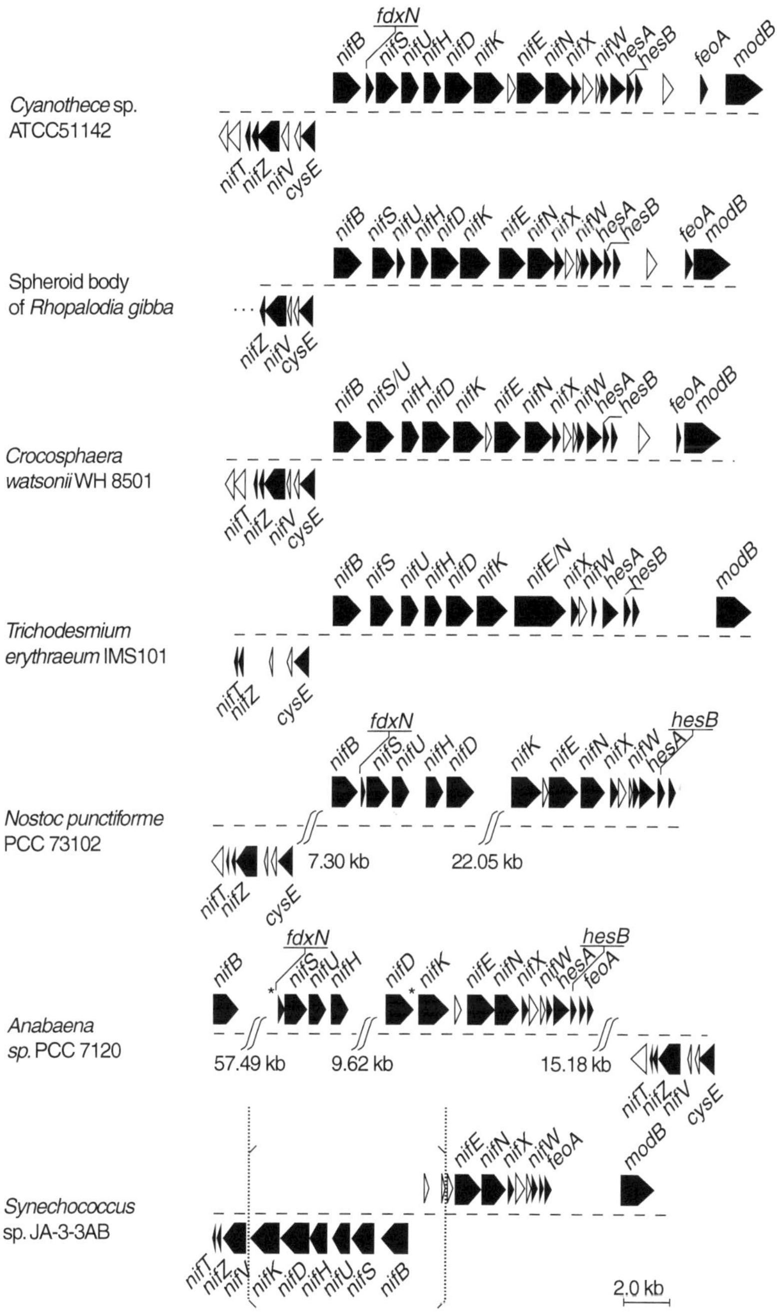

Fig. 16.1 (continued) AADV02000024; *Trichodesmium erythraeum*IMS101, CP000393; *Nostoc punctiforme*PCC 73102, CP001037; *Anabaena*sp. PCC 7120, BA000019; and *Synechococcus*sp. JA-3–3Ab, CP000239. Reproduced from PNAS 2008, vol 105:15094 with permission of the publisher

was acquired in a single ancient event, this single contiguous cluster may resemble that seen in the original ancestor. The smaller cluster present in the thermophilic *Synechococcus* strains either underwent significant gene loss and rearrangement from a single common ancestor common to all diazotrophic cyanobacteria or was acquired in a separate event by lateral gene transfer from a non-cyanobacterial organism. In either case, based on the phylogenetic evidence, nitrogen fixation was apparently present early in the evolutionary history of cyanobacteria and was subsequently lost in the non-diazotrophic β-cyanobacterial strains.

16.2.3 Metabolic Compartmentalization

Intracellular compartmentalization provides a strategy for *Cyanothece* cells to tightly regulate storage of metabolic products. The *Cyanothece* genome data confirmed the details of the pathways of storage granule accumulation and degradation by elucidation of the genes involved, provided insights into interconnections between different pathways, and highlighted the central role of nitrogen fixation in these organisms (Welsh et al. 2008). Certain filamentous nitrogen-fixing cyanobacterial strains separate photosynthesis and nitrogen fixation spatially by the differentiation of a subset of cells into heterocysts, which fix nitrogen and do not perform oxygenic photosynthesis (Haselkorn 1978; Wolk 1996). However, non-heterocyst-forming unicellular nitrogen-fixing cyanobacteria, such as *Cyanothece*, must separate these processes temporally by performing photosynthesis during the day and fixing nitrogen during the night (Sherman et al. 1998). Nitrogen fixation is an energy-intensive process requiring the use of 16 ATP molecules per molecule of N_2 converted to ammonia (Dean, Bolin and Zheng 1993). A respiratory burst at the beginning of the dark period provides some of the energy required for nitrogen fixation and serves to further deplete cellular O_2 which would otherwise inhibit nitrogen fixation. Photosynthetic capacity is also at a minimum during this time (Sherman et al. 1998), which further protects nitrogenase from oxygen damage. Fixed nitrogen is stored in cyanophycin granules, which are fully depleted during the next light period. Carbohydrates produced from photosynthesis are stored during the day in glycogen granules, which are rapidly consumed early in the dark period as a substrate for respiration (Sherman et al. 1998). Interestingly, two distinctly different glycogen debranching enzymes are present in the *Cyanothece* genome, one that is found only in the β-cyanobacterial clade (*glgP*) and another (*glgX*) that is found mainly in the α-cyanobacteria. The presence of these two enzymes suggests that granule accumulation and degradation is important and carefully regulated. The circadian cycle in *Cyanothece* 51142 cells, once entrained, persists in continuous light (Colón-López and Sherman 1998) or continuous darkness (Schneegurt et al. 2000). The temporal regulation of metabolic processes is crucial to *Cyanothece* 51142, as each is involved in a different, but interrelated, aspect of cellular metabolism.

The *Cyanothece* genome contains all of the genes required for fermentation, including the production of ethanol and hydrogen, both of which require a low or anoxic environment. *Cyanothece* 51142 creates such an intracellular environment during the early dark period in order to fix nitrogen. Stal and Moezelaar (1997) have proposed that 6-phosphogluconate dehydrogenase, present in the pentose phosphate pathway of *Cyanothece*, is primarily involved in fermentation in cyanobacteria and that its presence indicates fermentative capability. Therefore, *Cyanothece* has the ability to ferment glucose to ethanol and to produce hydrogen, and may carry out these processes early in the dark period in order to generate additional ATP molecules for nitrogen fixation. However, this same intracellular environment is generated in LL conditions (Toepel et al. 2008) and this may provide additional opportunities for N_2fixation and H_2 evolution.

16.2.4 Genomic Sequencing of Six Additional Cyanothece *Strains*

An additional six *Cyanothece* strains were sequenced by the Department of Energy Joint Genome Initiative as a first step toward understanding the nature of the *Cyanothece* genus (Table 16.1). The strains were isolated in different environments around the world and the size and shape of the organisms, as well as other important properties, can be seen in Table 16.1. All strains are capable of nitrogen fixation as well as mixotrophic growth on different carbon sources. The genome size of the strains varies from 4.6 to 6.4 Mb and none of these strains appear to have a linear chromosome.

The phylogenetic analysis of a broad grouping of cyanobacteria, including all of the sequenced *Cyanothece* strains, is shown in Fig. 16.2. It is evident that all

Table 16.1 Genomic information of six species of the genus *Cyanothece* sequenced by the Department of Energy Joint Genome Initiative. Reproduced from Welsh et al. (2008) by permission

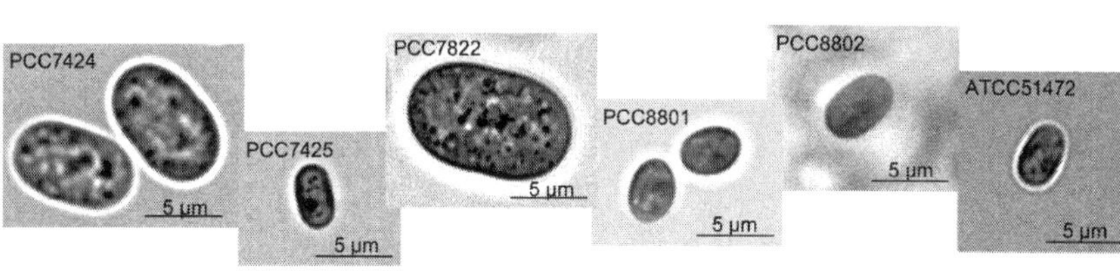

STRAIN	PCC7424	PCC7425	PCC7822	PCC8801	PCC8802	ATCC51472
ORIGIN	Senegal	Senegal	India	Taiwan	Taiwan	Texas, USA
SIZE	7-8 µm	3-4 µm	10 µm	3-4 µm	4-5 µm	4-5 µm
MIXOTROPHIC GROWTH	yes	yes	yes	yes	yes	yes
PHYCOERYTHRIN	yes	no	yes	yes	no	no
N_2 – FIXATION	yes	yes	yes	yes	yes	yes
% GC	39%	51%	40%	40%	40%	40%*
GENOME SIZE	6.4 Mb	5.7 Mb	5.7 Mb	4.6 Mb	4.7 Mb	5.5 Mb*
GENE NUMBER	6107	5574	5266	4436	4681	*
COG GENES	3432	3345	3006	2705	2866	*

*Estimated or not yet determined

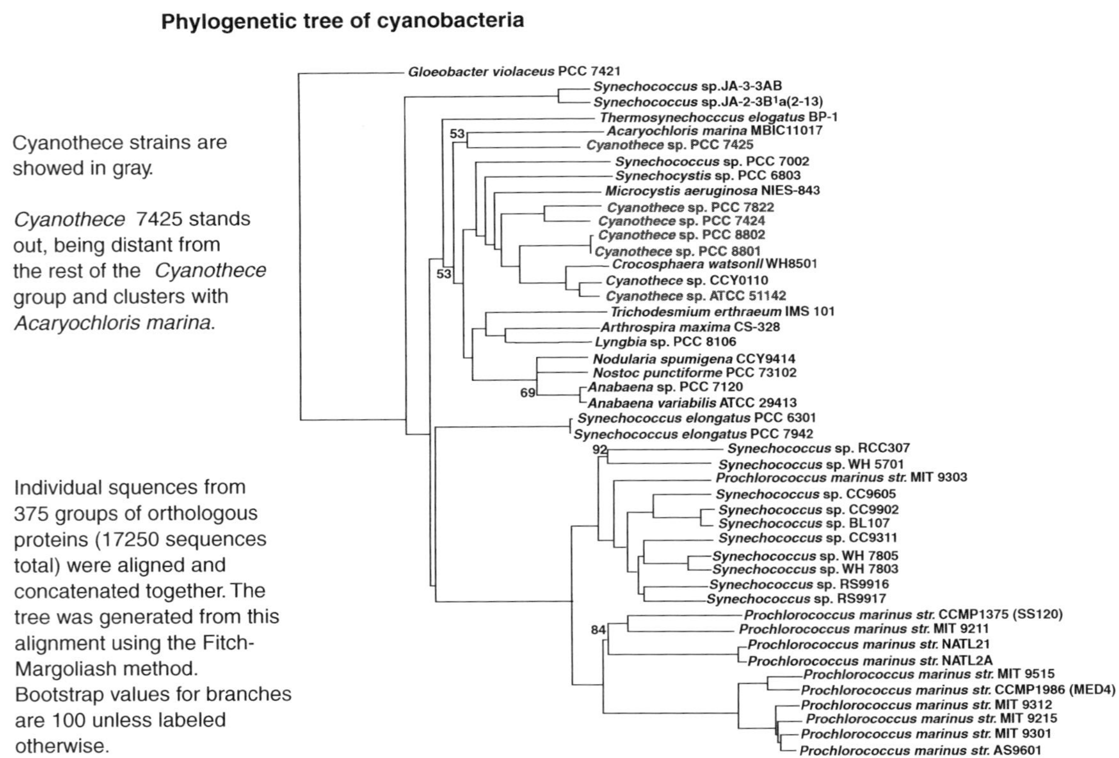

Fig. 16.2 Phylogenetic tree of cyanobacteria with sequenced genomes, including five of the recently sequenced strains of *Cyanothece*. Individual sequences from 375 groups of orthologous proteins (17250 sequences total) were aligned and concatenated together. The tree was generated from this alignment using the Fitch–Margoliash method. Bootstrap values for branches are 100 unless labeled otherwise. See Welsh et al. (2008) for details

of the *Cyanothece* strains, except for *Cyanothece* sp. PCC 7425, cluster closely together along with *Crocosphaera watsonii* WH8501. *Cyanothece* sp. PCC 7425 branches somewhat differently and is closer to *Acaryochloris marina* MBIC11017. Interestingly, this broad cluster includes a few non-nitrogen-fixing strains, such as *Synechocystis* sp. PCC 6803, which has many genes that show sequence similarity to those in *Cyanothece* sp. ATCC 51142. Such data again raise the question of whether or not *Synechocystis* once had nitrogen fixation genes that were subsequently lost or if *Cyanothece* acquired the genes at some later date. All of the *Cyanothece* strains so far sequenced have large clusters of nitrogenase genes, although not as complete as those in *Cyanothece* 51142. Once again, *Cyanothece* 7425 is the most divergent with the nitrogenase genes split into two medium-sized clusters. In addition, all the strains have clusters of the bidirectional hydrogenase genes (*hox*) (Fig. 16.3). *Cyanothece* 51142 has the most compact operon, whereas *Cyanothece* 7425 has the largest number of hypothetical genes within a larger cluster.

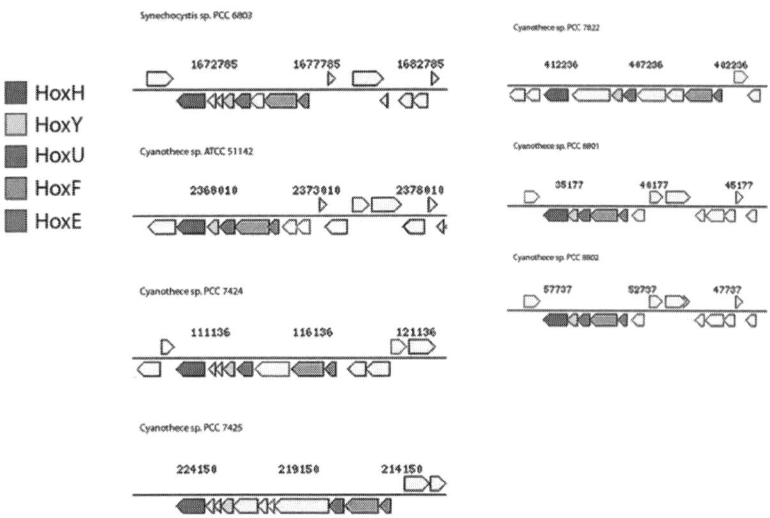

Fig. 16.3 Chromosomal organization of *hox* operons in seven strains of cyanobacteria, including *Synechocystis* sp. PCC 6803 and six *Cyanothece* strains. In all cases, the *hox* cluster encodes all of the proteins needed to produce a functional bidirectional hydrogenase

16.2.5 Genomics and Hydrogen Production

One objective for sequencing these genomes was to determine their capabilities for biofuels production. We were specifically interested in analyzing hydrogen production and to determine if all strains contain the genes for the three types of hydrogenases found in diazotrophic cyanobacteria: nitrogenase, uptake hydrogenase, and bidirectional hydrogenase. It has been demonstrated in *Anabaena* that mutants that lack the uptake hydrogenase evolve more hydrogen than the wild type

Table 16.2 Presence or absence of the three genes capable of hydrogen production in select cyanobacterial strains

Cyanobacterial strain	Nitrogenase	Bidirectional hydrogenase	Uptake hydrogenase
Cyanothece sp. ATCC 51142	+	+	+
Cyanothece sp. PCC 7425	+	+	–
Cyanothece sp. PCC 7822	+	+	+
Cyanothece sp. PCC 7424	+	+	+
Cyanothece sp. PCC 8801	+	+	+
Cyanothece sp. PCC 8802	+	+	+
Anabaena variabilis ATCC 29413	+	+	+
Anabaena sp. PCC 7120	+	+	+
Nostoc punctiforme PCC 73102	+	–	+
Synechocystis sp. PCC 6803	–	+	–
Trichodesmium erythraeum IMS101	+	–	+

(Dutta et al. 2005). Thus, we carefully analyzed the genomic sequences to determine if any of the strains were lacking an uptake hydrogenase system. In fact, all strains except *Cyanothece* 7425 had the uptake hydrogenase (Table 16.2). From this data, we were hopeful that *Cyanothece* 7425 might be a particularly important strain for hydrogen production. Of course, *Synechocystis* 6803 lacks an uptake hydrogenase, but it also lacks nitrogenase and produces relatively low levels of H_2 from the bidirectional hydrogenase. Thus, the presence or absence of the uptake hydrogenase is only one factor in determining the ultimate levels of hydrogen that can be produced.

16.2.6 Transcription and Translation of Hydrogen Production Genes

Another objective has been to understand transcription and transcriptional regulation of genes in *Cyanothece* 51142 under a variety of different environmental conditions. We are particularly interested in transcriptional regulation of N_2-fixing cells when grown under a variety of different light regimes. The most complete work to date has been on *Cyanothece* 51142, first using Northern blots for individual genes, and then using microarrays for full genome transcriptional analyses (Colón-López and Sherman 1998; Stöckel et al. 2008; Toepel et al. 2008, 2009). We have demonstrated that gene transcription is highly synchronized during 12 h LD conditions and that the large nitrogenase cluster is coordinately transcribed only in the early part of the dark period. As we varied the light regime to include LL and 6 h LD periods, we were able to discriminate between circadian and diurnal regulation of various genes (Toepel et al. 2009). The nitrogenase and uptake hydrogenase genes are regulated in a circadian fashion and occur only in the dark under 12 or 6 h LD conditions. However, once the cultures have been adapted to LD growth and then transferred to LL, they are capable of transcribing the nitrogenase genes at a high level (Toepel et al. 2009). On the other hand, the bidirectional hydrogenase typically follows a more diurnal pattern and is transcribed only in the dark when grown under 6 h LD periods.

In addition, there are substantial differences in transcript level between the nitrogenase and uptake hydrogenase on the one hand and the bidirectional hydrogenase on the other hand (Toepel et al. 2008, 2009). Nitrogenase and the uptake hydrogenase genes are usually transcribed in a similar fashion and to very high transcript levels. On the other hand, the bidirectional hydrogenase is always transcribed at quite modest levels in all conditions so far tested and with a peak transcript level that is typically only 1% that of the nitrogenase. On the protein level, we had previously shown that nitrogenase is very prominent during the first few hours of nitrogen fixation in the dark and this was verified by careful proteomic analysis over the 24 h period when grown under 12 h LD cycles. Once again, we could show that peak levels of nitrogenase were far greater than peak levels of the Hox proteins. These are critical factors as we consider overall productivity of hydrogen.

The transcriptomic data for *Cyanothece* 51142 have been published (Stöckel et al. 2008; Toepel et al. 2008, 2009), but the proteomic data are available only now for thorough analysis. The accurate mass and time (AMT) approach has identified 3,616 proteins with high confidence. This includes 70% coverage of gene products from the circular chromosome, but only 48% from the linear chromosome. This is consistent with our transcriptomics results that have indicated little or no expression for many of the genes on the linear chromosome under the conditions so far tested.

This information has been valuable for the overall annotation of the genome, as well as in conjunction with the transcriptomics results to determine if specific proteins are present or not under different growth conditions. Since one of our main objectives is to determine if *Cyanothece* strains are capable of H_2 production, we were particularly interested in protein levels of the three enzymes capable of producing hydrogen: nitrogenase, the uptake hydrogenase, and bidirectional hydrogenase (Table 16.2). The proteomic data are only from cells grown under 12 h LD conditions, but along with previous Western blots can provide a substantial amount of information as to protein levels under different LD regimes. Under nitrogen-fixing conditions, the nitrogenase proteins were the most cyclic proteins in the cell and showed peaks and valleys very similar to that of the gene transcription data – the Nif proteins were present only in the dark. Nonetheless, cells do grow under nitrogen-fixing conditions in LL, and we have shown that the nitrogenase genes are strongly induced in a circadian fashion when grown under LL conditions (Toepel et al. 2008). Additionally, we obtained reasonably high rates of nitrogenase activity in continuous light and we have demonstrated high Nif levels in LL by using Western blots with Nif antibodies (Colon-Lopez et al. 1998). On the other hand, we have shown that the *hupLS* genes are strongly up-regulated in the dark, but are expressed at much lower levels in continuous light (Toepel et al. 2008). Proteomic analysis of the HupL protein indicates that it is resent in high levels in the dark, but only at extremely low levels in the light (Fig. 16.4). This is consistent with our expression

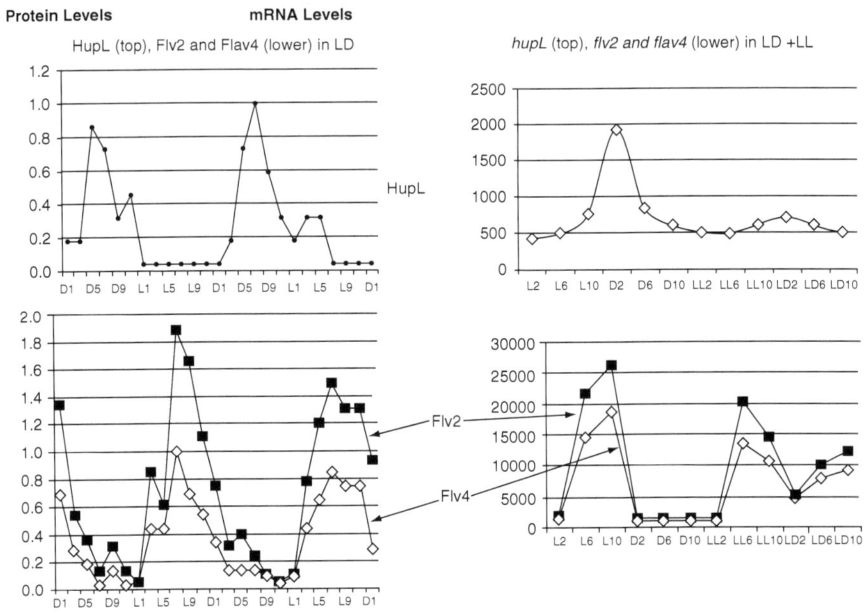

Fig. 16.4 Analysis of protein levels and transcript levels for the HupL proteins, representative of the uptake hydrogenase, and the Flv2 and Flv4 proteins

data which indicated that *hupLS* genes were up-regulated in a circadian fashion in the dark, coincident with the nitrogenase genes (Toepel et al. 2009). Based on these results, it would appear that the HupLS proteins are present, if at all, at very low levels during continuous light. This was the first suggestion that incubating cells under continuous light conditions might be beneficial for hydrogen production. Finally, the *hox* genes were expressed at very low levels in the dark under 12 h LD conditions. When grown under 6 h LD conditions, the *hox* genes were expressed at peaks in the dark but without a substantial trough in the intervening 6 h light period (Toepel et al. 2009). Consistent with the low transcript levels, very low levels of the Hox proteins were identified by proteomics either in the light or in the dark. We are left with the major conclusion that the nitrogenase enzyme is the only hydrogen-producing enzyme present in reasonable quantities during LL growth.

Another type of protein that might be important in permitting hydrogen production is the flavoprotein. Helman et al. (2003) have shown that flavoproteins are essential for photoreduction of O_2 in cyanobacteria via the Mehler reaction. We had determined that an operon containing two flavoproteins was induced under low-O_2 conditions in *Synechocystis* 6803 as well as in *Cyanothece* 51142 (Summerfield, Toepel and Sherman 2008); the data are summarized in Table 16.3. The *flv2* and *flv4* genes are inducible under low-oxygen conditions and we have checked levels of transcript and protein under different light conditions. As shown in Fig. 16.4, the transcript and protein levels of *flv2* and *flv4* were significantly higher in the light than

Table 16.3 Induction of flavoprpteins *flv2* and *flv4* under low O_2 conditions

	Synechocystis genes	Low O_2 induction	*Cyanothece* genes	Low O_2 induction
		t = 1, 2, 6		t = 1, 2, 6
flv1	*sll1521*	1 1 1	cce-2580	1 1 1
flv3	*sll0550*	1 1 1	cce-3635	1 2 1
Flv2	*sll0217*	3 3 3	cce-3835	6 4 2
Flv4	*sll0219*	2 2 2	cce-3833	4 3 2

in the dark, whereas the other two flavoproteins showed little change in expression under these conditions (data not shown). We suggest that these flavoproteins are utilized in the Mehler reaction to reduce oxygen under the appropriate conditions (see Helman et al. 2003). These flavoproteins are in an operon that includes a hypothetical membrane protein with four membrane-spanning regions. Although Flv2 and Flv4 appear to lack ideal $NADP^+$binding sites, the two proteins are likely tethered to the membrane by this membrane protein in such a way as to provide the appropriate activity. This system may help keep the intracellular concentration of oxygen low, thus permitting nitrogenase activity even in the presence of continuous light. These data represented a background as we began our detailed hydrogen production experiments.

16.2.7 Hydrogen Production in Cyanothece *Strains*

We have begun a detailed process of determining hydrogen production in the *Cyanothece* strains under a variety of conditions. Although the most complete work has so far been performed for *Cyanothece* 51142, we have measured hydrogen production under both nitrogen-sufficient and nitrogen-fixing conditions in all of the strains. Importantly, the analysis above led us to incubate N_2-fixing cells in LL under argon to provide a very low-O_2environment. All of the strains produced approximately 2–5 μmol H_2/mg Chl/h when grown under nitrogen-sufficient conditions. We obtained hydrogen production when cells were incubated in the light or dark, but there was an enhancement caused by light incubation. On the other hand, we get significantly more hydrogen produced when cultures are grown under nitrogen-fixing conditions. Strains *Cyanothece* 7424, 7425, 8801, and 8802 all produced ~30–40 μmol H_2/mg Chl/h, whereas *Cyanothece* 7822 and 51142 could produce significantly more. We have not performed as many experiments with *Cyanothece* 7822, but we have consistently obtained extremely high rates (over 150 μmol H_2/mg Chl/h) of hydrogen production with *Cyanothece* 51142 when grown either in the presence or absence of glycerol as a carbon source. The best results are obtained when cells are grown for 3 days under LL and then incubated under argon in LL. A critical feature appears to be the need to protect nitrogenase from oxygen inactivation, and incubation in argon provides that environment. From these results, it is evident that nitrogenase can produce far more hydrogen than the bidirectional hydrogenase, even when the *hupLS* genes are present. Of course, our previous results suggest that very little HupLS was present under LL incubation. Surprisingly, *Cyanothece* 7425 produced hydrogen at levels no higher than that of many of the other strains that contain an uptake hydrogenase. Thus, the presence or absence of HupLS may not be the most critical factor for H_2 production in all diazotrophic cyanobacteria.

16.3 Discussion

Strains of the cyanobacterial genus *Cyanothece* have become among the best studied of all cyanobacteria. We now have genome sequences for six strains and a detailed proteome for *Cyanothece* 51142. In addition, proteomes for the other sequenced strains are being developed as a way of initiating a process of providing proteomes for many bacterial genomes (D. Koppenaal, personal communication). Proteomics and transcriptomics have provided a great deal of high-throughput data that can be used to analyze metabolic processes in *Cyanothece*. In the current study, we have determined the best conditions under which to measure hydrogen production by *Cyanothece* 51142. At the current time, we have obtained the highest levels of hydrogen in *Cyanothece* 51142 when incubated under argon under LL conditions. This was somewhat surprising, but we believe that it has helped to identify the major metabolic issues concerned with hydrogen production in this strain. It is obvious that nitrogenase is the enzyme that produces the hydrogen and we have also demonstrated that this enzyme is produced continuously under these conditions even though cells do not grow (data not shown). The key energetic features for hydrogen production include plentiful energy (provided by photosynthesis and respiration), reducing power, and protection of the nitrogenase from oxygen (provided by argon and the Mehler reaction mediated by the flavoproteins). We also know from previous results that PS I is highly expressed at the same time as the nitrogenase (Colon-Lopez and Sherman 1998). It is highly likely that PS I is being used both in a cyclic fashion for ATP production and along with the flavoproteins in the Mehler reaction.

Overall, the amount of hydrogen produced by these *Cyanothece* species is quite significant based on previous results with cyanobacteria (Dutta et al. 2005). The high-throughput data have provided background information to help us try various physiological conditions for growth and hydrogen production and such experiments are continuing. These rates have so far been with wild-type strains without any genetic or molecular manipulations as the development of such resources are still in progress. Nonetheless, the initial results are promising and demonstrate the benefit of studying novel photosynthetic microbes as a source of alternative energy sources.

Acknowledgments We would like to thank the many people who have been involved with this work, including those at the Washington University Genome Center and the W. R. Wiley Environmental Molecular Science Laboratory (EMSL) for work on proteomics, especially Jon M. Jacobs and Richard D. Smith. Special thanks goes to Eric Welsh of the Pakrasi Lab for all his efforts on the annotation of the *Cyanothece* 51142 genome. We would like to thank the following members of the Pakrasi Lab for figures: Michelle Liberton (Fig. 16.1), Eric Welsh (Fig. 16.2), and Anindita Banerjee (Fig. 16.3).

This work was supported by the Danforth Foundation at Washington University, a Department of Energy grant on hydrogen production, and by the Joint Genome Initiative. This work is also part of a Membrane Biology Scientific Grand Challenge project at the W. R. Wiley Environmental Molecular Science Laboratory, a national scientific user facility sponsored by the U.S. Department of Energy's Office of Biological and Environmental Research program (Pacific Northwest National Laboratory).

References

Allardet-Servent A, Michaux-Charachon S, Jumas-Bilak E et al (1993) Presence of one linear and one circular chromosome in the *Agrobacterium tumefaciens* C58 genome. J Bacteriol 142:7869–7874

Colón-López MS, Sherman DM, Sherman LA (1997) Transcriptional and translational regulation of nitrogenase in light-dark- and continuous-light-grown cultures of the unicellular cyanobacterium *Cyanothece* sp. strain ATCC 51142. J Bacteriol 179:4319–4327

Colón-López MS, Sherman LA (1998) Transcriptional and translational regulation of photosystem I and II genes in light-dark- and continuous-light-grown cultures of the unicellular cyanobacterium Cyanothece sp. strain ATCC 51142. J Bacteriol 180:519–526

Colón-López MS, Tang H, Tucker DL et al (1999) Analysis of the *nifHDK* operon and structure of the NifH protein from the unicellular, diazotrophic cyanobacterium, *Cyanothece* strain sp. ATCC 51142. Biochim Biophys Acta 1473:363–375

Dean DR, Bolin JT, Zheng L (1993) Nitrogenase metalloclusters: structures, organization, and synthesis. J Bacteriol 175:6737–6744

Dutta D, De D, Chaudhuri S et al (2005) Hydrogen production by cyanobacteria. Microb Cell Fact 4:36

Ferdows MS, Barbour AG (1989) Megabase-sized linear DNA in the bacterium *Borrelia burgdorferi*, the Lyme disease agent. Proc Natl Acad Sci U S A 86:5969–5973

Haselkorn R (1978) Heterocysts. Annu Rev Plant Physiol 29:319–344

Helman Y, Tchernov D, Reinhold L et al (2003) Genes encoding A-type flavoproteins are essential for photoreduction of O_2 in cyanobacteria. Curr Biol 13:230–235

Kinashi H, Shimaji-Murayama M, Hanafusa T (1992) Integration of SCP1, a giant linear plasmid, into the *Streptomyces coelicolor* chromosome. Gene 115:35–41

Lipton MS, Pasa-Tolic L, Anderson GA et al (2002) Global analysis of the *Deinococcus radiodurans* proteome by using accurate mass tags. Proc Natl Acad Sci USA 99:11049–11054

Meunier PC, Colón-López MS and Sherman LA (1998) Photosystem II cyclic heterogeneity and photoactivation in the diazotrophic, unicellular cyanobacterium *Cyanothece* species ATCC 51142. Plant Physiol 116:1551–1562

Michael TP, Mockler TC, Breton G et al (2008) Network discovery pipeline elucidates conserved time-of-day-specific cis-regulatory modules. PLoS Genet 4:e14

Reddy KJ, Haskell JB, Sherman DM et al (1993) Unicellular, aerobic nitrogen-fixing cyanobacteria of the genus *Cyanothece*. J Bacteriol 175:1284–1292

Schneegurt MA, Sherman DM, Nayar S et al (1994) Oscillating behavior of carbohydrate granule formation and dinitrogen fixation in the cyanobacterium *Cyanothece* sp. strain ATCC 51142. J Bacteriol 176:1586–1597

Schneegurt MA, Sherman DM, Sherman LA (1997a) Composition of the carbohydrate granules of the cyanobacterium, *Cyanothece* sp. strain ATCC 51142. Arch Microbiol 167:89–98

Schneegurt MA, Sherman DM, Sherman LA (1997b) Growth, physiology, and ultrastructure of a diazotrophic cyanobacterium, *Cyanothece* sp. strain ATCC 51142, in mixotrophic and chemoheterotrophic cultures. J Phycol 33:632–642

Schneegurt MA, Tucker DL, Ondr JK et al (2000) Metabolic rhythms of a diazotrophic cyanobacterium, *Cyanothece* sp. strain ATCC 51142, heterotrophically grown in continuous dark. J Phycol 36:107–117

Sherman LA, Meunier P, Colón-López MS (1998) Diurnal rhythms in metabolism: a day in the life of a unicellular diazotrophic cyanobacterium. Photosynth Res 58:25–42

Stal LJ, Moezelaar R (1997) Fermentation in cyanobacteria. FEMS Microbiol Rev 21:179–211

Stöckel J, Welsh EA, Liberton M et al (2008) Global transcriptomic analysis of *Cyanothece* 51142 reveals robust diurnal oscillation of central metabolic processes. Proc Natl Acad Sci USA 105:6156–6161

Summerfield TC, Toepel J, Sherman LA (2008) Low-oxygen induction of normally cryptic *psbA* genes in cyanobacteria. Biochemistry 47:12939–12941

Toepel J, McDermott JE, Summerfield TC et al (2009) Transcriptional analysis of the unicellular diazotrophic cyanobacterium *Cyanothece* sp. ATCC 51142 grown under short day/night cycles. J Phycol 45:610–620

Toepel J, Welsh E, Summerfield TC et al (2008) Differential transcriptional analysis of the cyanobacterium *Cyanothece* sp. strain ATCC 51142 during light-dark and continuous-light growth. J Bacteriol 190:3904–3913

Volff JN, Altenbuchner J (2000) A new beginning with new ends: linearisation of circular chromosomes during bacterial evolution. FEMS Microbiol Lett 186:143–150

Welsh EA, Liberton M, Stöckel L et al (2008) The genome *Cyanothece* 51142, a unicellular diazotrophic cyanobacterium important in the marine nitrogen cycle. Proc Natl Acad Sci U S A 105:15094–15099

Wolk CP (1996) Heterocyst formation. Annu Rev Genet 30:59–78

Chapter 17
A Feasibility Study of Large-Scale Photobiological Hydrogen Production Utilizing Mariculture-Raised Cyanobacteria

Hidehiro Sakurai, Hajime Masukawa, Masaharu Kitashima, and Kazuhito Inoue

Abstract In order to decrease CO_2 emissions from the burning of fossil fuels, the development of new renewable energy sources sufficiently large in quantity is essential. To meet this need, we propose large-scale H_2 production on the sea surface utilizing cyanobacteria. Although many of the relevant technologies are in the early stage of development, this chapter briefly examines the feasibility of such H_2 production, in order to illustrate that under certain conditions large-scale photobiological H_2 production can be viable. Assuming that solar energy is converted to H_2 at 1.2% efficiency, the future cost of H_2 can be estimated to be about 11 (pipelines) and 26.4 (compression and marine transportation) cents kWh^{-1}, respectively.

17.1 Our Need for Research and Development of Large-Scale Production of Renewable Energy

By 2005 the global atmospheric concentration of the greenhouse gas CO_2 had increased from a pre-industrial value of about 280 to 379 ppm (IPCC 2007). In order to mitigate global warming, the development of renewable non-polluting energy alternatives to fossil fuels on a worldwide scale is urgently needed (see Sakurai and Masukawa 2007; Sakurai et al. in press).

The amount of solar energy received on the earth's surface is vast (about $2{,}700{,}000 \times 10^{18}$ J/yr) and exceeds the present use of fossil fuel energy (404×10^{18} J/yr, in 2006) by more than 6,000 times. The technical challenge that must be overcome for solar energy to be an economically feasible alternative is the low intensity at which it is received on the earth's surface (about 1,500 kWh m^{-2} yr^{-1}, at the middle latitudes). If we are able to convert solar energy into a usable form of energy

H. Sakurai (✉)
Research Institute for Photobiological Hydrogen Production, Kanagawa University, 2946 Tsuchiya, Hiratsuka, Kanagawa 259-1293, Japan
e-mail: sakurai@waseda.jp

This chapter is dedicated to Dr. Anthony San Pietro who passed away on September 13, 2008.

P.C. Hallenbeck (ed.), *Recent Advances in Phototrophic Prokaryotes*, Advances in Experimental Medicine and Biology 675, DOI 10.1007/978-1-4419-1528-3_17,

at 1 and 2% efficiency, about 15 and 30 kWh m^{-2} yr^{-1}, respectively, of renewable energy will be acquired in this region. If photobiological conversion of solar energy is to substitute for or supplement fossil fuels, economical energy production is essential. Considering that the amount of energy in foods accounts for only about 5% of the anthropogenic primary energy use (Sakurai and Masukawa 2007), we cannot expect large amounts of additional energy to be produced from land biomass, and thus we have proposed large-scale H_2 production utilizing mariculture-raised cyanobacteria. In proposing the system described below, we do not intend to criticize other systems such as the hydrogenase-based H_2 production and algal fuels (biodiesel).

17.2 Nitrogenase-Based Photobiological Hydrogen Production by Cyanobacteria

17.2.1 Hydrogenase and Nitrogenase as Hydrogen-Producing Enzymes

If large-scale H_2 production by mariculture is to be practical, the candidate photosynthetic organisms must use H_2O as the electron donor, thus narrowing the possibilities to cyanobacteria and eukaryotic microalgae. Both hydrogenase and nitrogenase are potential candidates as H_2-producing enzyme (review: Rao and Cammack 2001; for cyanobacteia: Tamagnini et al. 2002). In terms of the theoretical maximum energy conversion efficiency, hydrogenase (32.9% vs. 550 nm light (single-stage process), 22% (two-stage process)) is superior to nitrogenase (13.9–16.5%) (cf. C3 photosynthesis: 27.6%, C4 photosynthesis: 20.7–24.5%) (Sakurai and Masukawa 2007). However, hydrogenase catalyzes a reversible reaction and absorbs H_2 in the presence of O_2, when storage metabolites are exhausted, during the night or when shady conditions prevail. Hydrogenase-based processes therefore require frequent harvesting of H_2 or some measures to restrict H_2 reabsorption.

17.2.2 Hydrogen Production by Nitrogenase

Nitrogenase catalyzes the reduction of nitrogen to ammonia with reduced ferredoxin/flavodoxin as electron donors and with H_2 as the inevitable by-product. The reaction is expressed under the optimal conditions for nitrogen fixation, as

$$N_2 + 8e^- + 8H^+ + 16\,ATP \rightarrow H_2 + 2\,NH_3 + 16(ADP + Pi) \qquad (17.1)$$

and more generally as

$$(1-n)N_2 + 8e^- + 8H^+ + 16\,ATP \rightarrow (1+3n)H_2 + 2(1-n)\,NH_3 + 16\,(ADP + Pi) \qquad (17.2)$$

Nitrogenases typically bind a MoFeS cluster (Mo type) as the catalytic center, but some bind V (V type) or Fe (Fe-only type) instead of Mo. The latter types of enzymes are less efficient in nitrogen fixation, in other words, more favorable than the Mo type for H_2 production in the presence of N_2.

In the absence of N_2 (e.g., under Ar), all the electrons are allocated to H_2 production:

$$2e^- + 2H^+ + 4\,ATP \rightarrow H_2 + 4\,(ADP + Pi) \qquad (17.3)$$

Although nitrogenase is less efficient in H_2 production than hydrogenase in terms of its theoretical maximum energy conversion efficiency as the reaction consumes large amounts of ATP, it has the merit of catalyzing a unidirectional production of H_2.

17.2.3 Heterocyst-Forming Cyanobacteria

There are several types of strategies adopted by cyanobacteria in order to protect O_2-sensitive nitrogenase from the potentially dangerous O_2-evolving photosynthesis. We are using heterocyst-forming cyanobacteria because they are amenable to genetic engineering (Elhai and Wolk 1988) and because the whole-genome sequence of *Nostoc/Anabaena* sp. PCC 7120 strain was the first to be determined among the nitrogen-fixing cyanobacteria groups.

17.2.4 Effects of Inactivation of Hydrogenase Activity by Genetic Engineering

The entire process of photoinduced H_2 production is depicted as (1) production of organic compounds by ordinary C3 photosynthesis accompanied by O_2 evolution in vegetative cells, (2) supply of organic compounds to cells specialized for N_2 fixation (heterocysts) that is devoid of O_2-evolving photosynthesis, (3) H_2 evolution (and N_2 fixation) by nitrogenase using organic compounds as electron donors. The presence of hydrogenases that reabsorb the H_2 is considered to be one of the major obstacles to achieving efficient solar energy conversion by a nitrogenase-based system, and a hydrogenase mutant of *Anabaena variabilis* ATCC 29413 generated by disrupting the *hup* gene was shown to have higher hydrogen-producing activity than the wild type (Happe et al. 2000). We have also created genetically defined hydrogenase-inactivated mutants of *Nostoc/Anabaena* sp. PCC 7120 and have shown that the mutants produced H_2 at four to seven times the wild-type rate (Masukawa et al. 2002).

Since our H_2 production system is based on photosynthesis and nitrogenase activities of cyanobacteria, we speculated that the wild-type strain with high nitrogenase activity under light might be a good candidate as the parent strain for further improved photobiological H_2 production through genetic engineering. *Nostoc* sp. PCC 7422 was chosen from 12 other heterocystous strains because it has the highest nitrogenase activity. We sequenced the uptake hydrogenase gene (*hup*) cluster from the strain and constructed a mutant (Δ*hupL*) by insertional disruption of the *hupL* gene (the wild-type cells of this strain showed almost no Hox activity). The Δ*hupL* cells could accumulate H_2 to about 29% (Yoshino et al. 2007) in several days, in the presence of O_2 production (Fig. 17.1).

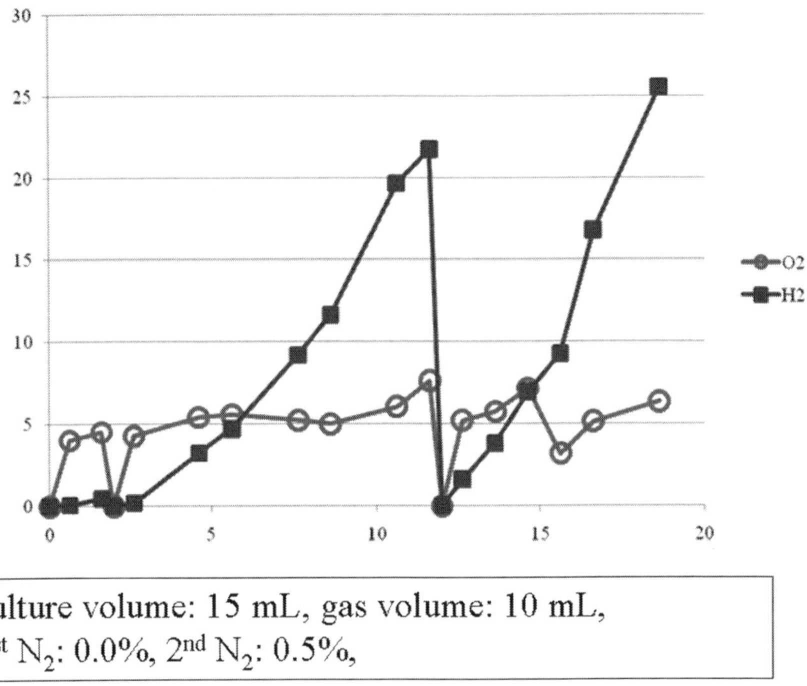

Fig. 17.1 Accumulation of H_2 by *Nostoc* sp. PCC 7422 Δ*hup* mutant in the presence of evolved O_2. A total of 15 ml of cells containing 30 μg chlorophyll *a* grown in $BG11_0$ for 2 days were transferred to 25-ml flasks, and the H_2 (■) and O_2 (○) concentrations in the gas phase were determined daily. Light: 12-hour light–12-hour dark cycle (Kitashima et al. unpublished)

17.3 Outline of the Process Design of Large-Scale Hydrogen Production in the Future Utilizing Mariculture-Raised Genetically Improved Cyanobacteria

One of the plausible economical large-scale H_2 production systems for the future may be growth of cyanobacteria in large bioreactor floating on the sea surface, production of H_2 and its repeated harvesting (followed by H_2 gas separation), and finally recycling of the waste cells as fish feed (Sakurai and Masukawa 2007). A plausible future process design is shown in Fig. 17.2.

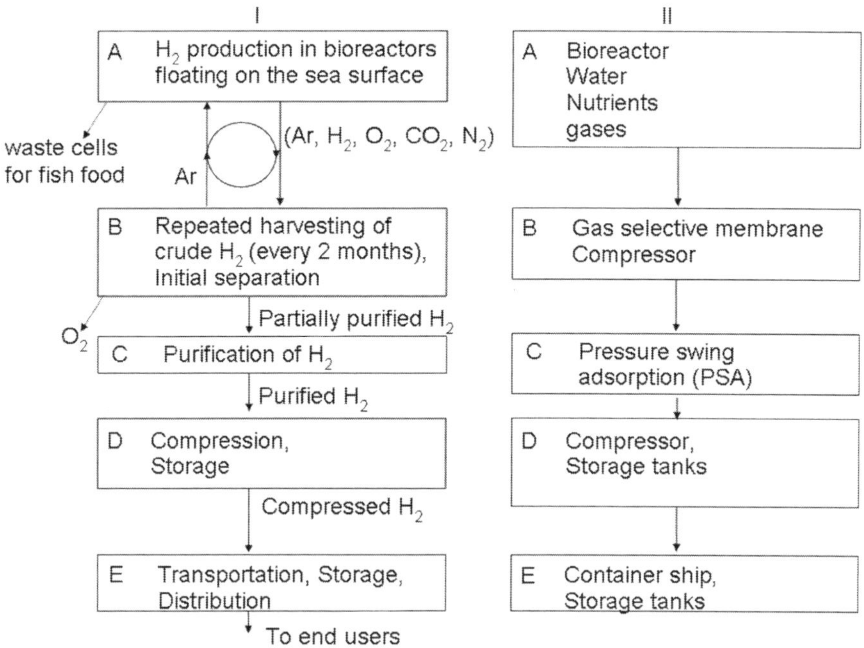

Fig. 17.2 Outline of photobiological H_2 production and transportation. (**a**) flowchart of process and (**b**) required process equipment. See text for (A–E)

17.3.1 H_2 Production in Bioreactors Floating on the Surface of the Sea

Cyanobacteria cells are first grown in a medium containing water and mineral nutrients under air plus CO_2 (e.g., 5%) fixing nitrogen in large plastic bioreactors consisting of several layers of plastic film, with at least one having low permeability to H_2. Each floating bag may be large (e.g., 25 m wide and 200 m long). Some areas of calm sea (such as inland seas) and ocean (e.g., the calm belts, the doldrums near the equator, and the horse latitudes of about 30° north or south) seem to be especially suitable for such large-scale mariculture in inexpensive plastic bags. If the medium is based on freshwater, the bioreactor would spread over the sea surface since the medium would have a lower density than the surrounding seawater. After a period of cell growth, simply decreasing the N_2 concentration (e.g., 1% N_2 in 5% CO_2 plus Ar) will prevent further growth while at the same time promote continuous H_2 production with concomitant evolution of O_2.

17.3.2 Repeated Harvesting of Crude H_2 and Initial Gas Separation

From the following assumptions, about 0.84 m^3 (STP) of H_2 m^{-2} of the bioreactor is produced in 2 months:

- Average solar energy received on the sea surface: 1,500 kWh m^{-2} $year^{-1}$.
- Energy conversion efficiency by cyanobacteria: 1.2% (solar energy into H_2).
- H2 produced in 2 months: 3 kWh m^{-2} = about 0.84 m^3 (STP) (with evolution of 0.42 m^3 O_2) m^{-2} of the bioreactor surface.
- If the volume of the initial gas phase is 0.5 m^3 (STP) m^{-2} of the bioreactor, then the final concentration of H_2 is about 48% (v/v) (0.84/(0.5 + 0.84 + 0.42)).

The gas mixture is harvested to a factory ship with hoses every 2 months with the assistance of working boats, and H_2 is initially separated from O_2 by gas-selective membranes (e.g., H_2 permeates a polychlorovinylidene film about 38 times faster than O_2, and the H_2 concentration can be increased to about 97% by a single operation). (This process of the initial separation is tentative.)

17.3.3 Further Purification of H_2

Contaminating O_2 (about 1.3%) is removed either by a second cycle of separation with gas-selective membranes or by using a catalyst (which would consume two volumes of H_2 for each volume of O_2). The H_2 is finally purified by pressure swing adsorption (PSA) on the factory ship.

17.3.4 Compression or Transformation to a Form Suitable for Transportation by Ship and Storage

For long-distance transportation of purified H_2 from the sea surface to the port, its volume should be greatly decreased by some means, possibly by compression (other possibilities: liquefaction, adsorption to alloy, etc). The H_2 is compressed into storage containers, transported by ships to final destination ports, unloaded, and stored awaiting final distribution.

17.4 Estimation of the Future Production Cost–Energy Balance

There are considerable uncertainties regarding the production processes and, therefore, the cost estimates of H_2 production are subject to change. Nevertheless, we present here an estimate so that the readers may understand the potential for such a

H_2 production system. We have detailed the costs of each item in the process separately so as to allow for the identification of the parts of the process that may be improved in order to reduce the total cost. With advances in relevant technologies, it will also be possible to recalculate the cost based on improved assumptions. In calculating the cost of H_2, the currency exchange rates assumed are US \$1 = 0.7 € = 95 ¥. As the value of energy of H_2, a high heating value (HHV, the oxidation product is condensed water) of 12.8 MJ m^{-3} is assumed (the low heating value (LHV, the product is vapor) is 10.8 MJ m^{-3}, about 84% of HHV).

17.4.1 H_2 Production in Bioreactors Floating on the Sea Surface

A number of assumptions need to be made to estimate the net energy yield of the initial process. These can be divided into four areas.

17.4.1.1 Energy Conversion Efficiency (in the Future)

Cyanobacteria photobiologically convert solar energy (1,500 kWh m^{-2} yr^{-1}, total radiation) into H_2 at 1.2% efficiency, resulting in 18 kWh or 64.8 MJ of H_2 m^{-2} yr^{-1}.

17.4.1.2 Photobioreactor

The bioreactor is composed of three layers of plastic bags, for a total of six layers (sunny side and shady side) of transparent plastic film. The innermost bag holds the cyanobacterial culture, the middle bag has very low permeability to H_2, and the outermost bag serves as mechanical protection for the inner bags. The thickness of each film is 0.08 mm, and therefore 480 cm^3 of plastic per m^2 of the bioreactor's sunny side surface is required. Assuming an average plastic price of \$2–4 kg^{-1} (or liter), the material cost is 96–192 cents m^{-2} of bioreactor. The used plastics can be recycled many times to regenerate plastic films at about half the price of the new materials. The above assumptions result in the cost of the bioreactor being about 48–96 cents m^{-2} of bioreactor surface per year assuming once-a-year renewal.

Note that plastic film of 480 cm^3 is assumed to be produced by consuming 360 ml of crude oil for processing, which is equivalent to 13.9 MJ (3.9 kWh) m^{-2} $year^{-1}$. The plastics can be recycled at an energy cost of 20% of the feedstocks (about 0.78 kWh, 4.3% of H_2 produced). The amount of energy in feedstocks derived from fossil fuels can be decreased further because currently H_2 generated from fossil fuels is used as a part of the feedstocks for plastic film production, and photobiologically produced H_2 can replace some part of it.

17.4.1.3 Culture Medium

Nitrogen-fixing cyanobacterial cells can grow in liquid media without combined nitrogen. Cyanobacteria are cultured in liquid medium 20 cm in depth (200 l m^{-2} or 0.2 ton m^{-2} of the bioreactor) utilizing freshwater. Potentially growth-limiting nutritional elements (especially the major ones, 18 mM K_2HPO_4, 0.03 mM $FeCl_3$) are added to the medium as "fertilizers" akin to agricultural practices (5–20 cents). Once grown, cyanobacteria continuously produce H_2 allowing repeated harvesting, and further addition of nutrients and CO_2 is not necessary. If the medium is renewed twice a year, the cost of chemicals is calculated to be about 10–40 cents m^{-2} of the bioreactor per year (Sakurai et al. 2009). The cost of water for the medium (0.2 ton m^{-2}) is calculated to be 1.5–16 cents m^{-2} of the bioreactor surface from a reference price of water for industrial use sold by local governments in Japan at about 7.5–80 cents ton^{-1}. If the medium is renewed twice a year, the cost of water is 3–32 cents m^{-2} of the bioreactor per year.

As the water substrate for H_2 production, 18 g of H_2O can generate 22.4 l (STP) of H_2, which corresponds to 1 kg of H_2O being converted to 1.24 m^3 of H_2, with an energy content equivalent to about 0.35 l of crude oil (3.3 m^3 H_2 is equivalent to about 1 l of crude oil in enthalpy). The cost of water substrate used as the electron donor is thus negligible. Using eutrophic water could further reduce the cost of chemicals in the culture media.

17.4.1.4 Cost of Culture Gases

The initial gas phase composition is 5% CO_2, 1% N_2, and 94% Ar (0.5 m^3 m^{-2} of bioreactor surface). The price of Ar is assumed to be $56 (a bulk rate), which leads to about 5 cents (2 cents with recycle, see below 17.4.2) m^{-2} of bioreactor. The costs of CO_2 and N_2 are small compared with Ar.

The sum of the Costs *A1-4* is calculated to be 63–170 (cents m^{-2}).

In addition to the Costs *A1-4*, the following costs will be incurred in the biological H_2 production stage: cyanobacteria growth costs, labor costs, the cost of ships, interest on capital goods, and the cost of marine transportation of production materials to the site of H_2 production.

17.4.2 Repeated Harvesting of Crude H_2 and Initial Separation

The gas mixture is harvested every 2 months (containing about 48% H_2) from bioreactors to a factory ship with the aid of a small group of boats. H_2 is partially purified in the initial separation process by gas-selective membranes, Ar is recycled to bioreactors, and O_2 is removed. We assume that 2% of energy in H_2 is lost in the initial separation process.

17.4.3 Further Purification of H_2

Contaminating O_2 (about 1.3%) is either removed by the catalyst consuming the two volumes of H_2 (2.6% of the energy). Thereafter, the H_2 is finally purified by pressure swing adsorption (PSA) on the factory ship with an overall energy efficiency of 85% with losses of 15% of energy. A subtotal of about 20% of energy is lost in 17.4.2 and 17.4.3.

17.4.4 Compression for Transportation by Ship

The purified H_2 is compressed to 35 MPa (about 15 kg/m^3, energy content: 2.1 GJ/kg).

In the presentation "Well-to-Wheels Analysis," Joseck and Wang (2007) estimated that 2,000 and 7,200 Btu (British thermal unit) of energy are required for compression (to about 35 MPa) and storage, respectively (about 8% in total), for the H_2 (116,000 Btu (100%)) originally generated by electrolysis powered by electricity from wind. By analogy, we assume that 8% of energy in 17.4.3 after PSA (80% energy yield) is lost in this process, which is equivalent to 6.4% of H_2 energy in the starting gas in *4.1A*.

17.4.5 Marine Transportation and Storage

H_2 purified and compressed on a factory ship is transported to final destination ports in storage tanks by container ships. The landed H_2 can be transported either by pipelines or by trucks to end users. The compressed H_2 in containers is transported to ports by a container ship and delivered to final users. If the distance between the marine area of H_2 production and the port is 2,000 km, we assume that the energy lost is about 4% of H_2.

17.5 Estimation of Net Energy Production

We assume that 18×10^6 kWh of H_2 (100%) is produced per km^2 per year and 6% of energy is lost in the process A (including bioreactors), resulting in the subtotal energy losses (A–E) of about 36%: A (6%), B (5%), C (15%), D (6%), and E (4%). In addition to the above losses, fuels for a factory ship and a group of working boats will be required (estimated to be 4%). As a total of about 40% of energy in photobiologically produced H_2 is lost, the net energy of H_2 at the port is 10.8×10^6 kWh (270 ton) km^{-2} year^{-1} (equivalent to about 930 tons of gasoline (11.6 kWh kg^{-1}) or 980 tons of crude oil (10.8 kWh kg^{-1})). This amount of energy will be more than enough to cover the energy cost required for manufacturing PSAs, compressors, storages, factory ships, cargo ships, etc., and therefore photobiological H_2 production would be able to produce a large quantity of net energy.

17.6 Estimation of Cost

17.6.1 Cost Analysis for Chlamydomonas-Based H_2 Production

The cost of hydrogenase-based photobiological H_2 production from the green alga *Chlamydomonas reinhardtii* was analyzed by Amos (2004). One of his assumed production systems is depicted roughly as follows: (1) growth of cells by ordinary photosynthesis in ponds, (2) transfer of cells to anaerobic bioreactor (\$1 m^{-2}, ponds covered with transparent plastic films) and H_2 production (about 39 kWh m^{-2} yr^{-1}, in Arizona; estimated solar radiation of about 2,600 kWh m^{-2} yr^{-1}), (3) harvesting and purification of H_2 (compressor and PSA), and (4) compression of H_2 to 20 MPa (storage compressor and high-pressure storage). Assuming that ongoing improvements in technology are successful, he estimated a H_2 sale price of \$8.97 kg^{-1} or 22.8 cents kWh^{-1}. The cost includes a 15% return on investment, and capital-related charges comprise about 90% of the cost. The largest cost arises from point 4, that is, compression and high-pressure storage (especially the latter), and is estimated to be \$7.75 kg^{-1} of H_2, about 86% of the sale price. If the reactor is more expensive, the estimated H_2 sale price rises to 34.4 and 1,110 cents kWh^{-1} for a reactor price of \$10 and 100 m^{-2}, respectively, indicating that reduction of reactor cost is very important in achieving economically viable production. By contrast, we are proposing a reactor consisting of three layers of plastic bags with a cost of about \$0.48–0.96 m^{-2}.

17.6.2 Estimation of the Cost of H_2 Production by Cyanobacteria

17.6.2.1 Comparison with Photobiological H_2 Production by *Chlamydomonas*

Overall, our cyanobacterial H_2 production system (System I) is rather similar to that of *Chlamydomonas* (System II) (Amos 2004) with some notable differences in H_2 production in the bioreactors.

Comparisons:

(1) *Cell culture.* The initial H_2 production costs from the bioreactors in System II are estimated to be about 1.45 and 9.34 cents kWh^{-1} of the total capital costs assuming the reactors cost \$1 and 10 m^{-2}, respectively. In System I, the reactor cost is estimated to be \$0.48–0.96. In System I, a single type of bioreactor is required, and combined nitrogen can be omitted from the culture medium, which is renewed twice a year. In System II, a system with two continuous-flow reactors are used. Therefore, System II requires much more water and nutrient, notably combined nitrogen, than System I. In System I, ships and boats are required. The amounts of H_2 produced are 18 and 39 kWh m^{-2} yr^{-1} in System I and II, respectively. Overall, we simply assume here that the cost of the biological H_2 production stage is about the same (1.5 cents kWh^{-1}).

(2) *Harvesting of $H_2$and initial separation.* In System I, initial separation of H_2 from O_2 is required, but not in System II. We tentatively assume a higher cost of 1 cent (about 4% of the final sale) kWh^{-1} of H_2 as the final commodity. In System I, the gas is harvested at any time (typically every 2 months), but in System II, gas must be frequently harvested almost everyday or at least every week so that a higher number of backup storage systems and more labor would be required. In System I, no such backup system is required because the produced H_2 just inflates the plastic bags (see point 4).

(3) *PSA and high-pressure storage.* They are required in both systems, but the initial concentration of H_2 differs: about 48% in System I and nearly 100% in System

II. The pressure is higher in System I (35 MPa) than in System II. We assume an additional cost of 0.8 cent (about 3% of the final sale) kWh^{-1} of the final commodity H_2 for PSA and compression. As 15% more H_2 energy is lost in PSA in System I, about 0.2 cents should be added (the cost of this process in System II is about 1.5 cents kWh^{-1} of H_2).

(4) *Marine transportation*. In System I, marine transportation of the product H_2 is required, but not in System II. The cost is estimated from the following assumptions: (4.1) The distance from the site of H_2 production to port is 2,000 km. If the speed of the freighter is 800 km day^{-1} (4,000 km of a round trip requires 5 days), and if it takes a half-day each for uploading and unloading the high-pressure storage, a total of 6 days will be required for one round trip. Therefore, a freighter may carry H_2 about 60 times a year. (4.2) The mass percent of compressed H_2 is assumed to be 8% (% in weight of $H_2/(H_2$ + high-pressure storage)). A 10,000 ton-class freighter carries 800 ton of H_2 (about 31.5×10^6 kWh) to a port at a time or 48×10^3 ton of H_2 (about 1.9×10^9 kWh) (collected from 180 km^2 of bioreactors) a year. The annual sale of 10,000 ton-class freighters is assumed to be $20 million, and the cost of H_2 transportation is calculated to be 1.1 cents kWh^{-1}.

Because of the marine transportation used in System I, a higher capacity storage system is required than for System II. However, as discussed in Section 17.6.1, point 2, System II requires a greater number of storage backup systems. We assume here that the total storage capacity is about the same for the two systems.

17.6.2.2 Estimation of the Price of H_2 Produced by Mariculture-Raised Cyanobacteria

Amos (2004) estimated a H_2 sale price of 22.8 cents kWh^{-1} ($8.97 kg^{-1}) assuming the reactor cost of $1 m^{-2}. From the above-described comparisons, the sale price of H_2 produced by cyanobacteria (System I) is calculated to be 25.9 (22.8 + 1 + 0.8 + 0.2 + 1.1) cents kWh^{-1} plus costs of the factory ship and working boats and labor costs thereof. The cost of the factory ship itself is calculated to be very small (about 0.01 cent kWh^{-1}) from the assumptions; the price of a factory ship of 40,000 DWT (dead weight ton) is $2 million, life: 40 years, annual interest: 5%, the ship produces 1.9×10^9 kWh of H_2 a year (see 17.6.2.1). The cost of working boats is also small. The labor of the crew is assumed to be 0.5 cent kWh^{-1}: 50 persons, annual salary of $100,000 per person (including the cost of management), divided by 1.9×10^9 kWh of H_2. From the above assumptions, the sale price of H_2 is calculated to be 26.4 cents kWh^{-1}.

In System II, the greatest cost arises from point 4, that is, compression and high-pressure storage (estimated to be 19.7 cents kWh^{-1} of H_2 ($7.75 kg^{-1}) of H_2). If this process can be omitted by directly connecting to H_2 pipelines, then the final price of H_2 produced by *Chlamydomonas* drops from 22.8 to 7.2 cents kWh^{-1} of H_2 (Amos 2004). With the cyanobacteria system, if the bioreactors are floated near land, and if a pipeline system is available, the final price of H_2 will drop to about 11 cents kWh^{-1} of H_2.

17.7 Improvements Required in Biological Research

By increasing the energy conversion efficiency of photobiological H_2 production, the H_2 selling price drops due to reduced bioreactor cost and reduced labor charges per unit amount of H_2. The theoretical maximum energy conversion efficiency of photobiological H_2 production is estimated to be 13.9–16.5% vs. 550-nm visible light (about 6.3–7.4% vs. total solar radiation assuming that visible light (photosynthetically active radiation, PAR) is 45% of the total solar radiation) (Sakurai and Masukawa 2007). Under laboratory conditions, the efficiency of around 3.8% (vs. visible light, which corresponds to about 1.7% vs. total solar radiation) was reported by several groups (e.g., Yoshino et al. 2007). These values apparently exceed our tentative target of 1.2%. However, these high efficiencies are only attained over a relatively short period (several hours) under low light intensities of about one twenty-fifth of full sunlight at the equator. Under outdoor conditions, a reported best efficiency over a relatively long period (days) is about 0.1% (Tsygankov et al. 2002). Thus, more research is needed to improve the long-term outdoor efficiency.

17.7.1 Potential Methods for Further Improvement in Efficiency

Potential methods for improvement of outdoor energy conversion include (1) reduction of antenna size, (2) improvement of nitrogenase (site-directed mutagenesis, use of V-type nitrogenase, reduced concentration of homocitrate essential for efficient nitrogen fixation; Masukawa et al. 2007), (3) improvement of culture conditions, (4) selection of promising wild-type strains followed by genetic engineering (e.g., Yoshino et al. 2007).

17.8 Conclusions

The future price of photobiologically produced H_2 at 1.2% energy conversion efficiency by mariculture-raised cyanobacteria is calculated to be 26.4 cents kWh^{-1} of H_2. If H_2 pipelines were available the price would drop to 11 cents kWh^{-1} of H_2. Although this is more expensive than the current price of crude oil, $50–150 per barrel (about 159 l), equivalent to 2.9–8.8 cents kWh^{-1}, and gasoline (the retail price of $1.5–4 per gallon is equivalent to about 4–11 cents kWh^{-1}), the price of H_2 could be further decreased by improving the light conversion efficiency and by advances in other relevant technologies. Research and development of photobiological renewable energy sources should be more earnestly pursued because photobiologically produced H_2 contributes to the reduction of the greenhouse gas CO_2 emission, and H_2 fuel cells are expected to be more energy efficient than internal combustion engines.

Acknowledgment This work is supported in part by JSPS Grant KAKENHI (B) 21380200 to HS and MEXT Grant (high-tech research center project) to KI.

References

Amos WA (2004) Updated cost analysis of photobiological hydrogen production from *Chlamydomonas reinhardtii* green algae – Milestone completion report. NREL/MP-560-35593. www.nrel.gov/docs/fy04osti/35593.pdf. Accessed 5 August 2008

Elhai J, Wolk CP (1988) Conjugal transfer of DNA to cyanobacteria. Methods Enzymol 167: 747–754

Happe T, Schütz K, Böhme H (2000) Transcriptional and mutational analysis of the uptake hydrogenase of the filamentous cyanobacterium *Anabaena variabilis* ATCC 29413. J Bacteriol 182:1624–1631

Joseck F, Wang M (2007) Well-to-Wheels Analysis. Presented to HTAC on July 31, 2007. www.hydrogen.energy.gov/pdfs/htacjuly07_well_to_wheels.pdf. Accessed 5 August

Masukawa H, Mochimaru M, Sakurai H (2002) Disruption of the uptake hydrogenase gene, but not of the bidirectional hydrogenase gene, leads to enhanced photobiological hydrogen production by nitrogen-fixing cyanobacterium *Anabaena* sp. PCC 7120. Appl Microbiol Biotechnol 58:618–624

Masukawa H, Inoue K, Sakurai H (2007) Effects of disruption of homocitrate synthase genes on photobiological hydrogen production and nitrogenase of *Nostoc* sp. PCC 7120. Appl Environ Microbiol 73:7562–7570

Rao KK, Cammack R (2001) Producing hydrogen as a fuel. In: Cammack R, Frey M, Robson R (eds) Hydrogen as a fuel – Learning from nature. Taylor & Francis, London and New York, pp 201–230

Sakurai H, Masukawa H (2007) Promoting R & D in photobiological hydrogen production utilizing mariculture-raised cyanobacteria. Mar Biotechnol 9:128–145

Sakurai H, Masukawa H, Inoue K (2009) A preliminary survey of the economical viability of large-scale photobiological hydrogen production utilizing mariculture-raised cyanobacteria. In: Gault PM, Marler HJ (eds) Handbook on cyanobacteria: biochemistry, biotechnology and applications. Nova Publishers, COMMACK, NY, pp 443–462

Tamagnini P, Axelsson R, Lindberg P, Oxelfelt F, Wünschiers R, Lindblad P (2002) Hydrogenases and hydrogen metabolism of cyanobacteria. Microbiol Mol Biol Rev 66:1–20

Tsygankov AA, Fedorov AS, Hisourov SN, Rao KK (2002) Hydrogen production by c cyanobacteria in an automated outdoor photobioreactor under aerobic conditions. Biotechnol Bioenbineer 80:777–783

Yoshino F, Ikeda H, Masukawa H, Sakurai H (2007) High photobiological hydrogen production activity of a *Nostoc* sp. PCC 7422 uptake hydrogenase-deficient mutant with high nitrogenase activity. Mar Biotechnol 9:101–112

Chapter 18
Hydrogenases and Hydrogen Metabolism in Photosynthetic Prokaryotes

Christoph Schwarz, Zach Poss, Doerte Hoffmann, and Jens Appel

Abstract Hydrogen plays important, different roles in a variety of photosynthetic prokaryotes. The variety of different hydrogenases, their catalytic sites, maturation, and genetic organization are reviewed in the context of what is known about these enzymes from model non-photosynthetic organisms. Examples from specific cyanobacteria and various anoxygenic photosynthetic bacteria are discussed in detail along with what is known about their metabolic role. The latest findings on transcriptional regulators and the metabolic conditions that regulate the expression of hydrogenases in various photosynthetic prokaryotes are emphasized.

18.1 Biological Impact of Hydrogen Metabolism

Hydrogen (H_2), with 0.5–0.6 ppm, is the second most abundant trace gas in the atmosphere (Novelli et al. 1999). Due to photochemical oxidation of hydrocarbons in the upper layers of the atmosphere, combustion of fossil fuels and biomass burning, it is constantly produced. Simultaneously, a large portion of this H_2 (75–77%) is consumed by microbial communities in the soil (Conrad 1996) and serve the energetic needs of these organisms. In the archaean atmosphere H_2 concentrations were certainly higher than at present and hydrogen likely played an important role in supporting life on the early earth. This is also valid for the first photosynthetic prokaryotes. They could have used hydrogen to provide reducing power for CO_2 fixation or as a convenient electron donor for their photosystems. Both processes are found in many of the contemporary anoxygenic photosynthetic bacteria. Hydrogen metabolism in photosynthetic prokaryotes may generally involve the action of two different enzymes, nitrogenase and hydrogenase.

C. Schwarz (✉)
School of Life Science, Arizona State University, POBox 874501, Tempe, AZ 85287, USA
e-mail: christoph.schwarz@asu.edu

P.C. Hallenbeck (ed.), *Recent Advances in Phototrophic Prokaryotes*, Advances in Experimental Medicine and Biology 675, DOI 10.1007/978-1-4419-1528-3_18,

Nitrogenase produces hydrogen as a by-product of the nitrogen fixation process according to the following reaction: 16 ATP + 16 H_2O + N_2 + 10 H^+ + $8e^-$ $\Rightarrow$ 16 ADP + 16 Pi + 2 NH_4 + H_2 (Rees et al. 2005). Additionally, in the absence of nitrogen or other substrates, nitrogenase catalysis continues, reducing protons and thereby releasing especially large amounts of hydrogen (Benemann and Weare 1974). The second enzyme participating in hydrogen cycling is hydrogenase, originally discovered and named by Stephenson and Stickland (1931). Generally hydrogenases are enzymes catalyzing the reversible oxidation or evolution of hydrogen according to the following reaction: $H_2 \Leftrightarrow 2H^+ + 2e^-$ (Adams and Stiefel 1998). Hydrogenases are widely distributed in biological systems (Vignais et al. 2001; Schwartz and Friedrich 2006; Vignais and Billoud 2007). Hydrogen is used as a metabolic exchange intermediate in many different microbial consortia. One of these is a well-known prokaryotic isolate, *Methanobacillus omelianskii*, which has later been shown to be a symbiotic association of two organisms that interact mainly via interspecies hydrogen transfer (Bryant et al. 1967). Likewise, certain anaerobic protozoa, which contain specialized organelles for fermentative energy generation, the so-called hydrogenosomes, maintain intracellular methanogens to consume the produced H_2 (van Bruggen et al. 1983; Finlay and Fenchel 1992). Knowledge about these tight connections, which depend on H_2, led to the hypothesis that the first eukaryotes arose from similar microbial interactions (Martin and Müller 1998).

However, hydrogen cycling is also an important process within a single microbial cell. Many fermentative processes release H_2, which serves to dispose of excess reducing equivalents. In addition nitrogenase constantly produces H_2 as a by-product of N_2 fixation. Although a certain amount of this H_2 will escape from the cell, being used by associated microbial H_2 oxidizers or ending up in the atmosphere, many prokaryotes, especially diazotrophs, including cyanobacteria and the majority of the anoxygenic phototrophs, are equipped with several hydrogenases. Some of these are solely consuming or uptake hydrogenases, which are linked to the respiratory chains. These enzymes enable the intracellular recycling of H_2 for energy generation. Importantly, the simultaneous consumption of H_2 keeps H_2-evolving processes feasible, since the accumulation of higher concentrations of H_2 would be thermodynamically unfavorable. However, certain ecological niches in coastal zones, especially microbial mats, have been reported to have very high photosynthetic activity and significant H_2 metabolism. The H_2 partial pressure in gas bubbles on the surface of *Lyngbya* mats, e.g., varies in the range of four orders of magnitude following a 24-h diel cycle. Maximum H_2 levels during night reach 10% (Hoehler et al. 2001, 2002). Since H_2 concentrations were observed to be highest in the upper 2 mm of the mat body, the observed H_2 evolution was attributed to the cyanobacteria which dominate these layers. Primary productivity of the mats is dependent on photosynthetic light harvesting during the day, but most of the H_2 evolved is thought to originate from nitrogen fixation and fermentation during the night phases (Fay 1992; Severin and Stal 2008). This review shall discuss hydrogen metabolism and the hydrogenase inventory of selected phototrophic prokaryotes with particular emphasis on transcriptional regulation.

18.2 Classification and Distribution of Hydrogenases

According to the metal content of their active sites hydrogenases are classified into three major groups, Fe, FeFe, and NiFe (including NiFeSe) hydrogenases.

18.2.1 NiFe Hydrogenases

NiFe hydrogenases can further be subdivided into four phylogenetically distinct groups: (1) the H_2 uptake hydrogenases, (2) cyanobacterial uptake- and sensor hydrogenases, (3) different soluble hydrogenases, and (4) membrane-bound, energy-converting, H_2-evolving hydrogenases. A detailed description of the various characteristics of hydrogenases within the specific groups has been covered by several reviews (Vignais and Colbeau 2004; Vignais and Billoud 2007; Vignais 2008) and shall therefore only be reviewed in brief.

Group 1 enzymes are membrane-bound, heterodimeric, and unidirectional hydrogen-consuming (uptake) hydrogenases which enable cells to grow on hydrogen as a substrate. Hydrogenase-mediated oxidation of hydrogen is coupled to a respiratory chain thereby leading to the generation of a proton motive force. Hydrogenase structural genes are typically associated with two downstream genes, one encoding a hydrogenase-specific protease, while the other encodes a membrane-bound b- or c-type cytochrome. The latter cytochromes link hydrogen consumption to the reduction of the quinone pool and finally respiration which besides oxygen mostly uses alternative electron acceptors like NO_3^-, SO_4^{2-}, or fumarate. The small subunit of group 1 enzymes contains an N-terminal targeting sequence which mediates membrane translocation via the TAT pathway. Among the group 1 enzymes there are hydrogenase 2 from *Escherichia coli*, Hyn from *Thiocapsa roseopersicina* (Rakhely et al. 1998), and HupSL enzymes found in *Rhodopseudomonas palustris*, *Roseovarius* sp. HTCC2601 as well as *Chlorobi*.

Hydrogenases in group 2 are very similar to group 1 enzymes but can be distinguished by the lack of an N-terminal signal peptide responsible for the membrane targeting. Among these cytoplasmic hydrogenases there are the cyanobacterial uptake hydrogenase (group 2a) and the regulatory hydrogen sensors (group 2b). Cyanobacterial uptake hydrogenases are exclusively found in nitrogen-fixing strains. They are membrane associated and thought to recover energy released in the form of hydrogen during the nitrogen fixation process (Appel and Schulz 1998; Tamagnini et al. 2007). In accordance with their physiological function gene expression is coregulated with nitrogenase. The second subclass within group 2 is the sensor hydrogenases. This type of hydrogenase is unable to perform normal catalytic hydrogen oxidation, rather they function as sensors of hydrogen availability and are the sensor component of a two-component signal transduction system. Examples of regulatory hydrogenases are HoxBC of *Ralstonia eutropha* or HupUV of *Rhodobacter capsulatus* (Friedrich et al. 2005; Vignais et al. 2005), which are discussed in further detail in Section 18.6.1.

A characteristic feature of group 3 hydrogenases is that they contain additional subunits, which allow binding of soluble cofactors like NAD, NADP, or F_{420} (8-hydroxy-5-deazaflavin). Examples of group 3 enzymes are the heteropentameric NAD(P)-reducing HoxEFUYH found in cyanobacteria, *T. roseopersicina*, *Allochromatium vinosum*, *R. eutropha* as well as the Fpo of *Methanosarcina mazei* (Tamagnini et al. 2007; Rakhely et al. 2004; Long et al. 2007; Schneider and Schlegel 1976; Baumer et al. 2000). Finally group 4 is represented by the hydrogenase 3 and 4 (Hyc and Hyf) of *E. coli*, the CO dehydrogenase complex of *Rhodospirillum rubrum*, Mbh of *Pyrococcus furiosus* (Sapra et al. 2000) as well as the Ech (energy-conserving hydrogenase) in *Methanosarcina barkeri* (Kunkel et al. 1998; Hedderich and Forzi 2005). Enzymes belonging to this group are membrane-bound multisubunit hydrogenases which couple H_2 evolution to energy conservation.

18.2.2 FeFe Hydrogenases

FeFe hydrogenases are frequently found in anaerobic prokaryotes like *Clostridia*, *Desulfovibrio*, *Shewanella*, or *Thermotoga* (Vignais and Billoud 2007) but also occur in chloroplasts of eukaryotic algae (*Chlamydomonas*, *Chlorella*, *Scenedesmus*) as well as in the hydrogenosomes of protozoan parasites (*Trichomonas*, *Giardia*, *Entamoeba*, *Nyctotherus*) and anaerobic chytridiomycota (*Neocallimastix*, *Piromyces*) (Happe et al. 2002; Hackstein et al. 1999). Generally FeFe hydrogenases may be separated into three families: (1) cytoplasmic or soluble, monomeric enzymes, for example, the FeFe hydrogenases of *Clostridium pasteurianum*; (2) periplasmatic, heterodimeric enzymes; and (3) soluble, monomeric enzymes, found in algal chloroplasts. Among the photosynthetic bacteria, FeFe hydrogenases are almost exclusively found only in the gram-positive heliobacteria (*Heliobacterium modesticaldum*, *Heliobacterium mobilis*). However, homologs of HydA are present in *R. rubrum* and *R. palustris* strains (Meyer 2007), but neither of the genomes of these entirely sequenced phototrophic strains harbor FeFe hydrogenase maturation genes. It still remains to be elucidated if these strains really synthesize active FeFe hydrogenases.

18.3 Hydrogenase Anatomy and Active Site Biochemistry

18.3.1 NiFe Hydrogenases

Initial progress in the understanding of hydrogenase active site biochemistry was achieved by a combination of spectroscopic analysis and X-ray crystallography (Armstrong and Albracht 2005). The elucidation of crystal structures of NiFe hydrogenases from *Desulfovibrio gigas* and *Desulfovibrio vulgaris* (Volbeda et al. 1995; Higuchi et al. 1997) revealed for the first time detailed insights into the molecular organization and active site composition (Fig. 18.1). NiFe hydrogenases consist of

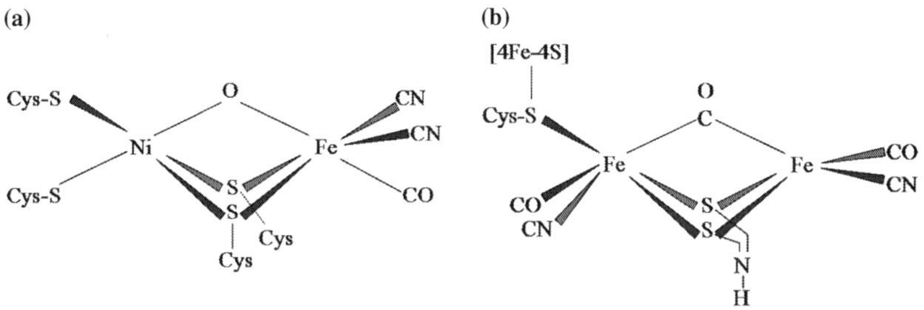

Fig. 18.1 Active site biochemistry of (**a**) a standard [NiFe] hydrogenase (based on the crystal structure of *D. gigas*) and (**b**) a [FeFe] hydrogenase (CpI of *C. pasteurianum*)

two core components which are found in all NiFe hydrogenases, the large (α, around 60 kDa) and the small subunit (β, about 30 kDa). The large subunit of the NiFe hydrogenases carries a bimetallic active site. The active site nickel atom is coordinated by four highly conserved cysteine thiolates, two of which are also involved in bridging to the iron ion. The iron ion additionally has three diatomic ligands, one carbon monoxide (CO) and two cyanide (CN^-) as was determined by Fourier transform infrared spectroscopy (Volbeda et al. 1996; Happe et al. 1997; Pierik et al. 1999). While the active site iron is Fe^{2+}, the nickel may be found in two oxidation states Ni^{2+} or Ni^{3+} (Armstrong 2004). The two metal atoms in the dimetallic center are linked by an additional bridging ligand, known as ligand "X," which shows varying composition according to the activation state of the enzyme and is discussed in Section 18.4.3. From the crystal structures it can be concluded that the active site can only be accessed via a hydrophobic channel (Fontecilla-Camps et al. 1997; Montet et al. 1997). This channel is to be considered the hydrogen as well as oxygen transport pathway to the deeply buried active site, while protons released during the oxidation of hydrogen are thought to reach the enzyme surface via four histidines and a glutamate, involving internal water molecules (Volbeda et al. 1995; Fontecilla-Camps et al. 1997). Large and small subunits of the hydrogenase assemble as a globular heterodimer, exhibiting a large contact surface (Volbeda et al. 1995; Higuchi et al. 1997; Garcin et al. 1999). The small subunit of the NiFe hydrogenases in *D. gigas* contains three FeS clusters, a [4Fe-4S] cluster in close proximity to the active site (13 Å), a [3Fe-4S] cluster within a distance of 12 Å, and another [4Fe-4S] cluster located at the enzyme surface, separated again by about 12 Å from the second (Volbeda et al. 1995). This linear arranged FeS cluster wire was proposed to mediate electron transfer from the active site to the enzyme surface, where they can be donated to acceptors, e.g., cytochrome *c* (Vignais et al. 2001).

18.3.2 FeFe Hydrogenases

FeFe hydrogenases are mostly composed of one or two subunits, although heterotrimeric and tetrameric forms are known as well. Similar to the NiFe hydrogenases the essential key information about the organization and the active site biochemistry of the H cluster were mainly derived from X-ray crystallographic data. The crystal structures of the FeFe enzymes from *C. pasteurianum* and *Desulfovibrio desulfuricans* (Peters et al. 1998; Nicolet et al. 1999) led to the discovery of an FeFe active site coordinated by five diatomic ligands: 1 CO and 1 CN^- per Fe plus another CO bridging the two metal ions (Fig. 18.1). While the bimetallic center was shown to be additionally bridged by a non-protein dithiolate (propane dithiolate, dithiomethylamine, or dithiomethylether), it is also bound to a [4Fe-4S] cluster via a bridging cysteine thiolate (Nicolet et al. 2001; Pandey et al. 2008). FTIR (Fourier transform infrared spectroscopy) experiments revealed reversible inhibition of FeFe enzymes by carbon monoxide (Lemon and Peters 1999).

18.4 Hydrogenase Maturation

18.4.1 NiFe Hydrogenases

The assembly of an intricate active site, like that of NiFe hydrogenases, that contains toxic ligands, requires several accessory proteins, which have been designated Hyp for "*hy*drogenase *p*leiotropic," due to their parallel effect on the activity of all hydrogenases in *E. coli.* Most of the current knowledge about the maturation of NiFe hydrogenases has been derived from early studies of the *E. coli* hydrogenase 3 (Hyc) by Böck and coworkers. The assembly and maturation of hydrogenase 3 in *E. coli* involves a minimal set of at least six Hyp proteins, HypA, HypB, HypC, HypD, HypE, HypF, and an endoprotease, HycI. Hydrogenase maturation constitutes a multistep process which is initiated by (a) the biosynthesis of the diatomic ligands (CO, CN^-) that are coordinated to an iron ion, followed by (b) nickel insertion and (c) endoproteolytic activation of the mature complex (Böck et al. 2006; Forzi and Sawers 2007).

The present model (derived from *E. coli*) proposes that initially iron is coordinated to a complex of HypC/HypD through an N-terminal thiolate on HypC (Blokesch and Böck 2002). Then the cyanide ligands are synthesized by HypF and HypE. During this process HypF acts as a carbamoyltransferase and transfers the carboxamido group of carbamoylphosphate to the thiolate of the C-terminal cysteine of HypE in an ATP-dependent reaction. This HypE thiocarboxamide intermediate is then dehydrated via another ATP-dependent phosphorylation and a subsequent dephosphorylation, leading to HypE thiocyanate (Paschos et al. 2002; Reissmann et al. 2003). The synthesis of HypE thiocyanate is followed by the transfer of the cyano group to iron which is still bound to the HypC/HypD complex (Reissmann et al. 2003). The origin of the CO ligand is not yet known. Generally acetate or CO_2 might serve as substrates (Roseboom et al. 2005). Finally the HypC/HypD ligand complex is thought to interact with the precursor of the large hydrogenase subunit and releases HypD (Blokesch et al. 2004). Besides playing a crucial role in cofactor assembly HypC additionally functions as a chaperone responsible for stabilizing an open conformation of the large subunit (Drapal and Bock 1998; Magalon and Böck 2000; Böck et al. 2006). HypC in *E. coli* specifically acts on hydrogenase 3 while another paralogue, HybG, is responsible for maturation of hydrogenases 1 and 2. Both are able to interact with HypD (Blokesch et al. 2001). In contrast both HypC homologs in *T. roseopersicina* are essential for and contribute equally to the maturation of all four hydrogenases (Maroti et al. 2003).

Once the hydrogenase large subunit precursor containing $Fe(CN)_2(CO)$ is assembled, nickel is to be introduced into the active site. In the presence of very high concentrations of nickel its incorporation can take place without the help of additional accessory proteins (Waugh and Boxer 1986; Hube et al. 2002; Hoffmann et al. 2006). Nevertheless, under physiological concentrations the nickel requirement for the hydrogenase active site is provided by a complex of the maturation factors HypA and HypB. While HypA is a nickel-binding metallochaperone (Hube et al. 2002; Atanassova and Zamble 2005), HypB displays GTPase activity and is thought to provide energy for the switch-dependent incorporation of nickel as well as the subsequent release of the maturation factors from the large subunit (Maier et al. 1993; Gasper et al. 2006). In *E. coli* HypA specifically serves hydrogenase 3 while another homolog, HybF, incorporates nickel into hydrogenases 1 and 2 (Atanassova and Zamble 2005; Hube et al. 2002; Blokesch et al. 2004). The HypB protein, on the other hand, carries both low- and high-affinity binding sites for nickel (Leach et al. 2005), which together might act as nickel storage/sequestering sites as suggested for HypB in *Bradyrhizobium japonicum* (Olson and Maier 2000). Recently an interaction partner of HybB, SlyD, a peptidylprolyl isomerase, was shown to stimulate the release of nickel ions from HypB, thereby fulfilling an essential function in hydrogenase maturation (Zhang et al. 2005; Leach et al. 2007). The final step in hydrogenase maturation requires the cleavage of a C-terminal peptide of the large subunit apoprotein complex, which is carried out by an endoprotease. In *E. coli*, which harbors four different hydrogenases, endoproteolytic cleavage at the C-terminus is performed by different, specific proteases, HyaD, HybD, HycI, in *E. coli* (Menon et al. 1991; Rossmann et al. 1995). Although a number of bioinformatics studies have been done on the putative cyanobacterial proteases (HoxW, HupW, e.g., Devine et al. 2009) only one study could show that the respective homolog in *Synechocystis* PCC 6803 is involved in hydrogenase maturation (Hoffmann et al. 2006).

It has been suggested that the presence of the C-terminal peptide keeps the hydrogenase apoprotein active site in a particular conformation accessible to metal insertion and may even serve as an interaction site between the large and the small subunit of the enzyme (Sawers et al. 2004; Magalon and Böck 2000). While purified hydrogenase endoproteases do not contain a metal cofactor, they were shown to cleave their substrate only when nickel has been incorporated (Fritsche et al. 1999). Crystal structures of HybD (Fritsche et al. 1999) and HycI (Kumarevel et al. 2009) revealed cadmium and calcium ions bound to specific sites within the proteases which likely constitute potential nickel recognition sites. Theodoratou et al. (2000, 2005) suggested that the protease might recognize the Ni ion on the large subunit precursor as a binding motif and this determines the cleavage site in a regiospecific manner. Mutational studies, replacing the cysteine residues in the hydrogenase large subunit relevant to the coordination of nickel, supported this proposal (Magalon and Böck 2000). Using NMR Yang et al. (2007) performed metal titration experiments and observed significant conformational changes of HycI upon nickel binding which activate its proteolytic function. During recent years, a large number of X-ray structures of hydrogenase maturation proteins have been determined (Dias et al. 2008, Watanabe et al. 2007; Rosano et al. 2002; Shomura et al. 2007; Rangarajan et al. 2008; Watanabe et al. 2009) which are a promising starting point to unfold the molecular details of the entire maturation process.

18.4.2 FeFe Hydrogenases

Surprisingly, the maturation of FeFe hydrogenases seems to rely on only three protein factors, HydE, HydF, and HydG. In eukaryotes HydE and HydF are found fused in one open reading frame. Initial conclusive results came from deletion and complementation experiments of the corresponding *Chlamydomonas reinhardtii* genes (Posewitz et al. 2004). Subsequent attempts to heterologously coexpress *C. reinhardtii* hydrogenase HydA1 along with maturation factor homologs from *Shewanella oneidensis* or *Clostridium acetobutylicum* in *E. coli* successfully yielded active enzyme (Posewitz et al. 2005; King et al. 2006). Similarly Boyer et al. (2008) were able to obtain functional algal as well as bacterial FeFe hydrogenases in *E. coli* cell-free lysates by coexpression with the maturation factors.

The accessory factors HydE and HydG belong to the radical SAM (*S*-adenosyl-L-methionine) superfamily. Radical SAM enzymes are known to catalyze chemically difficult reactions such as C–H and C–S bond formation, for example, during biotin and lipoic acid metabolism (Fontecave et al. 2003). Both proteins, in their reduced form, show SAM cleavage activity. UV and EPR spectroscopic investigations of the *Thermotoga maritima* factors indicate that HydE binds two 4Fe-4S clusters while HydG binds at least one and possibly another one (Rubach et al. 2005). The crystal structure of HydE shows similarity to the triose phosphate isomerase barrel structure of the biotin synthase while having unique characteristics such as a significantly larger interior cavity (Nicolet et al. 2008). The X-ray data display three anion binding sites, one of which binds thiocyanate but not cyanate with high affinity. Based on these results Pilet et al. (2009) postulated that HydE might be responsible for synthesis of the CO and CN ligands while HydG on the other hand is involved in synthesis of the dithiolate ligand. The latter hypothesis is supported by the observation of HydG-dependent cleavage of tyrosine to *p*-cresol and led to the suggestion of a reaction mechanism for the synthesis of the dithiomethylamine bridging ligand (Pilet et al. 2009). On the other hand HydF from *T. maritima* was shown to hydrolyze GTP and bind a 4Fe-4S cluster (Brazzolotto et al. 2006). Attempts to mature inactive FeFe hydrogenase in vitro using HydF purified from an overexpressing E coli strain expression demonstrated that HydF alone was able to activate the FeFe enzyme (McGlynn et al. 2008). Based on these results the authors suggested HydF might catalyze the terminal step in FeFe hydrogenase maturation by functioning as a scaffold protein.

The maturation of hydrogenases in photosynthetic prokaryotes has been studied only to a limited extent using mainly mutational analysis. Complementation of deletion mutants of *Synechocystis* sp. PCC 6803 verified the involvement of *hypA1, B1, C, D, E, F* and *hoxW* in the maturation of the large subunit, HoxH. The function of HypA1 and HypB1 in the Ni insertion into the active site could be shown by hydrogenase activity restorage via supplementation of the respective deletion mutants with nickel (Hoffmann et al. 2006). Additionally, the requirement of *hypC1, C2, D, E,* and *F* for the maturation of the functional enzyme is similarly clarified (Kovacs et al. 2005).

18.4.3 Oxygen Sensitivity of NiFe Hydrogenases and the Potential Function of HypX

Hydrogenases are immediately inactivated in the presence of oxygen (George et al. 2004; Kurkin et al. 2004). In general, inactivation is observed via Ni in two different EPR recognizable states known as Ni_u-A ("unready") and Ni_r-B ("ready"). In the Ni_r-B state the bridging ligand X has been identified as mono-oxo (hydroxo) while the Ni_u-A state probably contains a hydroperoxide (Ogata et al. 2005; Volbeda et al. 2005). This specific bridging ligand is removed during reductive activation of hydrogenase leaving the enzyme in one of its active states (Ni_a-S), now capable of binding H_2 (Fernandez et al. 1985; Ogata et al. 2005). $^{17}O_2$ labeling has demonstrated that the bridging oxygen originates from O_2 (Van der Zwaan et al. 1990), which would avoid immediate reaction of hydrogen with oxygen in the form of a Knallgas reaction, with consequent irreversible destruction of the enzyme.

One of the proposed means of conferring oxygen tolerance to hydrogenases is based on gas diffusion selectivity caused by variations in the architecture of the gas channel. Narrowing the gas channel by targeted amino acid replacement conferred increased oxygen tolerance to the generally oxygen-sensitive enzyme in *Desulfovibrio fructosovorans* (Dementin et al. 2009). As well, other studies using the oxygen-tolerant regulatory hydrogenases HupUV (*R. capsulatus*) and HoxBC (*R. eutropha*), or the membrane-bound hydrogenase of *R. eutropha*, have shown that exchanging the existing amino acids to those of more oxygen-sensitive enzymes increased oxygen sensitivity (Duche et al. 2005; Buhrke et al. 2005; Ludwig et al. 2009). In addition, these investigations also indicate that a more sophisticated selectivity filter is used that seems to rely on the overall architecture of the protein as well as the electronic structure of the active site.

Some bacteria that metabolize H_2 under aerobic conditions harbor an additional *hyp* gene, *hypX*. In *R. eutropha*, the HypX protein has been proposed to play a crucial role in recruiting the diatomic ligands CO and/or CN^- in a tetrahydrofolate-coupled reaction (Rey et al. 1996). Primary structure alignments of HypX reveal two typical domains: an N-terminal tetrahydrofolate binding motif and a C-terminal signature similar to the enoyl-CoA hydratases/isomerases (Rey et al. 1996). Deletion of *hypX* in *R. eutropha* strains harboring only the soluble hydrogenase abolishes growth under aerobic conditions with H_2 as sole energy source (Bleijlevens et al. 2004). Evidence has been presented that *R. eutropha* SH contains additional CN ligands at the active site bound to the nickel and iron atoms that render the enzyme oxygen tolerant (Bleijlevens et al. 2004; van der Linden et al. 2004).

18.5 Phototrophic Prokaryotes: Hydrogenase Inventory in a Physiological Context

18.5.1 Proteobacteria: R. rubrum

The purple non-sulfur bacterium *R. rubrum* is equipped with a versatile hydrogen metabolism. It harbors two H_2-evolving hydrogenases: a CO-dependent enzyme and a formate-dependent enzyme which functions in fermentation. As well, *R. rubrum* has been shown to contain a H_2-consuming uptake hydrogenase (Van Praag et al. 2000; Maness and Weaver 2001). Accordingly, the genome of *R. rubrum* ATCC 11170 (Copeland et al. unpublished) carries three different genomic loci encoding hydrogenase structural and accessory genes (Fig.18.2).

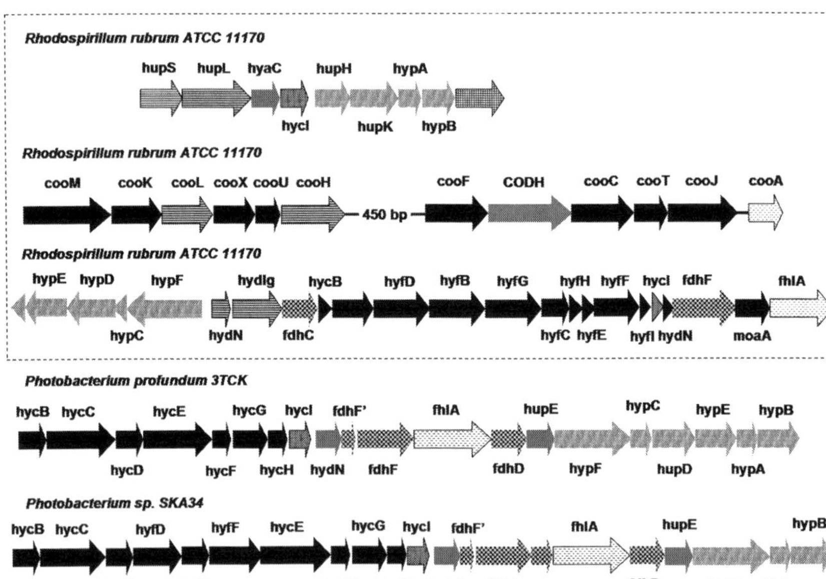

Fig. 18.2 Genetic inventory of *R. rubrum* ATCC 11170 with respect to its hydrogen metabolism. *R. rubrum* encodes an uptake hydrogenase (*hupSL*), a CO dehydrogenase (*cooM-H*, *cooF-A*) as well as the formate hydrogenlyase gene cluster (*hyc*, *hyf*). Genes encoding homologs to the formate dehydrogenase are highlighted in gray (*plaid*), hydrogenase core components in striped, structural complex genes in black, associated maturation genes *hypFCDE* and *hupHK-hypAB* in dark gray (*shaded*), and open reading frames without known function or homology in light gray. The formate and CO sensor/regulator genes *cooA* and *fhlA*, respectively, are indicated by stippled. For comparison two additional formate hydrogenlyase operons from non-photosynthetic prokaryotes are depicted

18.5.1.1 CO Dehydrogenase

CO-dependent H_2 evolution, also known as the biological water gas shift reaction $CO + H_2O \Rightarrow CO_2 + H_2$ $\Delta G = -20$ kJ/mol, enables *R. rubrum* to grow on CO as sole energy source under dark conditions (Kerby et al. 1995). The CO-induced hydrogenase in *R. rubrum*, like the formate hydrogenlyase, belongs to the energy-conserving membrane-bound NiFe hydrogenases (group 4; Ech). The Coo-complex consists of multiple subunits (Fig.18.3) encoded by two operons, the *cooMLXUH* and the *cooFSCTJ* (Shelver et al. 1997). The *cooFSCTJ* operon encodes the CO-oxidizing carbon monoxide dehydrogenase (CooS), an FeS electron transfer protein (CooF), and accessory proteins required for maturation (Ni insertion) of CooS (Ensign and Ludden 1991; Kerby et al. 1992). The *cooMKLXUH* operon, on the other hand, codes for the hydrogenase structural proteins, two of which, CooL and CooH, resemble the small and large hydrogenase subunits conserved in all NiFe hydrogenases, respectively (Fox et al. 1996). Nevertheless, the core subunits CooH and CooL show much higher overall homology to NADH:quinone:oxidoreductase (complex I) than to standard NiFe hydrogenases. Since the four additional subunits CooMKXU also show significant homology to complex I, Fox et al. (1996) proposed that the CODH complex might couple CO-dependent H_2 evolution to the generation of a proton gradient. In the photosynthetic bacterium *Rubrivivax gelatinosus*, which possesses a similar CODH complex, CO-dependent ATP production has been demonstrated (Maness et al. 2005).

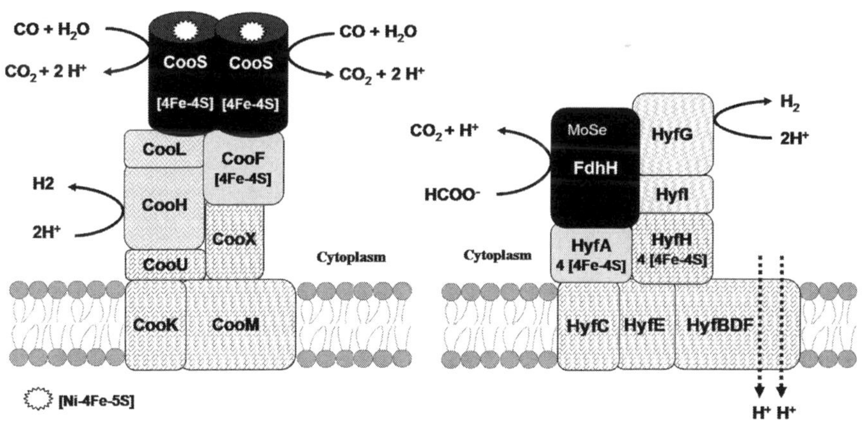

Fig. 18.3 Schematic model of the two group 4 NiFe hydrogenase complexes found in *R. rubrum*. The hypothetical subunit structure and subcellular localization of the formate hydrogenlyase and the CO dehydrogenase complexes have been drawn according to Hedderich and Forzi (2005) and Andrews et al. (1997), respectively. The figure does not consider a potential function of the FeFe hydrogenase homolog, encoded upstream of the FHL cluster

CODH has been solubilized from membranes of *R. rubrum* and purified either in its monomeric form or as a complex with the ferredoxin-like electron transfer protein CooF (Bonam and Ludden 1987; Ensign and Ludden 1991). Subsequently, its crystal structure was solved and revealed a homodimeric CooS with an unusual [Ni4Fe-5S] cluster, which constitutes the catalytic center of CO oxidation (Svetlitchnyi et al. 2001; Drennan et al. 2001). Mutation of amino acids near the CODH active site, Cys531→Ala or His265→Val, results in the conversion of the enzyme into either an uptake hydrogenase or a hydroxylamine reductase (Heo et al. 2002). The catalytic activity of CO dehydrogenase is redox dependent, being mostly in its inactive form at redox potentials above –300 mV in the absence of CO (Heo et al. 2001). The *R. rubrum* CO-dependent hydrogenase is exceptional among the NiFe hydrogenases in being CO tolerant. In an atmosphere of 100% CO, hydrogenase activity only decreased about 60%, while 50% inhibition was observed at 35% CO (Bonam et al. 1989; Fox et al. 1996).

18.5.1.2 Formate Hydrogenlyase

The second enzyme involved in the hydrogen metabolism of *R. rubrum* is a putative formate hydrogenlyase (FHL) (Fig. 18.3). The best characterized FHL so far is hydrogenase 3 of *E. coli* (HycBCDEFGI), which has been extensively studied in the group of Böck (Bohm et al. 1990; Sauter et al. 1992; Böck et al. 2006). However, *E. coli* also synthesizes a second 10 subunit FHL isoenzyme encoded by the *hyfABCDEFGHIJR* operon. In comparison with hydrogenase 3, Hyf, also known as hydrogenase 4, possesses three additional transmembrane subunits (Hyf D, E, F) sharing sequence homology to complex I subunits (Andrews et al. 1997). The same authors suggested that Hyf might perform proton translocation.

Hydrogenase 3, as well as probably hydrogenase 4, forms a complex with the selenocysteine- and molybdenum-containing formate dehydrogenase FdhH (encoded by *fdhF*). This complex is relevant to the mixed acid fermentation pathway and catalyzes formate hydrogenlyase activity, according to the overall reaction: $HCOO^- + H_2O \rightarrow HCO^- + H_2$ (Thauer et al. 1977; Bagramyan and Trchounian 2003). Hydrogenase 3 couples the oxidation of formate to the reduction of protons to H_2, which under low H_2 partial pressure and low pH is an exergonic (–20 kJ/mol) reaction (Böck and Sawers 1996). HycE and HycG, as well as their counterparts in hydrogenase 4, HyfG and HyfI, are the large and small hydrogenase subunits. Communication of the hydrogenase with the formate dehydrogenase subunits is thought to be mediated by HycB, which contains four [4Fe-4S] clusters and, as CooF of the CO dehydrogenase complex, might function in electron transfer to the hydrogenase.

Electron donation to the FHL-catalyzed reaction is performed by formate dehydrogenase. *E. coli* synthesizes three different membrane-associated formate dehydrogenase isoenzymes. While FDH-N and FDH-O face the periplasm, FDH-H is located at the cytoplasmic side of the membrane (Jormakka et al. 2002; Sawers et al. 1991). FDH-H is characterized by a low substrate affinity and thus is only active at high formate concentrations (Berks et al. 1995). FHL-dependent H_2 evolution in *E. coli* has been shown to be sensitive to the protonophore CCCP (carbonylcyanide-*m*-chlorophenylhydrazone) and the F_0F_1-ATPase inhibitor DCCD

(*N,N'*-dicyclohexylcarbodiimide) indicating its involvement in the formation of a transmembrane potential. Investigations by Bagramyan and Trchounian (2003) demonstrated a coupling of formate-dependent proton translocation and H_2 evolution to $2H^+–K^+$ exchange via the low-affinity TrkA system. H_2 evolution and $2H^+/K^+$ exchange were abolished in deletion mutants of hyf genes (Bagramyan et al. 2001). Similarly both H_2 evolution and $2H^+/K^+$ exchange were lost in ATP synthase mutants.

Growth on organic carbon sources generally makes use of one of three different biochemical pathways: the Emden–Meyerhof–Parnas (Glycolysis), the oxidative pentose phosphate, or the Entner–Doudoroff (KDPG, 2-keto-3-deoxy-6-phosphogluconate). All of them lead to the formation of the intermediate pyruvate which may further be metabolized to generate reducing equivalents for oxidative phosphorylation via the TCA (*tricarboxylic acid*) cycle or is channeled into fermentation, depending on the availability of oxygen. Under aerobic conditions pyruvate is oxidatively decarboxylated to CO_2 and acetyl-CoA. This reaction is catalyzed by the pyruvate dehydrogenase (PDH) complex and constitutes the entry step into the TCA cycle. Under fermentative growth regimes, when oxygen is absent, pyruvate:ferredoxin/flavodoxin:oxidoreductase (PFOR) or pyruvate formate lyase (PFL) replaces PDH. While PFOR catalyzes the disposal of reducing equivalents via ferredoxin (pyruvate + CoA + 2 ferredoxin$_{ox}$ $\rightarrow$ acetyl-CoA + CO_2 + 2 ferredoxin$_{red}$ + 2 H+) PFL uses radical chemistry to oxidize pyruvate and CoA to formate and acetyl-CoA. Formate then serves as substrate for the FHL-catalyzed reaction. Early investigations of anaerobic growth capabilities of *R. rubrum* revealed typical products of the mixed acid fermentation pathway: succinate, acetate, propionate, formate, hydrogen, and carbon dioxide (Schultz and Weaver 1982). Based on the genetic organization of the formate hydrogenlyase operon in *R. rubrum* (Fig. 18.3) as well as primary sequence homologies of its subunits we propose that the formate-dependent hydrogenase in this organism resembles hydrogenase 4 of *E. coli*. The genetic organization of hydrogenase operons in two other purple non-sulfur photosynthetic bacteria is shown in Fig. 18.4.

18.5.2 Proteobacteria: T. roseopersicina

T. roseopersicina is a member of the purple sulfur bacteria (Chromatiaceae) which was originally isolated from the North Sea (Bogorov 1974). It is able to perform anoxygenic photosynthesis using reduced sulfur compounds such as S_2^-, S_0, $S_2O_3^-$ as electron donors (Laurinavichene et al. 2007). Besides a regulatory hydrogenase (HupUV) and the adjacent two-component signal transduction chain HupRT, *T. roseopersicina* is equipped with three different hydrogenases: the membrane-bound HynSL and HupSL and the soluble HoxEFUYH. Both small subunits, HynS and HupS, appear to have signal sequences for membrane translocation via the TAT pathway. The heteropentameric Hox is a soluble cytoplasmic protein, whereas both HupSL and HynSL are localized at the periplasmic side of the CM (cytoplasmic membrane). While the soluble heteropentameric Hox and Hyn are known to

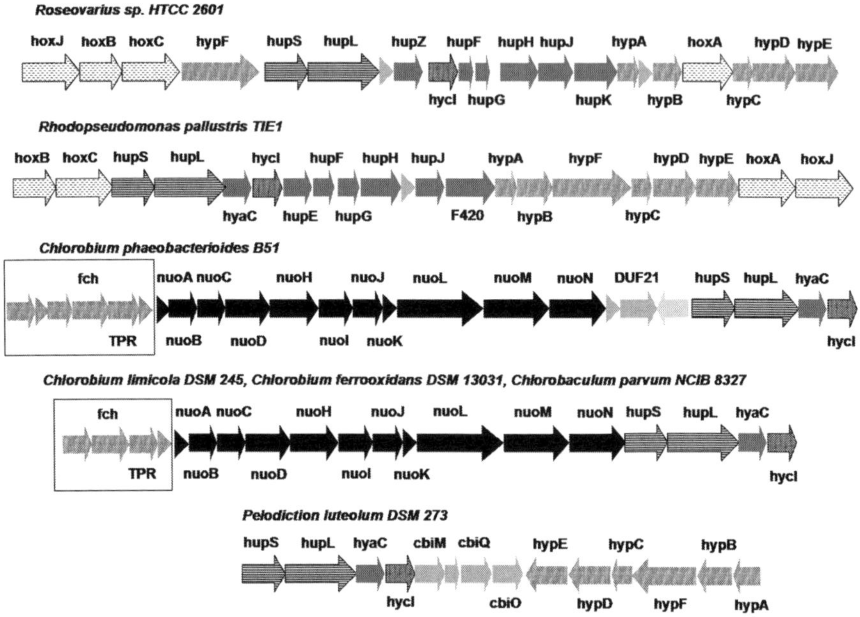

Fig. 18.4 Genetic organization of the uptake hydrogenase encoding operons in the purple non-sulfur bacteria *Roseovarius sp.* and *R. palustris* as well as the green sulfur bacteria *Chlorobium limicola*, *C. ferrooxidans*, *Chlorobaculum parvum*, and *Pelodiction luteolum*. Hydrogenase structural genes (*hupSL*) and associated primary electron acceptor (*hupC* homologs) are depicted as striped arrows, whereas accessory hyp genes are in dark *gray*, patterned and open reading frames without known function or homology are shown in light *gray*. Genes encoding the regulatory hydrogenase (*hoxBC*) and the corresponding two-component signaling cascade (*hoxJA*) were named according to the nomenclature used in *R. eutropha*. For simplicity, and based on their initial discovery and description, accessory genes (hyps, *gray* patterned) were depicted using the nomenclature for *E. coli* hydrogenase 3. The genes *cbiMQO* in *P. luteolum* are homologs of an ABC-type metal transport system, according to its chromosomal localization, likely encoding a nickel transporter

work bidirectionally, the uptake hydrogenase HupSL encoded by the *hupSLCDHIR* operon (Fig. 18.5) is an exclusively H_2-consuming hydrogenase believed to recycle H_2 generated during nitrogen fixation (Colbeau et al. 1994, Colbeau et al. 1980). Since this hydrogenase is anchored to the periplasmic side of the CM, oxidation of H_2 leads to the formation of a proton gradient across the CM (Kovacs et al. 1983; Kovacs and Bagyinka 1990).

Hyn was shown to be an exceptionally stable enzyme with relatively high O_2 tolerance (Gogotov et al. 1978). Catalytic activity reaches a maximum around 70°C, whereas inactivation occurs above 78°C. The structural genes *hynSL* of the stable hydrogenase are arranged in an operon (*hynS-isp1-isp2-hynL*) together with two additional genes which show significant homology to genes encoding subunits DsrK and DsrM of the dissimilatory sulfite reductase (Rakhely et al. 1998). A similar genetic organization of Hyn homologous hydrogenases is only found

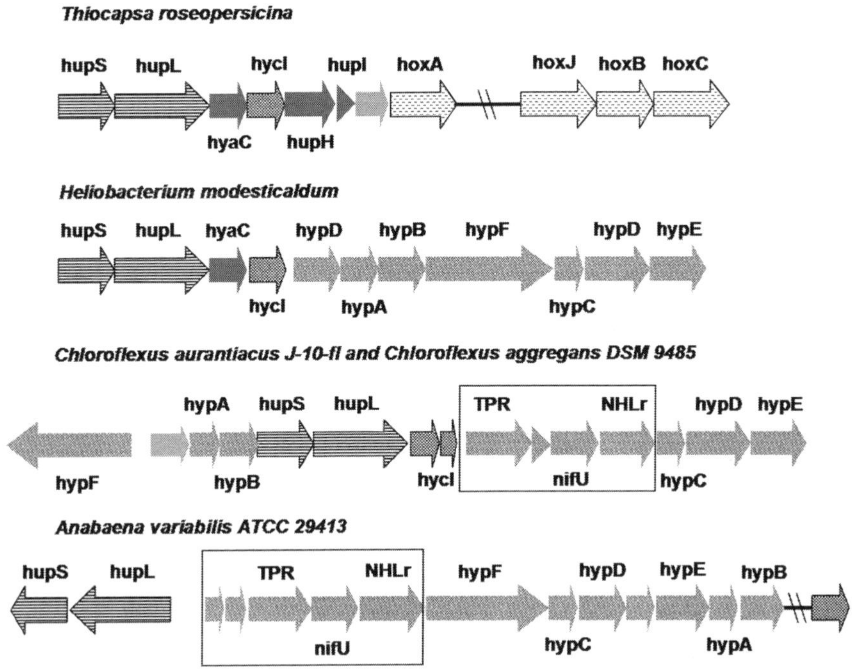

Fig. 18.5 Genetic organization of the uptake hydrogenase encoding operons in *T. roseopersicina*, *H. modesticaldum*, *C. aurantiacus J-10-fl*, and the cyanobacterium *A. variabilis* ATCC 29413. Gene nomenclature has been used as described in Fig. 18.4. Framed area marks additional *hyp* genes, present in all uptake hydrogenase expressing cyanobacteria as well as the *Chloroflexi* and, although without higher primary sequence homology, the *Chlorobi*

in *Chromatium vinosum*, *Aquifex aeolicus*, or the archaeum *Acidianus ambivalens* (Dahl et al. 1999). Investigations to clarify the function and selectivity of electron mediators in electron transport from and to these hydrogenases confirmed the bioinformatics predictions. Isp1 is a transmembrane electron carrier containing a binding site for a b-type heme and is required for activity of HynSL. Deletion of *isp1* and *isp2* in a *ΔhoxH*/*ΔhupSL* background results in disappearance of H_2 evolution as well as a decrease in H_2 uptake activity despite the retention of the ability to catalyze in vitro activity with artificial electron acceptors. Similarly a dependence of HupSL activity on the membrane-bound b-type cytochrome HupC (named HyaC in our figures according to the nomenclature in *E. coli*) could be demonstrated. *Thiocapsa* uses H_2S, SO_4^{2-}, or S^0 as electron donors for anoxygenic photosynthesis, generating NADH, with NAD being regenerated by the Hox hydrogenase. This leads to light-dependent H_2 evolution in the presence of thiosulfate. However, Hox is additionally responsible for dark fermentative H_2 production under conditions where nitrogenase expression is repressed. In contrast to the cyanobacteria, light-dependent H_2 evolution is not temporary but may last for several days (Kovács et al. 2006). This difference might be explained by higher amounts of NADH, continuously generated via reverse electron transport by complex I, and, especially, the absence of PSII-dependent O_2 evolution. Reverse electron transport involving NADH dehydrogenases homologous to complex I seems to be a common theme to generate reduced pyridine nucleotides (Zannoni 1995).

Light limitation is favorable for H_2 evolution in *T. roseopersicina* (Rakhely et al. 2007). The directionality of the hydrogenase-catalyzed reaction is, similar to the CODH in *R. rubrum*, regulated by the redox state. Activation in cell-free extracts occurs at approximately –365 mV, favorable for H_2 consumption, while below –420 mV Hox catalyzes H_2 evolution. While the pentameric Hox was initially partially purified using classical techniques, recent investigations used affinity chromatography to isolate active protein (Rakhely et al. 2004; Palagyi-Meszaros et al. 2009). In vitro reduced ferredoxins from *Synechococcus* PCC7942 as well as the green alga *Scenedesmus obliquus* were capable of driving hydrogenase-mediated H_2 evolution. Light-dependent H_2 uptake by Hox was stimulated by CO_2 and high concentrations of thiosulfate. Nevertheless, and unlike Hup, Hox is unable to catalyze oxygen-dependent H_2 uptake (Rakhely et al. 2007). Purple sulfur bacteria are known to make use of H_2 for photoautotrophic CO_2 fixation. In contrast to the bidirectional hydrogenase of *R. eutropha* $HoxFUYHI_2$, the Hox hydrogenases of cyanobacteria and anoxygenic photosynthetic bacteria contain another, possibly electron-mediating, subunit, HoxE. Deletion of *hoxE* in *Thiocapsa* results in a phenotype consisting of impaired in vivo hydrogenase activity. Nevertheless, the truncated hydrogenase complex HoxFUYH still shows unaltered H_2-dependent NAD+ reduction in vitro (Rakhely et al. 2004).

18.5.3 Green Sulfur and Filamentous Anoxygenic Phototrophs

The green sulfur bacteria (GSB or *Chlorobi*) and the green non-sulfur bacteria now called filamentous anoxygenic phototrophs (FAPs) both contain chlorosomes for light harvesting but seem to be quite different in many other aspects. Whereas the former do have a PSI-type reaction center the latter use a PSII-type reaction center. *Chlorobi* fix CO_2 via the reverse TCA cycle, are strictly anaerobic, and are obligately photoautotrophic whereas the FAPs use the hydroxypropionate pathway. So far, dark respiration or solely fermentative growth has not been reported for *Chlorobi*. The latter contain a bifunctional (NADP) hydrogenase formerly also called sulfhydrogenase. This type of hydrogenase has been described in *P. furiosus* where it is thought to dispose of reducing equivalents during fermentation (Ma et al. 1993; Chou et al. 2008). The *Chlorobi* have either obtained the sulfhydrogenase, which we name HydBGDA (or $\beta\gamma\delta\alpha$) according to the nomenclature used in *P. furiosus* (Pedroni et al. 1995), by lateral gene transfer from the archaea or vice versa (Ludwig et al. 2006).

This Hyd consists of at least four subunits, encoded in an operon (Fig. 18.6). HydB and HydG show significant homology to sulfite reductases, for example, to *Salmonella typhimurium* AsrAB (Huang and Barret 1991). Like AsrAB, HydG contains NADH- and FAD-binding sites and likely also catalyses the reduction of sulfite or S_0 to H_2S. Based on their homology, HydD and HydA constitute the small and

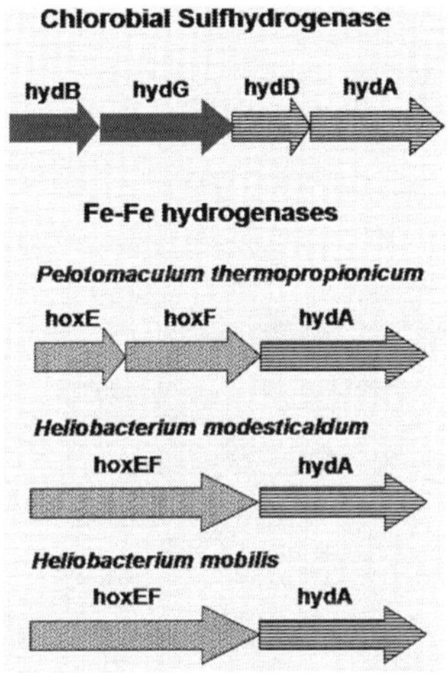

Fig. 18.6 Structural organization of the sulfhydrogenase operon found in *Chlorobi* as well as the FeFe hydrogenase encoding operons in *H. modesticaldum* and *H. mobilis* and the non-photosynthetic bacterium *P. thermopropionicum*. Genes encoding the catalytic FeFe hydrogenase (*hydA*) and the NiFe hydrogenase small and large subunits (*hydD* and *hydA*) are shown as striped arrows, genes encoding NADH dehydrogenase subunit homologous to the cyanobacterial *hoxE* and *hoxF* are in *gray* (*dotted*) while the sulfite oxidoreductase homologous genes of the sulfhydrogenase operon are depicted in black

large hydrogenase subunits. Nevertheless, none of the four subunits has targeting motifs for membrane translocation via the TAT pathway, which indicates that this Hyd is a soluble, cytoplasmic enzyme or perhaps membrane associated. In addition to this group 3 enzyme, the *Chlorobi* also contain a group 1 uptake hydrogenase. Although *Chlorobaculum tepidum* (the former *Chlorobium tepidum*) contains only a truncated version of the gene of the large subunit (*hupL*) and does not harbor a small subunit gene (*hupS*), there are several other *Chlorobi* that contain a full complement of this hydrogenase including the b-type cytochrome (*hupC*) that connects it to the membrane. Noticeably, the *hupSLC* genes (*hupC* is named *hyaC* in our figures) in all sequenced chlorobial genomes are located downstream of an operon encoding the NADH dehydrogenase (NDH-1) complex (Fig. 18.4). Similar to the cyanobacteria, the NDH-1 in *Chlorobi* is composed of only 11 of the typically 14 core subunits of the prokaryotic respiratory complex I. If the genetic proximity of *ndh* and *hup* genes results from a structural or functional relationship remains to be determined. The presence of NDH-1 in *Chlorobi* led to the suggestion of a possible function in reversed electron transport (Eisen et al. 2002). A functional interaction of both hydrogenases present in *Chlorobi* with NDH-1 as well as a structural role for the uptake hydrogenase HupSL in the complex is likely.

The thermophilic FAP *Chloroflexus aurantiacus* was originally isolated from hot spring microbial mats (Pierson and Castenholz 1974) (Fig. 18.7). In their natural

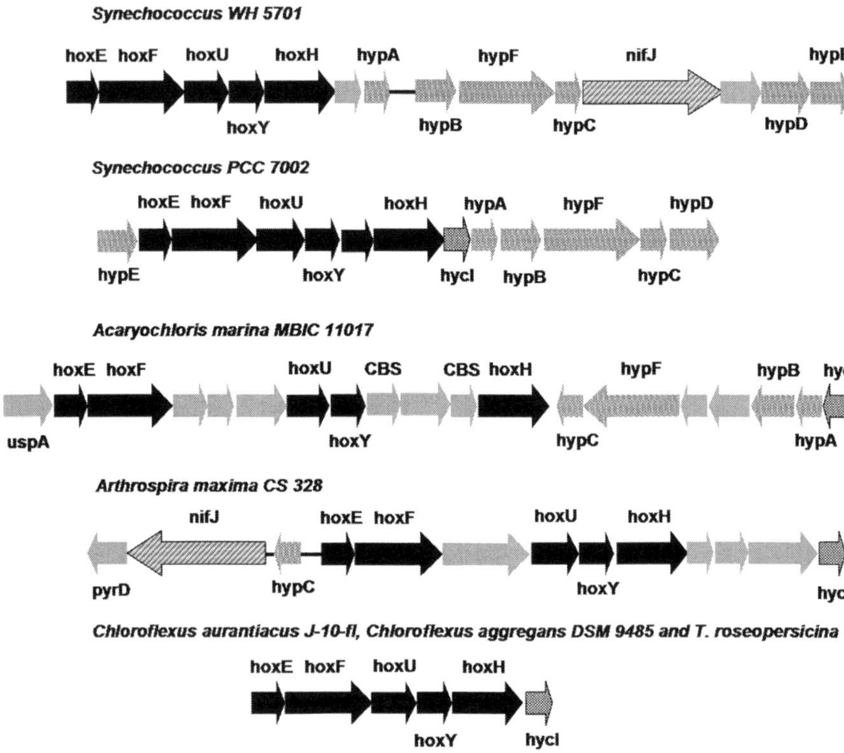

Fig. 18.7 Genetic organization of gene clusters encoding the bidirectional NAD(P)-reducing hydrogenases in the cyanobacteria *Synechococcus* WH 5701, *Synechococcus* PCC 7002, *Acaryochloris marina* MBIC 11017, *Arthrospira maxima* CS 328 as well as the filamentous green non-sulfur bacterium *C. aurantiacus* and the purple sulfur bacterium *T. roseopersicina*. Hydrogenase structural genes (*hoxEFUYH*) are depicted in *black*, *nifJ* encodes a potential pyruvate ferredoxin/flavodoxin oxidoreductase. *Hyp* genes and open reading frames without clear homology or assignment are shown as already described in Figs. 18.2 and 18.4. Additional *hyp* genes not encoded in the *hox* clusters, but located elsewhere in the bacterial chromosomes, are not shown

habitats *Chloroflexi* mostly grow photo- or chemoheterotrophically, but may also perform photoautotrophic growth at the expense of S^{2-} or H_2. *Chloroflexi* may also respire in the presence of oxygen. *C. aurantiacus* was reported to express a membrane-bound hydrogenase, which is repressed by sulfide, but induced by the presence of nickel and hydrogen (Drutschmann and Klemme 1985). The availability of the entire genomic sequence of *C. aurantiacus* revealed the presence of an uptake hydrogenase (HupSL), likely accounting for the observed activity, as well as a NAD(P)-dependent bidirectional hydrogenase (HoxE-H) (Fig. 18.8). The two enzymes found in *Chloroflexus* belong to the group 2, cyanobacterial-like uptake hydrogenases and the group 3 bidirectional NADP-linked hydrogenases,

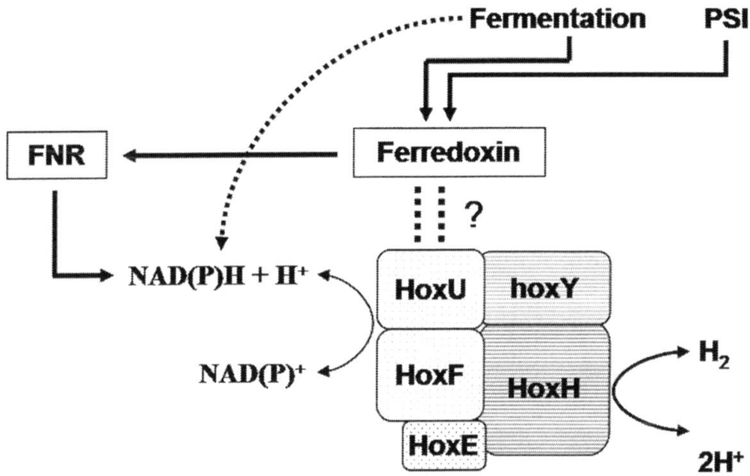

Fig. 18.8 Hypothetical model of organization of the NAD(P)-reducing, bidirectional hydrogenase (HoxEFUYH) in cyanobacteria as well as its connection to photosynthesis and fermentation, modified according to Houchins (1984). The reversible Hox serves as an acceptor for an excess of reducing equivalents produced during fermentation. An increase in the cellular NAD(P)H, which is generated during fermentation or photosynthesis (involving the ferredoxin:NAD(P)H oxidoreductase), is thought to be the source of the observed photo- and dark H_2 evolution. If ferredoxin can also serve as a direct electron donor has not yet been clarified

respectively. Phylogenetic investigations revealed that both share a common ancestor with hydrogenases of cyanobacteria (Ludwig et al. 2006). This suggests a close evolutionary connection between these two phototrophic groups. Since the *Chloroflexi* are also known to use hydrogen as electron donor during photosynthesis (REF) both of them might well function in hydrogen uptake. Although this hypothesis is still awaiting experimental proof, it is likely that *Chloroflexus* performs reversed electron flow to reduce NAD. Assuming some physiological similarity to *T. roseopersicina* photosynthesis with S^{2-}, S_0, $S_2O_3^-$, *Chloroflexus* could make use of the Hox hydrogenase as an acceptor leading to H_2 evolution. Surprisingly, *Roseiflexus castenholzii* also contains a group 1 uptake hydrogenase and *Roseiflexus* sp. RS-1 does not contain the cyanobacterial-like uptake hydrogenase but a group 1 uptake hydrogenase.

18.5.4 Firmicutes: Heliobacteria

Heliobacteria belong to the phylum firmicutes. They are the only gram-positive members among the anoxygenic phototrophs. Whereas most anoxygenic photosynthetic bacteria grow photoautotrophically, heliobacteria can only grow photoheterotrophically or by fermentation (Madigan and Ormerod 1995). Heliobacterial photosynthesis is based on a type I reaction center using bacteriochlorophyll g as a major antenna pigment. In contrast to the occurrence of specialized membrane structures like the chlorosomes in *Chlorobi* or the lamellae in *Chromatiales* and *Rhodospirillales*, the heliobacterial photosynthesis machinery is located in the cytoplasmic membrane. The recently published genome sequence of *H. modesticaldum* (Sattley et al. 2008) revealed that hydrogen metabolism in this organism uses three NiFe and one FeFe hydrogenase. According to the genomic information, the FeFe enzyme in heliobacteria consists of a single subunit hydrogenase HydA, but, similar to the cyanobacterial reversible hydrogenase, contains the diaphorase subunits HoxEF (NuoEF homologs). Considering the absence of homologs to known FeFe hydrogenase accessory genes in *Heliobacterium* (Fig. 18.6), the question arises as to how maturation of this enzyme is performed. In contrast to the situation in some obligate anaerobic bacteria, such as *Pelotomaculum thermopropionicum*, or cyanobacteria, *hoxE* and *hoxF* in *H. mobilis* and *H. modesticaldum* were found to be fused in one open reading frame (Ludwig et al. 2006, Sattley et al. 2008). The FeFe hydrogenase in *H. modesticaldum* is likely to function in the fermentative removal of reducing equivalents (NADH or $\text{Ferredoxin}_{\text{red}}$), thus mainly catalyzing H_2 evolution. In the presence of appropriate terminal electron acceptors or under light, hydrogen, which accumulates during fermentative growth or under nitrogen fixation conditions, can be used via two membrane-bound uptake hydrogenases.

18.5.5 Cyanobacterial Hydrogen Metabolism

18.5.5.1 The Bidirectional Hydrogenase

The reversible or bidirectional hydrogenase HoxEFUYH in *Synechocystis* PCC6803 constitutes a heteropentamer (Schmitz et al. 2002; Germer et al. 2009) which is able to use both NADH (*K*m, pH 6 = 12 μM) and NADPH (K_m, pH 6 = 100 μM) as an electron donor for H_2 evolution in vitro (Cournac et al. 2004). Both substrates were shown to effectively reactivate the catalytic center in the absence of oxygen. In contrast to many NiFe hydrogenases studied in anaerobes where reactivation of O_2-inactivated enzyme takes place in the range of 1 h (Fernandez et al. 1984; Lissolo et al. 1984), the cyanobacterial bidirectional hydrogenase is rapidly activated (Appel et al. 2000; Cournac et al. 2004; Germer et al. 2009). However, attempts to couple hydrogenase to plant ferredoxin or the major photosynthetic cyanobacterial ferredoxin failed to demonstrate physiologically relevant redox reactions (Serebryakova et al. 1996; Schmitz et al. 2002). It remains to be seen if in catalyzing in vivo hydrogen production Hox is only accepting NAD(P)H as an electron donor or if it can react simultaneously with both ferredoxin and NAD(P)H, as described for the *T. maritima* FeFe hydrogenase (Schut and Adams 2009). The diaphorase subunits, HoxE, HoxF, and HoxU, show striking homology to the NADH dehydrogenase type I (complex I) subunits NuoE, NuoF, and NuoG. Since the cyanobacterial complex I (NDH-1) and the NADH dehydrogenases in chloroplasts lack homologs to those subunits Appel and Schulz (1996) suggested that the bidirectional hydrogenase could function as the NADH dehydrogenase part of complex I (Fig. 18.9).

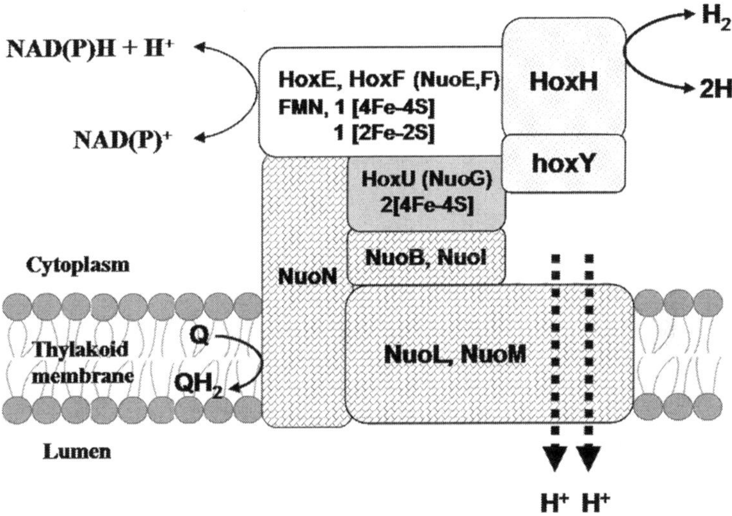

Fig. 18.9 Hypothetical model of the putative interaction of the NAD(P)-reducing, bidirectional hydrogenase in *Synechocystis* PCC6803 with the complex I homologous donor:quinine:oxidoreductase. Subunits of the NDH-1 (NADH dehydrogenase type I) were named according to the nomenclature of their homolog, *E. coli* Nuo. The diaphorase subunits of the hydrogenase HoxE, HoxF, and HoxU have been suggested to substitute for the missing homologs of NuoE,F,G under certain environmental growth conditions (Appel and Schulz 1996)

18.5.5.2 Fermentative Hydrogen Evolution

In many cyanobacteria, e.g., *Spirulina platensis* (Aoyama et al. 1997), *Oscillatoria limnetica*, or *Anabaena variabilis*, nitrogenase-independent hydrogen evolution can be observed under anaerobic conditions in the dark. This H_2 evolution can clearly be assigned to fermentative metabolism (Troshina et al. 2002; Gutthann et al. 2007). Bacterial fermentation uses hydrogenase to convert reduced NAD(P)H or ferredoxin into hydrogen, thereby removing an excess of reducing equivalents (Van der Oost and Cox 1988; Hallenbeck 2005). This process regenerates $NAD(P)^+$ or oxidized ferredoxin and enables the cell to subsequently recycle hydrogen by H_2-dependent respiration (Fig. 18.8). While many cyanobacteria like *Synechococcus* PCC6716 can only ferment endogenous metabolites (polyhydroxybutyrate, glycogen), *Microcystis* PCC7806 is also able to use exogenous substrates. For a detailed overview of fermentative capabilities and pathways in cyanobacteria, see Stal and Moezelaar (1997).

18.5.5.3 Photoreduction of CO_2

Since the early work of Frenkel et al. (1950), several reports have appeared that indicate an additional role of the bidirectional hydrogenase in photosynthetic electron transport under anoxic conditions. Initially Frenkel et al. (1950) measured H_2-driven photoreduction in *Synechococcus elongatus* as well as in *Chroococcus* sp. Similar results were observed in *Anacystis nidulans* (*Synechococcus* PCC 6301), which showed DCMU-insensitive anoxygenic $^{14}CO_2$ photoreduction using sulfide, sulfate, or hydrogen as electron donor (Peschek 1979). Later Belkin and Padan (1978) confirmed CO_2 fixation in *O. limnetica* and *Aphanothece halophytica* driven by anoxygenic photosynthesis with H_2 or sulfide as electron source. More recently, investigations by Ludwig et al. (2006) demonstrated the absence of any other hydrogenase except the bidirectional in the latter species, thus supporting the conclusion that this enzyme is involved in H_2 uptake driven by photosynthesis. Since anoxygenic photoreduction of CO_2 involves only photosystem I it is considered a cyanobacterial variant of anoxygenic photosynthesis (Oren et al. 1977). Despite many attempts, chemolithotrophic or anoxygenic phototrophic growth on hydrogen could not be demonstrated (Houchins 1984).

18.5.5.4 Photohydrogen Evolution

Abdel-Basset and Bader (1998) studied hydrogen metabolism in *Synechocystis* PCC6803 using membrane inlet mass spectrometry. The authors were able to demonstrate photohydrogen evolution after a dark to light switch. This photoevolution, which only leads to temporary net accumulation, normally lasts for about 30 s in WT cells (Appel et al. 2000; Cournac et al. 2004). Nevertheless, prolonged photohydrogen evolution analogous to the effects shown for FeFe hydrogenases in green algae has only been reported for *A. variabilis* (Laczko 1986) and *O. limnetica* (Belkin and Padan 1978). More recently, Cournac et al. (2004) and Gutthann et al. (2007) observed increased and prolonged H_2 photoevolution in a *ΔM55* mutant which is devoid of type I NADH dehydrogenase activity. The *ΔM55* mutant is characterized by reduced respiration and cyclic electron transport. According to Cooley and Vermaas (2001), the reduction of the NADPH pool in this mutant reaches a level of 100% while it is normally around 50% in WT cells. Photohydrogen evolution was found to last for at least 5 min in the mutant and could further be extended to close to 30 min by addition of glucose (Cournac et al. 2004). Similar effects, although less pronounced, were found in mutants impaired in respiratory oxidases, especially the quinol oxidase (Gutthann et al. 2007). The *ΔhoxH* mutant, in addition to having altered transcript levels of some photosynthetic genes (*psbA*, *psaA*, *petB*), showed increased photosystem II fluorescence and was impaired in photosystem I oxidation (Appel et al. 2000). It was concluded that the bidirectional hydrogenase in *Synechocystis* might function as a redox valve under changing light/dark regimes, especially when cells shift from fermentation under anaerobic conditions in the dark to photosynthesis. As already observed in green algae, photohydrogen evolution competes with nitrite-, nitrate-, and sulfide reduction as well as with CO_2

fixation and respiration (Aparicio et al. 1985). Similarly, Gutthann et al. (2007) were able to show that NO_3^- is a major electron acceptor of the H_2 uptake reaction in the absence of oxygen, as concluded using tungsten inhibition of nitrate reductase in *Synechocystis* PCC6803.

18.5.5.5 Cyanobacterial Uptake Hydrogenase

Competition experiments in *Anabaena* PCC 7120 did not result in the wild-type cells overgrowing the uptake hydrogenase-free mutant. At low light intensities the mutant grew as well as the wild type, whereas at higher light intensities its fraction of the population decreased but never became extinct (Lindblad et al. 2002), thus demonstrating that this type of hydrogenase does not confer a selective advantage under low-light conditions. The discovery of diazotrophic cyanobacteria without a H_2-consuming enzyme (Ludwig et al. 2006) further indicates that the cyanobacterial Hup is not essential for protection of the nitrogenase from oxygen inactivation. Nevertheless, the recycling of H_2 would seem to be bioenergetically favorable for the cell. In all cyanobacteria containing uptake hydrogenases the gene clusters encoding the maturation functions (Hyp) are extended for five additional genes (Agervald et al. 2008). All proteobacteria branching from the cyanobacterial clade also share these putative additional accessory genes (*asr0689*, *asr0690*, *alr0691*, *alr0692*, and *alr0693* according to the gene nomenclature in *Nostoc* sp. 7120). Besides a tricopeptide repeat protein (TPR) and a ferrochelatase homolog (named Fch in our figures), the five genes encode a NifU-like protein (Alr0692) and two further proteins without homologies to known functions. Despite lacking experimental evidence the additional "*hyp*" genes very likely encode functions required for the insertion of FeS clusters (Johnson et al. 2005) into the small hydrogenase subunit. As indicated in Fig. 18.5 the *hup* operon in *C. aurantiacus* also contains additional maturation genes. Since the *hupSL* operons in contrast to many other proteobacterial group 1 uptake hydrogenase operons do not encode any cytochromes, the primary electron acceptor of the cyanobacterial uptake hydrogenase is still not known. While *hupSL* forms an operon structure together with a gene encoding a putative hydrogenase-specific protease HupW in *Trichodesmium erythraeum* and *Gloeothece* ATCC it is found at a different location in the genomes of *A. variabilis* and *Anabaena* PCC7120.

18.6 Transcriptional Regulation of Hydrogenases

18.6.1 Hydrogen-Sensing Regulatory Systems

The various types of hydrogenases discussed earlier in Section 18.2 fulfill different physiological functions in their host organisms. It is therefore not surprising that diverse control mechanisms for regulation of their expression have arisen during evolution. Expression of the membrane-bound uptake hydrogenases in aerobic chemolithotrophs, for example, is induced by the presence of their substrate, hydrogen. This type of regulatory circuit has been studied in *R. capsulatus* (Dischert et al. 1999), *R. palustris* (Rey et al. 2006) as well as the Knallgas bacterium *R. eutropha* and *Alcaligenes hydrogenophilus* (Lenz et al. 1997) and the nitrogen-fixing legume symbiont *B. japonicum* (Durmowicz and Maier 1997). The mechanism of this signal transduction pathway relies on the heterodimeric sensor module HoxBC which communicates with a specific two-component system, HoxA/HoxJ (HupR/HupT in *R. capsulatus*). HoxBC (HupUV in *R. capsulatus*) which structurally resembles the catalytic subunits of the hydrogenase initially senses available concentrations of hydrogen. Despite very low turnover rates, this so-called regulatory hydrogenase (RH) has been shown to be capable of true hydrogenase activity as could be demonstrated by H^+/D_2 exchange as well as by dye-mediated H_2 consumption and evolution measurements (Lenz and Friedrich 1998; Bernhard et al. 2001).

Among hydrogenases, RHs separate into a distinct group showing unique features such as insensitivity toward oxygen, CO, and acetylene (Bernhard et al. 2001); no requirement for reductive activation of the active site; and no proteolytic cleavage during the maturation of the large subunit. The RH heterodimer is tightly attached to a tetrameric histidine kinase, HoxJ. A highly conserved 50–60 amino acid extension at the C-terminus of HoxB, not present in other NiFe hydrogenases, mediates communication with the N-terminal PAS domain of the cytoplasmic histidine kinase HoxJ (Buhrke et al. 2004). In the absence of hydrogen, HoxJ autophosphorylates followed by a phosphotransfer from His220 of the kinase to Asp55 of the NtrC-type response regulator HoxA (Lenz et al. 2002). Phosphorylation of HoxA leads to its inactivation and dissociation from its respective target gene promoters. Generally, NtrC-type DNA-binding proteins are known to initiate transcription of σ_{54} (*RpoN*)-dependent promoters. Upon receiving a phosphorylation signal most response regulators in this family activate transcription by interacting with RNA polymerase at the expense of ATP. Nevertheless, HoxA induces gene expression solely in its non-phosphorylated form in the presence of hydrogen.

While transcription of hydrogenase structural genes in *R. eutrophus*, *T. roseopersicina* (Kovacs et al. 2005) as well as in *B. japonicum* (Black and Maier 1995) and *Rhizobium leguminosarum* (Brito et al. 1997) is σ_{54} (*RpoN*) dependent, expression of the *hupSL* operon in *R. capsulatus* has been proven to rely on *rpoD* (σ_{70}). Recent transcriptomic studies in *R. palustris* indicate that the hydrogen-sensing signaling chain in this organism might also be responsible for differential expression of an additional set of genes unrelated to hydrogen metabolism (Rey et al. 2006). In a comparison of a *hupV* mutant to the WT the authors found significant changes in expression patterns of genes coding for glutamine synthetase (*glnA*), a formate- and a putative dicarboxylic acid transporter. Nevertheless, the H_2-sensing regulatory circuits found in *T. roseopersicina* as well as in *R. palustris* do not respond to the presence of hydrogen since they have been found to be non-functional due to alterations (frame shift mutation) in *hupV* (Rey et al. 2006; Kovacs et al. 2005). On the other hand, an additional histidine kinase (RPA0981), which is encoded within the *R. palustris hup* cluster, was found to affect *hupSL* gene expression under photoheterotrophic growth in the presence of hydrogen.

18.6.2 Regulation of Four Hydrogenases in E. coli

E. coli remains one of the best characterized model systems for the transcriptional regulation of hydrogenases. The *E. coli* genome harbors four different hydrogenases all of which are only induced under micro- or anaerobic growth conditions. Hydrogenases 1 (HyaA-F) and 2 (HybO-G) were shown to be transcribed as polycistronic messages from σ_{70}-dependent promoters. Both operons are subject to control by the two-component system ArcB–ArcA as well as the anaerobic regulator FNR (Vignais and Toussaint 1994; Richard et al. 1999). FNR (fumarate nitrate reduction regulator) is a global regulator of the FNR/CRP family. Besides a HTH DNA-binding domain at the C-terminus, FNR carries four essential cysteines (Cys20, 23, 29, 122) in its N-terminal sensor domain which coordinate a FeS cluster. Under anaerobic conditions and with the help of cysteine desulfurylase, FNR forms a $[4Fe\text{-}4S]^{2+}$ cluster, which promotes dimerization thereby activating high-affinity DNA-binding activity (Schwartz et al. 2000; Moore and Kiley 2001). In the presence of oxygen, the $[4Fe\text{-}4S]^{2+}$ cluster is converted to a $[2Fe\text{-}2S]^{2+}$ which subsequently leads to dissociation of the FNR dimer and to the inactivation of its ability to bind DNA (Green et al. 1996; Lazazzera et al. 1996). ArcB–ArcA is a homolog of the ResD–ResE system in *Bacillus subtilis*, which is known to sense and transmit redox signals in addition to being subject to metabolic control (Bauer et al. 1999). The *hyb* operon, e.g., is upregulated about 13-fold under anaerobic, fermentative conditions. Additionally, hydrogenases 1 and 2 are subject to additional NO_3-dependent repression mediated by the two-component signal transduction chains of NarL/NarX and NarP/NarQ (Richard et al. 1999). While *hya* transcription is enhanced by the action of the stress sigma factor $\sigma 38$ as well as ArcA, when the pH in the medium decreases the opposite effect can be observed for the *hyb* operon (King and Przybyla 1999; Brondsted and Atlung 1996; Hayes et al. 2006). Hya expression additionally responds to AppY which belongs to the family of AraC-type transcriptional activators and is itself the target of another two-component system, DpiAB (Richard et al. 1999; Atlung et al. 1997). Although recent investigations discovered negative regulation of *dpiAB* by a new small regulatory RNA, *rybC* (Mandin and Gottesman 2009), the AppY–DpiAB signaling chain is not sufficiently understood to date. The regulatory network controlling expression of hydrogenases in bacteria is quite diverse. This is demonstrated by the action of TyrR (regulator of aromatic amino acid metabolism) and CRP (catabolite repressor protein) on the expression of the HyaB homolog in the closely related enterobacterium *Salmonella* (Park et al. 1999).

On the other hand, the expression of hydrogenase complexes 3 (*hyc*) and 4 (*hyf*) in *E. coli* is induced by their substrate, formate (Sauter et al. 1992; Andrews et al. 1997). Despite not being able to completely complement each other, two σ_{54}-dependent transcriptional regulators, FhlA and HyfR, which share 46% homology, sense the substrate formate. In the presence of formate, FhlA and HyfR exhibit increased DNA-binding activity and activate transcription by directly interacting with RNA polymerase. While HyfR exclusively regulates the *hyf* operon, FhlA binds and activates the *hyf* operon as well as the *hyc* operon and the formate dehydrogenase gene, *fdhH* (Skibinski et al. 2002). An additional cysteine-rich segment

(-C-X6-H-C-X-C-X-P-X-C-X-P) found in the A3 domain of HyfR likely serves as a docking site for a FeS cluster or metallocofactor, as in FNR, IRP, or SoxR (Kiley and Beinert 1998; Hidalgo et al. 1995), and might act as redox sensor (Andrews et al. 1997). Since formate dehydrogenase contains molybdenum it was not surprising when Self et al. (1999) discovered a molybdate-dependent control of the formate hydrogenlyase regulon which relies on the metal-sensing repressor ModE.

18.6.3 Regulation of the CO Dehydrogenase in R. rubrum

The CO dehydrogenase complex in *R. rubrum*, encoded by *cooFSCTJ* and *cooMKLXUH* operons, is also under control of a substrate-sensing transcriptional factor. CooA, a member of the CRP/FNR family of DNA-binding proteins, employs a heme cofactor to fulfill a dual function: redox- and CO sensing (Roberts et al. 2005). The CooA heme holoprotein is easily oxidized from the reduced ferrous to the ferric form in the presence of oxygen. Only reduced CooA is able to respond to CO, and only CO-bound CooA displays DNA binding and thereby activates transcription (Shelver et al. 1997). The mid-point potential for this reduction is approximately –300 mV (Nakajima and Aono 1999). This is quite reasonable since CO dehydrogenase is only functional below that redox potential (Heo et al. 2001). Lanzilotta et al. (2000) reported the crystal structure of reduced CooA, which provides new insights into ligand/heme binding. Using a comparative approach, the authors were able to develop a reasonable model which describes the allosteric switch mechanism in CooA as well as CAP (catabolite activator protein) and FNR in *E. coli*.

18.6.4 Redox Regulation in Diazotrophs

In nitrogen-fixing bacteria, uptake hydrogenase is often coregulated with nitrogenase gene expression. Nitrogen fixation is an energetically expensive process which is only required under NO_3/NH_{4+}-limited growth conditions. Furthermore, since nitrogenase is very sensitive to oxygen inactivation, nature has developed control mechanisms to tightly regulate its gene expression. For example, the *R. capsulatus* uptake hydrogenase is controlled by the two-component RegB–RegA sensor kinase/response regulator pair (Elsen et al. 2000). RegB–RegA coregulates genes encoding essential functions in nitrogenase biosynthesis, photosynthesis, CO_2 fixation, and denitrification along with components of energy generation like the uptake hydrogenase and respiratory oxidases (Elsen et al. 2000; Swem et al. 2001; see Chapter 15). The isolation of *regA–regB* deletion mutants, which were unable to induce photosynthetic gene clusters (*puf*, *puh*, *puc*) under anaerobiosis, initially led to the assumption that it directly sensed oxygen. However, it was later found that the transmembrane sensor kinase RegB senses the cellular redox state. In addition to harboring a redox-responsive cysteine (Cys265) in its cytoplasmic dimerization domain, a RegB periplasmic loop was shown to bind quinone (Swem et al. 2003,

2006). Binding of oxidized quinone likely modulates RegB conformation leading to inhibition of its autophosphorylation function (Wu and Bauer 2008). In contrast to *Rhodobacter* species, redox- and oxygen-dependent regulation in *T. roseopersicina* and *R. leguminosarum* biovar *viciae* UPM791 is under the control of FNR homologs (Kovacs et al. 2005; Gutierrez et al. 1997).

R. leguminosarum and *B. japonicum* apply different systems for the control of hydrogenase expression. *hupSL* transcription in these strains is under control of the σ_{54}-dependent NifA–NifL (Brito et al. 1997, 2008) and the heme-based FixL/FixJ-FixK/FixT regulatory network, respectively (Palacios et al. 1990; Durmowicz and Maier 1998). All the four mentioned regulatory systems are able to control the aerobic/anaerobic adaptation of global gene expression, albeit based on different sensor chemistries. While FNR uses FeS clusters to respond to oxygen, FixL exploits heme to sense oxygen. In contrast RegB is thought to bind quinone as well as sense cytoplasmic redox levels. NifL, on the other hand, uses FAD as a redox-responsive cofactor in addition to sensing the nitrogen status, either by direct interaction with the PII protein, GlnK, or by the antagonistic action of 2-oxoglutarate binding to NifA (Martinez-Argudo et al. 2005).

Similar to eukaryotic cells where DNA binding by histones and HMG (high mobility group) proteins is important, bacterial gene regulation can also be influenced by chromosomal DNA-binding proteins (Sandman et al. 1998). Interaction of IHF (integration host factor) with *hupSL* promoters has been demonstrated for both *B. japonicum* and *R. capsulatus*, with IHF binding leading to repression of hydrogenase gene expression (Black and Maier 1995; Toussaint et al. 1991). Moreover, the promoter of *appY*, whose gene product is involved in *hya* expression in *E. coli*, shows AT-rich binding sites for the histone-like protein H-NS (Atlung et al. 1996).

18.6.5 Transcriptional Regulation of Cyanobacterial Hydrogenases

The *hox* operons encoding the cyanobacterial bidirectional hydrogenase are mostly organized as single gene clusters; as in *A. variabilis* (Boison et al. 2000), *Nodularia spumigena*, *Synechococcus* sp. PCC 7002, *Synechococcus* sp. WH 5701, and *Synechocystis* PCC 6803 (Appel and Schulz 1996). Exceptions to this rule are *A. nidulans* and *Anabaena* PCC7120 where the chromosomal location for the genes encoding the NADH dehydrogenase subunits HoxEF is separated by 16 and 9.5 kb from the remainder of the hydrogenase operon, respectively. While the *hox* operons in *Synechocystis* PCC 6803 (Gutekunst et al. 2005), *A. variabilis* (Boison et al. 2000), and *Lyngbya majuscula* CCAP1446/4 (Ferreira et al. 2009) are transcribed as polycistronic messages, *Synechococcus* PCC7942 *strain R2* forms a dicistronic *hoxEF* and a polycistronic *hoxUYH*, initiated from two separate promoters, as demonstrated by qualitative RT-PCR and primer extension analysis. Schmitz et al. (2001) found differing expression levels of *hoxEF* and *hoxUYH* in *Synechococcus* PCC7942 and also showed for the first time that hydrogenase gene expression is subject to oscillations that follow circadian rhythmicity, an effect also observed in global microarray profiling of *Synechocystis* (Kucho et al. 2005).

In cyanobacteria, like the heterocystous *Anabaena* sp. PCC 7120, *A. cylindrica*, and *A. variabilis*, activity of the bidirectional hydrogenase HoxEFUYH is absent during standard photoautotrophic growth, but is induced under dark anaerobic conditions or in the light when photosynthetic water splitting is inhibited by the PSII inhibitor DCMU (Houchins and Burris 1981; Laczko 1984; Serebryakova et al. 1992, 1994). In contrast, a number of unicellular strains like *Synechocystis* sp. PCC 6803, *Gloeocapsa alpicola*, and *A. halophytica* constitutively express this hydrogenase even under aerobic photoautotrophic growth conditions (Serebryakova et al. 1998; Appel et al. 2000). However, even in those strains transcription is subject to further induction when cells are exposed to microaerobic or anoxic conditions (Summerfield and Sherman 2008; Kiss et al. 2009). Appel and coworkers (unpublished results) have detected an increase in total hydrogenase activity under nitrogen limitation that was dependent on the light intensity. Transcript levels in *Synechocystis* PCC 6803 were reported to correlate well with hydrogenase activities, a redox-dependent rather than an oxygen-dependent control mechanism (Antal et al. 2006). In order to obtain better understanding of this particular question, Gutthann et al. (2007) analyzed mutants impaired in photosynthetic or respiratory electron transport and found upregulation of hydrogenase activity in a quinol oxidase mutant as well as severe downregulation in a mutant defective in the NADH dehydrogenase (NDH) complex. These results correspond well with the idea of redox-regulated expression of the bidirectional hydrogenase.

Although the key players responsible have not yet been identified, recent investigations by Kiss et al. (2009) have verified the light- and redox-driven regulatory effects on *hox* transcript level. While the entire operon *hoxEFUYH* was found to be upregulated about 5- to 6-fold after adjusting to a microaerobic environment, *hoxEF* (not *hoxUYH*) transcription was further increased about 10- to 12-fold when cells were shifted from anoxic growth in the light to anoxic dark conditions (Kiss et al. 2009). These results were further substantiated by examining the effect of inhibitors of photosynthetic and respiratory electron transport. Glycolaldehyde simulated the effects during a shift from light to dark in the presence of oxygen. DBMIB, on the other hand, led to the same results which were observed after a light-to-dark transition under anoxic conditions. Although only one transcriptional start site, located at 168 bp upstream of the *hoxE* translational start, was found in *Synechocystis*, the occurrence of additional internal start sites cannot be excluded.

18.6.5.1 Regulation via LexA

So far there are two transcription factors known to be involved in the regulation of the bidirectional hydrogenase in *Synechocystis* PCC6803: LexA (Gutekunst et al. 2005, Oliveira and Lindblad 2005) and the AbrB-like regulator Sll0359 (Oliveira and Lindblad 2008). Since transcription of LexA and the AbrB-like protein does not correlate with hydrogenase expression (Zhang et al. 2008; Kiss et al. 2009), one or both factors might be posttranslationally modified or modulated as part of a signal transduction pathway. Alternatively, another unknown regulator might account for the observed redox/O_2-dependent effects. While LexA, together with RecA, is known to be a crucial key regulator of the SOS response in *E. coli* and *B. subtilis*, microarray experiments of a *lexA* deletion mutant in *Synechocystis* PCC6803 instead indicate a role for it under inorganic carbon starvation conditions (Domain et al. 2004). The damage-inducible LexA in *E. coli* functions as a repressor. Derepression of the LexA regulon depends on an autoproteolytic intramolecular cleavage reaction. The *Synechocystis* LexA lacks the catalytic serine as well as the conserved alanine–glycine cleavage bond which is known to be essential for LexA degradation (Little 1984; Little 1991; Giese et al. 2008). Patterson-Fortin and coworkers suggested that LexA might mediate the regulation of the intracellular redox state (Patterson-Fortin et al. 2006). *Synechocystis* LexA, which generally binds to a 12 bp direct repeat motif [CTA-N9-CT(A/T)] on DNA, was found to be a negative regulator of *crhR* expression (Patterson-Fortin et al. 2006). However, LexA occupies multiple binding sites in the *hoxE-H* promoter (Gutekunst et al. 2005; Oliveira and Lindblad 2005) where it seems on the contrary to activate hydrogenase transcription as judged by *hoxEFUYH* downregulation in a *ΔlexA* mutant background (Gutekunst et al. 2005). Subsequently, a number of studies in other cyanobacteria have also shown LexA binding upstream of *hyp* and *hox* genes (Ferreira et al. 2007; Sjöholm et al. 2007).

18.6.5.2 Regulation via AbrB

Similar to the discovery of LexA, AbrB, another transcription factor known to directly regulate the bidirectional hydrogenase, has been isolated and identified based on its DNA-binding affinity to the *hox* promoter (Oliveira and Lindblad 2008). AbrB of *B. subtilis* is a prototype of the AbrB family. In *B. subtilis* it belongs to the transition state regulators which are known to control gene expression during entry into the stationary phase. It directly controls gene expression of more than 40 genes involved in spore formation, competence, and biofilm development (Phillips and Strauch 2002). Despite several crystal structures which reveal a unique swapped hairpin β-barrel (ββαββ) arrangement of the N-terminal DNA-binding domain (Bobay et al. 2005; Coles et al. 2005, Asen et al. 2009), the detailed mode of transcriptional activation remains unclear. Furthermore, no consensus recognition sequences could be determined although more than 80 high-affinity binding sites were selected in the upstream regions of its target promoters (Xu and Strauch 1996). It was suggested that AbrBs recognize a three-dimensional structure in the DNA helix (Bobay et al. 2004). Interestingly, it is able to bind to some DNA as a dimer or tetramer, but only as tetramer to other promoters (Vaughn et al. 2000).

Both AbrB-like proteins in *Synechocystis* PCC6803 autoregulate their own synthesis but activate transcription of their main target genes. While *sll0359* seems to be essential, as judged by the inability to obtain a fully segregated mutant, *sll0822* could be deleted, and its phenotype was studied using microarray analysis (Ishii and Hihara 2008). The results indicate a role of *sll0822* in activation of nitrogen-regulated genes, including the *hox* operon. Nevertheless, direct interaction of the regulator with the hydrogenase promoter remains to be addressed. It should be noted that investigations by Kaplan and coworkers discovered posttranslational modifications (N-acetylation, methylation) of an AbrB-like protein in *Aphanizomenon ovalisporum*, which significantly changed its DNA-binding capacity (Shalev-Malul et al. 2008). During the analysis of the *sbtA* promoter in *Synechocystis* the same group observed changes in the extent of electrophoretic mobility shifts when applying different combinations of the three transcription factors LexA, Sll0359, and Sll0822 (Lieman-Hurwitz et al. 2009). Using promoter/reporter measurements, Gutekunst et al. (2006) found an additional iron dependency of *hox* transcription in *Synechocystis*, which reached maximum expression levels under iron-deplete conditions at concentrations in the range of 0–5 μM Fe^{3+}. Whether or not the observed regulation is mediated by one of the Fur (*F*erric *U*ptake *R*egulators) homologs present in *Synechocystis* is still under investigation.

18.6.6 Regulation of the Cyanobacterial Uptake Hydrogenase

The cyanobacterial uptake hydrogenase HupSL is not expressed under non-nitrogen-fixing growth conditions. Maximum induction of this enzyme in *Nostoc muscorum* and *Anabaena cylindrica* could be observed after several days of incubation under a hydrogen atmosphere (Eisbrenner et al. 1978). Similarly, *A. variabilis* and *N. muscorum* showed a 5- to 20-fold increase in total activity when grown under an atmosphere of H_2:N_2:CO_2 (20:75:5) (Tel-Or et al. 1977). Much later these results were confirmed on the transcriptional level (Axelsson et al. 1999; Happe et al. 2000; Axelsson and Lindblad 2002; Oxelfelt et al. 1995). In non-heterocystous cyanobacteria like *L. majuscula* or the unicellular *Gloeothece* ATCC27152 and *Cyanothece sp* ATCC 51142, *hupSL* transcription follows light/dark changes with maximum expression in the dark phases (Oliveira et al. 2004; Leitao et al. 2005; Toepel et al. 2008).

Using electrophoretic mobility shift assays (EMSA) Oliveira et al. (2004) could demonstrate the interaction of the *hupSL* promoter in *Gleothece* with the global nitrogen regulator NtcA. NtcA, but not hydrogen or anoxic conditions, was found to be solely responsible for induction of *hupSL* after a shift to nitrogen-fixing growth (Weyman et al. 2008), although previous investigations reported H_2-dependent upregulation in *N. muscorum* and *Nostoc punctiforme* (Axelsson and Lindblad 2002, Stensjo et al. 2007). NtcA belongs to the cAMP receptor protein family and is highly conserved among cyanobacteria. It is known to be an autoregulatory transcriptional activator. In the absence of ammonia, it positively regulates target genes involved in nitrogen metabolism, such as *glnA* (glutamate synthase), *narB* (nitrate reductase), the *nir* operon (ferredoxin:nitrite reductase), as well as some genes not involved in the nitrogen assimilation process, such as *icd* (isocitrate dehydrogenase), *rbcLS* (Rubisco), the histone-like chromosomal DNA-binding protein HU, and the alternative sigma factor $RpoD_2$-V (Herrero et al. 2001).

NtcA is able to sense two different environmental signals. While it monitors the C/N ratio via the concentration of intracellular 2-oxoglutarate (Muro-Pastor et al. 2001), an additional role in thiol-dependent redox sensing has been suggested (Herrero et al. 2001; Jiang et al. 1997; Alfonso et al. 2001). In certain cases, DNA binding by NtcA is modulated by 2-oxoglutarate-dependent interaction with PipX and involves communication with the signal transducer PII (Espinosa et al. 2006; Forchhammer 2007). NtcA-dependent promoters contain a DNA recognition site with the consensus sequence (GTA-N_8-TAC). Although most NtcA binding sites are located at around –40 bp relative to the transcriptional start sites, sequential deletions of the *hupSL* promoter in *N. punctiforme* ATC29133 demonstrated that multiple binding sites might occur which could be located in the non-translated mRNA leader region as well (Holmqvist et al. 2009). A summary of the current knowledge is depicted in Fig. 18.10.

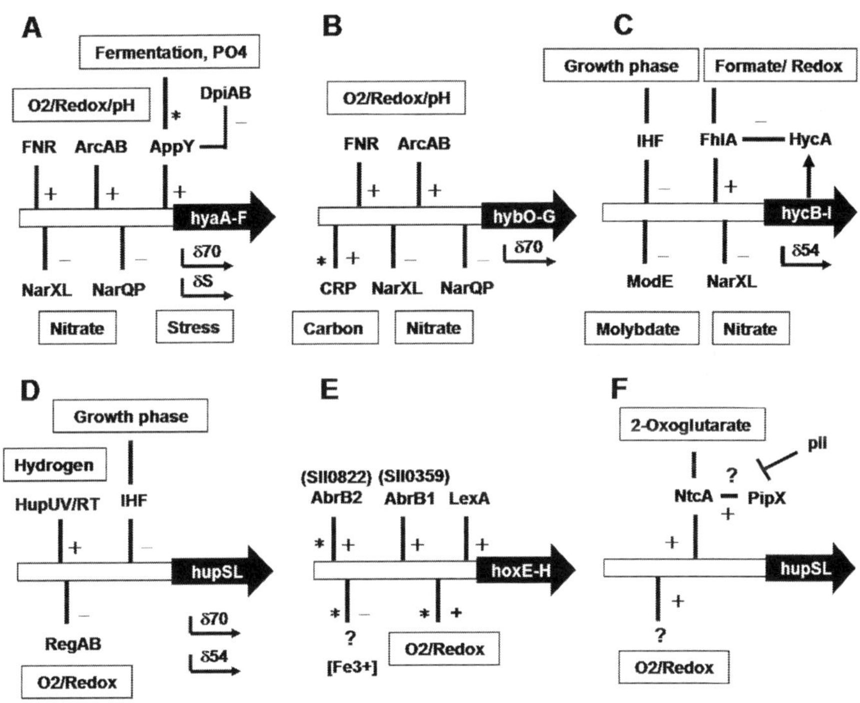

Fig. 18.10 Graphical overview representing the current knowledge about differential expression and transcriptional regulation of selected hydrogenases in model organisms. (**a**) hya operon, (**b**) *hyb* operon, and (**c**) *hyc* operon in *E. coli* (Richard et al. 1999, Atlung 1997, King and Przybyla 1999); (**d**) hupSLC in *R. capsulatus* (Dischert et al. 1999, Elsen et al. 2000, Touissant et al. 1997); as well as (**e**) *hoxEFUYH* in *Synechocystis* PCC6803 (Gutekunst et al. 2005, Oliveira et al. 2005, Gutekunst et al. 2006) and (**f**) the cyanobacterial *hupSLW* (+/–) indicate positive or negative regulation while * indicates indirect effects on hydrogenase gene expression

References

Abdel-Basset R, Bader KP (1998) Physiological analyses of the hydrogen gas exchange in cyanobacteria. J Photochem Photobiol 43:146–151

Adams MW,Stiefel EI (1998) Biological hydrogen production: not so elementary. Science 282:1842–1843

Agervald A, Stensjo K, Holmqvist M et al (2008) Transcription of the extended *hyp*-operon in *Nostoc* sp. strain PCC 7120. BMC Microbiol 8:69

Alfonso M, Perewoska I, Kirilovsky D (2001) Redox control of *ntcA* gene expression in *Synechocystis* sp. PCC 6803. Nitrogen availability and electron transport regulate the levels of the NtcA protein. Plant Physiol 125:969–981

Andrews SC, Berks BC, McClay J et al (1997) A 12-cistron *Escherichia coli* operon (*hyf*) encoding a putative proton-translocating formate hydrogenlyase system. Microbiology 143 3633–3647

Antal TK, Oliveira P Lindblad P (2006) The bidirectional hydrogenase in the cyanobacterium *Synechocystis* sp. strain PCC 6803. Int J Hydr Energ 31:1439–1444

Aoyama K, Uemura I, Miyake J et al (1997) Fermentative metabolism to produce hydrogen gas and organic compounds in a cyanobacterium, *Spirulina platensis*. J Ferment Bioeng 83: 17–20

Aparicio PJ, Azuara MP, Ballesteros A et al (1985) Effects of light intensity and oxidized nitrogen sources on hydrogen production by *Chlamydomonas reinhardtii*. Plant Physiol 78: 803–806

Appel J, Phunpruch S, Steinmuller K et al (2000) The bidirectional hydrogenase of *Synechocystis* sp. PCC 6803 works as an electron valve during photosynthesis. Arch Microbiol 173: 333–338

Appel J, Schulz R (1996) Sequence analysis of an operon of a NAD(P)-reducing nickel hydrogenase from the cyanobacterium *Synechocystis* sp. PCC 6803 gives additional evidence for direct coupling of the enzyme to NAD(P)H-dehydrogenase (complex I). Biochim Biophys Acta 1298:141–147

Appel J, Schulz R (1998) Hydrogen metabolism in organisms with oxygenic photosynthesis: hydrogenases as important regulatory devices for a proper redox poising? J Photochem Photobiol B: Biology 47:1–11

Armstrong F A (2004) Hydrogenases: active site puzzles and progress. Curr Opin Chem Biol 8:133–140

Armstrong FA, Albracht SP (2005) [NiFe]-hydrogenases: spectroscopic and electrochemical definition of reactions and intermediates. Philos Transact A Math Phys Eng Sci 363:937–954; discussion 1035–1040

Asen I, Djuranovic S, Lupas AN et al (2009) Crystal structure of SpoVT, the final modulator of gene expression during spore development in *Bacillus subtilis*. J Mol Biol 386:962–975

Atanassova A, Zamble DB (2005) *Escherichia coli* HypA is a zinc metalloprotein with a weak affinity for nickel. J Bacteriol 187:4689–4697

Atlung T, Knudsen K, Heerfordt L et al (1997) Effects of sigmaS and the transcriptional activator AppY on induction of the *Escherichia coli hya* and *cbdAB-appA* operons in response to carbon and phosphate starvation. J Bacteriol 179:2141–2146

Atlung T, Sund S, Olesen K et al (1996) The histone-like protein H-NS acts as a transcriptional repressor for expression of the anaerobic and growth phase activator AppY of *Escherichia coli*. J Bacteriol 178:3418–3425

Axelsson R, Lindblad P (2002) Transcriptional regulation of *Nostoc* hydrogenases: effects of oxygen, hydrogen, and nickel. Appl Environ Microbiol 68:444–447

Axelsson R, Oxelfelt F, Lindblad P (1999) Transcriptional regulation of *Nostoc* uptake hydrogenase. FEMS Microbiol Lett 170:77–81

Bagramyan K, Trchounian A (2003) Structural and functional features of formate hydrogen lyase, an enzyme of mixed-acid fermentation from *Escherichia coli*. Biochemistry (Mosc) 68: 1159–1170

Bagramyan K, Vassilian A, Mnatsakanyan N et al (2001) Participation of *hyf*-encoded hydrogenase 4 in molecular hydrogen release coupled with proton-potassium exchange in *Escherichia coli*. Membr Cell Biol 14:749–763

Bauer CE, Elsen S, Bird TH (1999) Mechanisms for redox control of gene expression. Annu Rev Microbiol 53:495–523

Baumer S, Ide T, Jacobi C et al (2000) The F420H2 dehydrogenase from *Methanosarcina mazei* is a redox-driven proton pump closely related to NADH dehydrogenases. J Biol Chem 275: 17968–17973

Belkin S, Padan E, (1978) Hydrogen metabolism in the facultative anoxygenic cyanobacteria (blue-green algae) *Oscillatoria limnetica* and *Aphanothece halophytica*. Arch Microbiol 116: 109–111

Benemann JR Weare NM(1974) Hydrogen evolution by nitrogen-fixing *Anabaena cylindrica* cultures. Science 184:174–175

Berks BC, Page MD, Richardson DJ et al (1995) Sequence analysis of subunits of the membrane-bound nitrate reductase from a denitrifying bacterium: the integral membrane subunit provides a prototype for the dihaem electron-carrying arm of a redox loop. Mol Microbiol 15: 319–331

Bernhard M, Buhrke T, Bleijlevens B et al (2001) The H_2 sensor of *Ralstonia eutropha*. Biochemical characteristics, spectroscopic properties, and its interaction with a histidine protein kinase. J Biol Chem 276:15592–15597

Black LK, Maier RJ, (1995) IHF- and RpoN-dependent regulation of hydrogenase expression in *Bradyrhizobium japonicum*. Mol Microbiol 16:405–413

Bleijlevens B, Buhrke T, van der Linden E et al (2004) The auxiliary protein HypX provides oxygen tolerance to the soluble [NiFe]-hydrogenase of *Ralstonia eutropha* H16 by way of a cyanide ligand to nickel. J Biol Chem 279:46686–46691

Blokesch M, Albracht SP, Matzanke BF et al (2004) The complex between hydrogenase-maturation proteins HypC and HypD is an intermediate in the supply of cyanide to the active site iron of [NiFe]-hydrogenases. J Mol Biol 344:155–167

Blokesch M, Bock A (2002) Maturation of [NiFe]-hydrogenases in *Escherichia coli*: the HypC cycle. J Mol Biol 324:287–296

Blokesch M, Magalon A, Bock A (2001) Interplay between the specific chaperone-like proteins HybG and HypC in maturation of hydrogenases 1, 2, and 3 from *Escherichia coli*. J Bacteriol 183:2817–2822

Bobay BG, Andreeva A, Mueller GA et al (2005) Revised structure of the AbrB N-terminal domain unifies a diverse superfamily of putative DNA-binding proteins. FEBS Lett 579:5669–5674

Bobay BG, Benson L, Naylor S et al (2004) Evaluation of the DNA binding tendencies of the transition state regulator AbrB. Biochemistry 43:16106–16118

Böck A, King PW, Blokesch M et al (2006) Maturation of hydrogenases. Adv Microb Physiol 51:1–71

Böck A, Sawers G (1996) Fermentation. ASM Press, Washington, DC

Bogorov LV (1974) The properties of *Thiocapsa roseopersicina*, strain BBS, isolated from an estuary of the white sea. Mikrobiololiia 43:326–332

Bohm R, Sauter M, Bock A (1990) Nucleotide sequence and expression of an operon in *Escherichia coli* coding for formate hydrogenlyase components. Mol Microbiol 4:231–243

Boison G, Bothe H, Schmitz O (2000) Transcriptional analysis of hydrogenase genes in the cyanobacteria *Anacystis nidulans* and *Anabaena variabilis* monitored by RT-PCR. Curr Microbiol 40:315–321

Bonam D, Lehman L, Roberts GP et al (1989) Regulation of carbon monoxide dehydrogenase and hydrogenase in *Rhodospirillum rubrum*: effects of CO and oxygen on synthesis and activity. J Bacteriol 171:3102–3107

Bonam D, Ludden PW (1987) Purification and characterization of carbon monoxide dehydrogenase, a nickel, zinc, iron-sulfur protein, from *Rhodospirillum rubrum*. J Biol Chem 262:2980–2987

Boyer ME, Stapleton JA, Kuchenreuther JM et al (2008) Cell-free synthesis and maturation of [FeFe] hydrogenases. Biotechnol Bioeng 99:59–67

Brazzolotto X, Rubach JK, Gaillard J et al (2006) The [Fe-Fe]-hydrogenase maturation protein HydF from *Thermotoga maritima* is a GTPase with an iron-sulfur cluster. J Biol Chem 281:769–774

Brito B, Martinez M, Fernandez D et al (1997) Hydrogenase genes from *Rhizobium leguminosarum* bv. viciae are controlled by the nitrogen fixation regulatory protein nifA. Proc Natl Acad Sci USA 94:6019–6024

Brito B, Toffanin A, Prieto RI et al (2008) Host-dependent expression of *Rhizobium leguminosarum* bv. viciae hydrogenase is controlled at transcriptional and post-transcriptional levels in legume nodules. Mol Plant Microbe Interact 21:597–604

Brondsted L, Atlung T (1996) Effect of growth conditions on expression of the acid phosphatase (*cyx-appA*) operon and the *appY* gene, which encodes a transcriptional activator of *Escherichia coli*. J Bacteriol 178:1556–1564

Bryant MP, Wolin EA, Wolin MJ et al (1967) *Methanobacillus omelianskii*, a symbiotic association of two species of bacteria. Arch Mikrobiol 59:20–31

Buhrke T, Lenz O, Krauss N et al (2005) Oxygen tolerance of the H_2-sensing [NiFe] hydrogenase from *Ralstonia eutropha* H16 is based on limited access of oxygen to the active site. J Biol Chem 280:23791–23796

Buhrke T, Lenz O, Porthun A et al (2004) The H_2-sensing complex of *Ralstonia eutropha*: interaction between a regulatory [NiFe] hydrogenase and a histidine protein kinase. Mol Microbiol 51:1677–1689

Chou CJ, Jenney FE, Jr, Adams MW et al (2008) Hydrogenesis in hyperthermophilic microorganisms: implications for biofuels. Metab Eng 10:394–404

Colbeau A, Kelley BC, Vignais PM (1980) Hydrogenase activity in *Rhodopseudomonas capsulata*: relationship with nitrogenase activity. J Bacteriol 144:141–148

Colbeau A, Kovacs KL, Chabert J et al (1994) Cloning and sequence of the structural (*hupSLC*) and accessory (*hupDHI*) genes for hydrogenase biosynthesis in *Thiocapsa roseopersicina*. Gene 140:25–31

Coles M, Djuranovic S, Soding J et al (2005) AbrB-like transcription factors assume a swapped hairpin fold that is evolutionarily related to double-psi beta barrels. Structure 13:919–928

Conrad R (1996) Soil microorganisms as controllers of atmospheric trace gases (H2, CO, CH_4, OCS, N_2O, and NO). Microbiol Rev 60:609–640

Cooley JW, Vermaas WF (2001) Succinate dehydrogenase and other respiratory pathways in thylakoid membranes of *Synechocystis* sp. strain PCC 6803: capacity comparisons and physiological function. J Bacteriol 183:4251–4258

Cournac L, Guedeney G, Peltier G et al (2004) Sustained photoevolution of molecular hydrogen in a mutant of *Synechocystis* sp. strain PCC 6803 deficient in the type I NADPH-dehydrogenase complex. J Bacteriol 186:1737–1746

Dahl C, Rakhely G, Pott-Sperling AS et al (1999) Genes involved in hydrogen and sulfur metabolism in phototrophic sulfur bacteria. FEMS Microbiol Lett 180:317–324

Dementin S, Leroux F, Cournac L et al (2009) Introduction of methionines in the gas channel makes [NiFe] hydrogenase aero-tolerant. J Am Chem Soc 131:10156–10164

Devine E, Holmqvist M, Stensjo K et al (2009) Diversity and transcription of proteases involved in the maturation of hydrogenases in *Nostoc punctiforme* ATCC 29133 and Nostoc sp. strain PCC 7120. BMC Microbiol 9:53

Dias AV, Mulvihill CM, Leach MR et al (2008) Structural and biological analysis of the metal sites of *Escherichia coli* hydrogenase accessory protein HypB. Biochemistry 47:11981–11991

Dischert W, Vignais PM, Colbeau A (1999) The synthesis of *Rhodobacter capsulatus* HupSL hydrogenase is regulated by the two-component HupT/HupR system. Mol Microbiol 34: 995–1006

Domain F, Houot L, Chauvat F et al (2004) Function and regulation of the cyanobacterial genes *lexA*, *recA* and *ruvB*: LexA is critical to the survival of cells facing inorganic carbon starvation. Mol Microbiol 53:65–80

Drapal N, Bock A (1998) Interaction of the hydrogenase accessory protein HypC with HycE, the large subunit of *Escherichia coli* hydrogenase 3 during enzyme maturation. Biochemistry 37:2941–2948

Drennan CL, Heo J, Sintchak MD et al (2001) Life on carbon monoxide: X-ray structure of *Rhodospirillum rubrum* Ni-Fe-S carbon monoxide dehydrogenase. Proc Natl Acad Sci USA 98:11973–11978

Drutschmann M, Klemme JH (1985) Sulfide repressed, membrane bound hydrogenase in the thermophilic facultative phototroph, *Chlorojiexus aurantiacus*. FEMS Microbiology Letters 28:231–235

Duche O, Elsen S, Cournac L et al (2005) Enlarging the gas access channel to the active site renders the regulatory hydrogenase HupUV of *Rhodobacter capsulatus* O_2 sensitive without affecting its transductory activity. Febs J 272:3899–3908

Durmowicz M, Maier R (1997) Roles of HoxX and HoxA in biosynthesis of hydrogenase in *Bradyrhizobium japonicum*. J Bacteriol 179:3676–3682

Durmowicz MC, Maier RJ (1998) The FixK2 protein is involved in regulation of symbiotic hydrogenase expression in *Bradyrhizobium japonicum*. J Bacteriol 180:3253–3256

Eisen JA, Karen EN, Paulsen IT et al (2002) The complete genome sequence of *Chlorobium tepidum* TLS, a photosynthetic, anaerobic, green-sulfur bacterium. Proc Natl Acad Sci USA. 99:9509–9514

Eisbrenner G, Distler E, Floener L et al (1978) The occurrence of the hydrogenase in cyanobacteria. Archives of Microbiology 118:177–184

Elsen S, Dischert W, Colbeau A et al (2000) Expression of uptake hydrogenase and molybdenum nitrogenase in *Rhodobacter capsulatus* is coregulated by the RegB-RegA two-component regulatory system. J Bacteriol 182:2831–2837

Ensign SA, Ludden PW (1991) Characterization of the CO oxidation/H_2 evolution system of Rhodospirillum rubrum. Role of a 22-kDa iron-sulfur protein in mediating electron transfer between carbon monoxide dehydrogenase and hydrogenase. J Biol Chem 266:18395–18403

Espinosa J, Forchhammer K, Burillo S et al (2006) Interaction network in cyanobacterial nitrogen regulation: PipX, a protein that interacts in a 2-oxoglutarate dependent manner with PII and NtcA. Mol Microbiol 61:457–469

Fay P (1992) Oxygen relations of nitrogen fixation in cyanobacteria. Microbiol Rev 56:340–373

Fernández VM, Aguirre R, Hatchikian EC (1984) Reductive activation and redox properties of hydrogenase from *Desulfovibrio gigas*. Biochim Biophys Acta 790:1–797

Fernandez VM, Hatchikian EC, Cammack R (1985) Properties and reactivation of two different deactivated forms of *Desulfovibrio gigas* hydrogenase. Biochim Biophys Acta 832:69–79

Ferreira D, Leitao E, Sjoholm J et al (2007) Transcription and regulation of the hydrogenase(s) accessory genes, *hypFCDEAB*, in the cyanobacterium *Lyngbya majuscula* CCAP 1446/4. Arch Microbiol 188:609–617

Ferreira D, Pinto F, Moradas-Ferreira P et al (2009) Transcription profiles of hydrogenases related genes in the cyanobacterium *Lyngbya majuscula* CCAP 1446/4. BMC Microbiol 9:67

Finlay BJ, Fenchel T (1992) An anaerobic ciliate as a natural chemostat for the growth of endosymbiotic methanogens. Eur J Protistol. 28:127–137

Fontecave M, Ollagnier-de-Choudens S, Mulliez E (2003) Biological radical sulfur insertion reactions. Chem Rev 103:2149–2166

Fontecilla-Camps JC, Frey M, Garcin E et al (1997) Hydrogenase: a hydrogen-metabolizing enzyme. What do the crystal structures tell us about its mode of action? Biochimie 79:661–666

Forchhammer K. (2007) Glutamine signalling in bacteria. Front Biosci 12:358–370

Forzi L, Sawers RG (2007) Maturation of [NiFe]-hydrogenases in *Escherichia coli*. Biometals 20:565–578

Fox JD, He Y, Shelver D et al (1996) Characterization of the region encoding the CO-induced hydrogenase of *Rhodospirillum rubrum*. J Bacteriol 178:6200–6208

Frenkel A, Gaffron H, Battley EH(1950) Photosynthesis and photoreduction by the blue green alga, *Synechococcus elongatus*, Nag. Biol Bull 99:157–162

Friedrich B, Buhrke T, Burgdorf T et al (2005) A hydrogen-sensing multiprotein complex controls aerobic hydrogen metabolism in *Ralstonia eutropha*. Biochem Soc Trans 33:97–101

Fritsche E, Paschos A, Beisel HG et al (1999) Crystal structure of the hydrogenase maturating endopeptidase HybD from *Escherichia coli*. J Mol Biol 288:989–998

Garcin E, Vernede X, Hatchikian EC et al (1999) The crystal structure of a reduced [NiFeSe] hydrogenase provides an image of the activated catalytic center. Structure 7:557–566

Gasper R, Scrima A, Wittinghofer A (2006) Structural insights into HypB, a GTP-binding protein that regulates metal binding. J Biol Chem 281:27492–27502

George SJ, Kurkin S, Thorneley RN et al (2004) Reactions of H_2, CO, and O_2 with active [NiFe]-hydrogenase from *Allochromatium vinosum*. A stopped-flow infrared study. Biochemistry 43:6808–6819

Germer F, Zebger I, Saggu M et al (2009) Overexpression, isolation and spectroscopic characterization of the bidirectional [NiFe]-hydrogenase from *Synechocystis* sp. PCC 6803. J Biol Chem 284:36462–36472

Giese KC, Michalowski CB, Little JW (2008) RecA-dependent cleavage of LexA dimers. J Mol Biol 377:148–161

Gogotov IN, Zorin NA, Serebriakova LT et al (1978) The properties of hydrogenase from *Thiocapsa roseopersicina*. Biochim Biophys Acta 523:335–343

Green J, Bennett B, Jordan P et al (1996) Reconstitution of the [4Fe-4S] cluster in FNR and demonstration of the aerobic-anaerobic transcription switch in vitro. Biochem J 316 (Pt 3): 887–892

Gutekunst K, Hoffmann D, Lommer M et al (2006) Metal dependence and intracellular regulation of the bidirectional NiFe–hydrogenase in *Synechocystis* sp. PCC 6803. IntJHydrEnerg 31:1452–1459

Gutekunst K, Phunpruch S, Schwarz C et al (2005) LexA regulates the bidirectional hydrogenase in the cyanobacterium *Synechocystis* sp. PCC 6803 as a transcription activator. Mol Microbiol 58:810–823

Gutierrez D, Hernando Y, Palacios JM et al (1997) FnrN controls symbiotic nitrogen fixation and hydrogenase activities in *Rhizobium leguminosarum* biovar *viciae* UPM791. J Bacteriol 179:5264–5270

Gutthann F, Egert M, Marques A et al (2007) Inhibition of respiration and nitrate assimilation enhances photohydrogen evolution under low oxygen concentrations in *Synechocystis* sp. PCC 6803. Biochim Biophys Acta 1767:161–169

Hackstein JH, Akhmanova A, Boxma B et al (1999) Hydrogenosomes: eukaryotic adaptations to anaerobic environments. Trends Microbiol 7:441–447

Hallenbeck PC (2005) Fundamentals of the fermentative production of hydrogen. Water Sci Technol 52:21–29

Happe RP, Roseboom W, Pierik AJ et al (1997) Biological activation of hydrogen. Nature 385:126

Happe T, Hemschemeier A, Winkler M et al (2002) Hydrogenases in green algae: do they save the algae's life and solve our energy problems? Trends Plant Sci 7:246–250

Happe T, Schutz K, Bohme H (2000) Transcriptional and mutational analysis of the uptake hydrogenase of the filamentous cyanobacterium *Anabaena variabilis* ATCC 29413. J Bacteriol 182:1624–1631

Hayes ET, Wilks JC, Sanfilippo P et al (2006) Oxygen limitation modulates pH regulation of catabolism and hydrogenases, multidrug transporters, and envelope composition in *Escherichia coli* K-12. BMC Microbiol 6:89

Hedderich R, Forzi L (2005) Energy-converting [NiFe] hydrogenases: more than just H_2 activation. J Mol Microbiol Biotechnol 10:92–104

Heo J, Halbleib CM Ludden PW (2001) Redox-dependent activation of CO dehydrogenase from *Rhodospirillum rubrum*. Proc Natl Acad Sci USA 98:7690–7693

Heo J, Wolfe MT, Staples CR et al (2002) Converting the NiFeS carbon monoxide dehydrogenase to a hydrogenase and a hydroxylamine reductase. J Bacteriol 184:5894–5897

Herrero A, Muro-Pastor AM,Flores E (2001) Nitrogen control in cyanobacteria. J Bacteriol 183:411–425

Hidalgo E, Bollinger JM, Jr, Bradley TM et al (1995) Binuclear [2Fe-2S] clusters in the *Escherichia coli* SoxR protein and role of the metal centers in transcription. J Biol Chem 270:20908–20914

Higuchi Y, Yagi T, Yasuoka N (1997) Unusual ligand structure in Ni-Fe active center and an additional Mg site in hydrogenase revealed by high resolution X-ray structure analysis. Structure 5:1671–1680

Hoehler TM, Albert DB, Alperin MJ et al (2002) Comparative ecology of H_2 cycling in sedimentary and phototrophic ecosystems. Antonie Van Leeuwenhoek 81:575–585

Hoehler TM, Bebout B.M, Des Marais DJ (2001) The role of microbial mats in the production of reduced gases on the early Earth. Nature 412:324–327

Hoffmann D, Gutekunst K, Klissenbauer M et al (2006) Mutagenesis of hydrogenase accessory genes of *Synechocystis* sp. PCC 6803. Additional homologues of *hypA* and *hypB* are not active in hydrogenase maturation. FEBS J 273:4516–4527

Holmqvist M, Stensjo K, Oliveira P et al (2009) Characterization of the *hupSL* promoter activity in *Nostoc punctiforme* ATCC 29133. BMC Microbiol 9:54

Houchins JP (1984) The physiology and biochemistry of hydrogen metabolism in cyanobacteria. Biochim Biophys Acta 768:227–255

Houchins JP, Burris RH (1981) Physiological Reactions of the Reversible Hydrogenase from *Anabaena* 7120. Plant Physiol 68:717–721

Huang CJ, 1. Barrett EL (1991) Sequence analysis and expression of the Salmonella typhimurium asr operon encoding production of hydrogen sulfide from sulfite. J Bacteriol 173:1544–1553

Hube M, Blokesch M, Bock A (2002) Network of hydrogenase maturation in *Escherichia coli*: role of accessory proteins HypA and HybF. J Bacteriol 184:3879–3885

Ishii A, Hihara Y (2008) An AbrB-like transcriptional regulator, Sll0822, is essential for the activation of nitrogen-regulated genes in *Synechocystis* sp. PCC 6803. Plant Physiol 148: 660–670

Jiang F, Mannervik B, Bergman B. (1997) Evidence for redox regulation of the transcription factor NtcA, acting both as an activator and a repressor, in the cyanobacterium *Anabaena* PCC 7120. Biochem J 327 (Pt 2): 513–517

Johnson DC, Dean DR, Smith AD et al (2005) Structure, function, and formation of biological iron-sulfur clusters. Annu Rev Biochem 74:247–281

Jormakka M, Tornroth S, Byrne B et al (2002) Molecular basis of proton motive force generation: structure of formate dehydrogenase-N. Science 295:1863–1868

Kerby RL, Hong SS, Ensign SA et al (1992) Genetic and physiological characterization of the *Rhodospirillum rubrum* carbon monoxide dehydrogenase system. J Bacteriol 174:5284–5294

Kerby RL, Ludden PW, Roberts GP (1995) Carbon monoxide-dependent growth of *Rhodospirillum rubrum*. J Bacteriol 177:2241–2244

Kiley PJ, Beinert H (1998) Oxygen sensing by the global regulator, FNR: the role of the iron-sulfur cluster. FEMS Microbiol Rev 22:341–352

King PW, Posewitz MC, Ghirardi ML et al (2006) Functional studies of [FeFe] hydrogenase maturation in an *Escherichia coli* biosynthetic system. J Bacteriol 188:2163–2172

King PW, Przybyla AE (1999) Response of *hya* expression to external pH in *Escherichia coli*. J Bacteriol 181:5250–5256

Kiss E, Kos PB, Vass I (2009) Transcriptional regulation of the bidirectional hydrogenase in the cyanobacterium *Synechocystis* 6803. J Biotechnol 142:31–37

Kovacs AT, Rakhely G, Balogh J et al (2005) Hydrogen independent expression of hupSL genes in *Thiocapsa roseopersicina* BBS. Febs J 272:4807–4816

Kovacs KL, Bagyinka C (1990) Structural properties, functional states and physiological roles of hydrogenase in photosynthetic bacteria. FEMS Microbiol. Rev. 87:407–412

Kovacs KL, Bagyinka C Serebriakova LT (1983) Distribution and orientation of hydrogenase in various photosynthetic bacteria. Current Microbiology 9:215–218

Kovács KL, Maróti G, Rákhely G (2006) A novel approach for biohydrogen production. Int J Hydrogen Energy 31:1460–1468

Kucho K, Okamoto K, Tsuchiya Y et al (2005) Global analysis of circadian expression in the cyanobacterium *Synechocystis* sp. strain PCC 6803. J Bacteriol 187:2190–2199

Kumarevel T, Tanaka T, Bessho Y et al (2009) Crystal structure of hydrogenase maturating endopeptidase HycI from *Escherichia coli*. Biochem Biophys Res Commun 389:310–314

Kunkel A, Vorholt JA, Thauer RK et al (1998) An *Escherichia coli* hydrogenase-3-type hydrogenase in methanogenic archaea. Eur J Biochem 252:467–476

Kurkin S, George SJ, Thorneley RN et al (2004) Hydrogen-induced activation of the [NiFe]-hydrogenase from *Allochromatium vinosum* as studied by stopped-flow infrared spectroscopy. Biochemistry 43:6820–6831

Laczkó H (1984) Protective mechanisms in photosynthesis of Anabaena cylindrica. Physiologia Plantarum 63:221–224

Laczko I (1986) Appearance of a reversible hydrogenase activity in *Anabaena cylindrica* grown in high light. Physiologia Plantarum 67:634–637

Lanzilotta WN, Schuller DJ, Thorsteinsson MV et al (2000) Structure of the CO sensing transcription activator CooA. Nat Struct Biol 7:876–80

Laurinavichene TV, Rakhely G, Kovacs KL et al (2007) The effect of sulfur compounds on H_2 evolution/consumption reactions, mediated by various hydrogenases, in the purple sulfur bacterium, *Thiocapsa roseopersicina*. Arch Microbiol 188:403–10

Lazazzera BA, Beinert H, Khoroshilova N et al (1996) DNA binding and dimerization of the Fe-S-containing FNR protein from *Escherichia coli* are regulated by oxygen. J Biol Chem 271: 2762–2768

Leach MR, Sandal S, Sun H et al (2005) Metal binding activity of the *Escherichia coli* hydrogenase maturation factor HypB. Biochemistry 44:12229–12238

Leach MR, Zhang JW, Zamble DB (2007) The role of complex formation between the *Escherichia coli* hydrogenase accessory factors HypB and SlyD. J Biol Chem 282:16177–16186

Leitao E, Oxelfelt F, Oliveira P et al (2005) Analysis of the *hupSL* operon of the nonheterocystous cyanobacterium *Lyngbya majuscula* CCAP 1446/4: regulation of transcription and expression under a light-dark regimen. Appl Environ Microbiol 71:4567–4576

Lemon BJ, Peters JW (1999) Binding of exogenously added carbon monoxide at the active site of the iron-only hydrogenase (CpI) from *Clostridium pasteurianum*. Biochemistry 38: 12969–12973

Lenz O, Bernhard M, Buhrke T et al (2002) The hydrogen-sensing apparatus in *Ralstonia eutropha*. J Mol Microbiol Biotechnol 4:255–262

Lenz O, Friedrich B (1998) A novel multicomponent regulatory system mediates H_2 sensing in *Alcaligenes eutrophus*. Proc Natl Acad Sci USA 95:12474–12479

Lenz O, Strack A, Tran-Betcke A et al (1997) A hydrogen-sensing system in transcriptional regulation of hydrogenase gene expression in *Alcaligenes* species. J Bacteriol 179:1655–1663

Lieman-Hurwitz J, Haimovich M, Shalev-Malul G et al (2009) A cyanobacterial AbrB-like protein affects the apparent photosynthetic affinity for CO_2 by modulating low-CO_2-induced gene expression. Environ Microbiol 11:927–936

Lindblad P, Christensson K, Lindberg P et al (2002) Photoproduction of H_2 by wildtype *Anabaena* PCC 7120 and a hydrogen uptake deficient mutant: from laboratory experiments to outdoor culture. Int J Hydrogen Energy:1271–1281

Lissolo T, Pulvin S, Thomas D (1984) Reactivation of the hydrogenase from *Desulfovibrio gigas* by hydrogen. Influence of redox potential. J Biol Chem 259:11725–11729

Little JW (1984) Autodigestion of *lexA* and phage lambda repressors. Proc Natl Acad Sci USA 81:1375–1379

Little JW (1991) Mechanism of specific LexA cleavage: autodigestion and the role of RecA coprotease. Biochimie 73:411–421

Long M, Liu J, Chen Z et al (2007) Characterization of a HoxEFUYH type of [NiFe] hydrogenase from Allochromatium vinosum and some EPR and IR properties of the hydrogenase module. J Biol Inorg Chem 12:62–78

Ludwig M, Cracknell JA, Vincent KA et al (2009) Oxygen-tolerant H_2 oxidation by membrane-bound [NiFe] hydrogenases of ralstonia species. Coping with low level H_2 in air. J Biol Chem 284:465–477

Ludwig M, Schulz-Friedrich R, Appel J (2006) Occurrence of hydrogenases in cyanobacteria and anoxygenic photosynthetic bacteria: implications for the phylogenetic origin of cyanobacterial and algal hydrogenases. J Mol Evol 63:758–768

Ma K, Schicho RN, Kelly RM et al (1993) Hydrogenase of the hyperthermophile *Pyrococcus furiosus* is an elemental sulfur reductase or sulfhydrogenase: evidence for a sulfur-reducing hydrogenase ancestor. Proc Natl Acad Sci USA 90:5341–5344

Madigan MT, Ormerod JG (eds) (1995) Taxonomy, physiology and ecology of heliobacteria. Anoxygenic Photosynthetic Bacteria. Kluwer, Dordrecht, The Netherlands

Magalon A, Bock A (2000) Analysis of the HypC-hycE complex, a key intermediate in the assembly of the metal center of the *Escherichia coli* hydrogenase 3. J Biol Chem 275: 21114–21120

Maier T, Jacobi A, Sauter M et al (1993) The product of the *hypB* gene, which is required for nickel incorporation into hydrogenases, is a novel guanine nucleotide-binding protein. J Bacteriol 175:630–635

Mandin P, Gottesman S (2009) A genetic approach for finding small RNAs regulators of genes of interest identifies RybC as regulating the DpiA/DpiB two-component system. Mol Microbiol 72:551–565

Maness PC, Huang J, Smolinski S et al (2005) Energy generation from the CO oxidation-hydrogen production pathway in *Rubrivivax gelatinosus*. Appl Environ Microbiol 71:2870–2874

Maness PC, Weaver PF (2001) Evidence for three distinct hydrogenase activities in *Rhodospirillum rubrum*. Appl Microbiol Biotechnol 57:751–756

Maroti G, Fodor BD, Rakhely G et al (2003) Accessory proteins functioning selectively and pleiotropically in the biosynthesis of [NiFe] hydrogenases in *Thiocapsa roseopersicina*. Eur J Biochem 270:2218–2227

Martin W, Muller M (1998) The hydrogen hypothesis for the first eukaryote. Nature 392:37–41

Martinez-Argudo I, Little R, Shearer N et al (2005) Nitrogen fixation: key genetic regulatory mechanisms. Biochem Soc Trans 33:152–166

McGlynn SE, Shepard EM, Winslow MA et al (2008) HydF as a scaffold protein in [FeFe] hydrogenase H-cluster biosynthesis. FEBS Lett 582:2183–2187

Menon NK, Robbins J, Wendt JC et al (1991) Mutational analysis and characterization of the *Escherichia coli hya* operon, which encodes [NiFe] hydrogenase 1. J Bacteriol 173:4851–4861

Meyer J (2007) [FeFe] hydrogenases and their evolution: a genomic perspective. Cell Mol Life Sci 64:1063–1084

Montet Y, Amara P, Volbeda A et al (1997) Gas access to the active site of Ni-Fe hydrogenases probed by X-ray crystallography and molecular dynamics. Nat Struct Biol 4:523–526

Moore LJ, Kiley PJ (2001) Characterization of the dimerization domain in the FNR transcription factor. J Biol Chem 276:45744–45750

Muro-Pastor AM, Herrero A, Flores E (2001) Nitrogen-regulated group 2 sigma factor from *Synechocystis* sp. strain PCC 6803 involved in survival under nitrogen stress. J Bacteriol 183:1090–1095

Nakajima H, Aono S (1999) Redox properties of the heme in the CO-sensing transcriptional activator CooA: Electrochemical evidence of the redox-controlled ligand exchange of the heme. Chem Lett 28:1233–1234

Nicolet Y, de Lacey AL, Vernede X et al (2001) Crystallographic and FTIR spectroscopic evidence of changes in Fe coordination upon reduction of the active site of the Fe-only hydrogenase from *Desulfovibrio desulfuricans*. J Am Chem Soc 123:1596–1601

Nicolet Y, Piras C, Legrand P et al (1999) *Desulfovibrio desulfuricans* iron hydrogenase: the structure shows unusual coordination to an active site Fe binuclear center. Structure 7: 13–23

Nicolet Y, Rubach JK, Posewitz MC et al (2008) X-ray structure of the [FeFe]-hydrogenase maturase HydE from *Thermotoga maritima*. J Biol Chem 283:18861–18872

Novelli PC, Lang PM, Masarie KA et al (1999) Molecular Hydrogen in the troposphere: Global distribution and budget. J Geophys Res 104:30427–30444

Ogata H, Hirota S, Nakahara A et al (2005) Activation process of [NiFe] hydrogenase elucidated by high-resolution X-ray analyses: conversion of the ready to the unready state. Structure 13: 1635–1642

Oliveira P, Leitao E, Tamagnini P et al (2004) Characterization and transcriptional analysis of *hupSLW* in *Gloeothece* sp. ATCC 27152: an uptake hydrogenase from a unicellular cyanobacterium. Microbiology 150:3647–3655

Oliveira P, Lindblad P (2005) LexA, a transcription regulator binding in the promoter region of the bidirectional hydrogenase in the cyanobacterium *Synechocystis* sp. PCC 6803. FEMS Microbiol Lett 251:59–66

Oliveira P Lindblad P (2008) An AbrB-Like protein regulates the expression of the bidirectional hydrogenase in *Synechocystis* sp. strain PCC 6803. J Bacteriol 190:1011–1019

Olson JW, Maier RJ (2000) Dual roles of *Bradyrhizobium japonicum* nickel protein in nickel storage and GTP-dependent Ni mobilization. J Bacteriol 182:1702–1705

Oren A, Padan E, Avron M (1977) Quantum yields for oxygenic and anoxygenic photosynthesis in the cyanobacterium *Oscillatoria limnetica*. Proc Natl Acad Sci USA 74:2152–2156

Oxelfelt F, Tamagnini P, Salema R et al (1995) Hydrogen uptake in *Nostoc* strain PCC 73102: effects of nickel, hydrogen, carbon and nitrogen. Plant Physiol. Biochem. 33:617–623

Palacios JM, Murillo J, Leyva A et al (1990) Differential expression of hydrogen uptake (hup) genes in vegetative and symbiotic cells of *Rhizobium leguminosarum*. Mol Gen Genet 221: 363–370

Palagyi-Meszaros LS, Maroti J, Latinovics D et al (2009) Electron-transfer subunits of the NiFe hydrogenases in *Thiocapsa roseopersicina* BBS. Febs J 276:164–174

Pandey AS, Harris TV, Giles LJ et al (2008) Dithiomethylether as a ligand in the hydrogenase h-cluster. J Am Chem Soc 130:4533–4540

Park KR, Giard JC, Eom JH et al (1999) Cyclic AMP receptor protein and TyrR are required for acid pH and anaerobic induction of *hyaB* and *aniC* in *Salmonella* typhimurium. J Bacteriol 181:689–694

Paschos A, Bauer A, Zimmermann A et al (2002) HypF, a carbamoyl phosphate-converting enzyme involved in [NiFe] hydrogenase maturation. J Biol Chem 277:49945–49951

Patterson-Fortin LM, Colvin KR, Owttrim GW (2006) A LexA-related protein regulates redox-sensitive expression of the cyanobacterial RNA helicase, *crhR*. Nucleic Acids Res 34: 3446–3454

Pedroni P, Della Volpe A, Galli G, Mura GM, Pratesi C, Grandi G (1995) Characterization of the locus encoding the [Ni-Fe] sulfhydrogenase from the archaeon *Pyrococcus furiosus*: evidence for a relationship to bacterial sulfite reductases. Microbiology 141:449–458

Peschek GA (1979) Anaerobic hydrogenase activity in *Anacystis nidulans*. H_2-dependent photoreduction and related reactions. Biochim Biophys Acta 548:187–202

Peters JW, Lanzilotta WN, Lemon BJ et al (1998) X-ray crystal structure of the Fe-only hydrogenase (CpI) from *Clostridium pasteurianum* to 1.8 angstrom resolution. Science 282:1853–1858

Phillips ZE, Strauch MA (2002) *Bacillus subtilis* sporulation and stationary phase gene expression. Cell Mol Life Sci 59:392–402

Pierik AJ, Roseboom W, Happe RP et al (1999) Carbon monoxide and cyanide as intrinsic ligands to iron in the active site of [NiFe]-hydrogenases. NiFe(CN)2CO, Biology's way to activate H_2. J Biol Chem 274:3331–3337

Pierson BK, Castenholz RW (1974) A phototrophic gliding filamentous bacterium of hot springs, *Chloroflexus aurantiacus*, gen. and sp. nov. Arch Microbiol 100:5–24*

Pilet E, Nicolet Y, Mathevon C et al (2009) The role of the maturase HydG in [FeFe]-hydrogenase active site synthesis and assembly. FEBS Lett 583:506–511

Posewitz MC, King PW, Smolinski SL et al (2005) Identification of genes required for hydrogenase activity in *Chlamydomonas reinhardtii*. Biochem Soc Trans 33:102–104

Posewitz MC, King PW, Smolinski SL et al (2004) Discovery of two novel radical S-adenosylmethionine proteins required for the assembly of an active [Fe] hydrogenase. J Biol Chem 279:25711–25720

Rakhely G, Colbeau A, Garin J et al (1998) Unusual organization of the genes coding for HydSL, the stable [NiFe]hydrogenase in the photosynthetic bacterium *Thiocapsa roseopersicina* BBS. J Bacteriol 180:1460–1465

Rakhely G, Kovacs AT, Maroti G et al (2004) Cyanobacterial-type, heteropentameric, NAD+-reducing NiFe hydrogenase in the purple sulfur photosynthetic bacterium *Thiocapsa roseopersicina*. Appl Environ Microbiol 70:722–728

Rakhely G, Laurinavichene TV, Tsygankov AA et al (2007) The role of Hox hydrogenase in the H_2 metabolism of *Thiocapsa roseopersicina*. Biochim Biophys Acta 1767:671–676

Rangarajan ES, Asinas A, Proteau A et al (2008) Structure of [NiFe] hydrogenase maturation protein HypE from *Escherichia coli* and its interaction with HypF. J Bacteriol 190: 1447–1458

Rees DC, Akif Tezcan F, Haynes CA et al (2005) Structural basis of biological nitrogen fixation. Philos Transact A Math Phys Eng Sci 363:971–984; discussion 1035–1040

Reissmann S, Hochleitner E, Wang H et al (2003) Taming of a poison: biosynthesis of the NiFe-hydrogenase cyanide ligands. Science 299:1067–1070

Rey FE, Oda Y, Harwood CS (2006) Regulation of uptake hydrogenase and effects of hydrogen utilization on gene expression in *Rhodopseudomonas palustris*. J Bacteriol 188: 6143–6152

Rey L, Fernandez D, Brito B et al (1996) The hydrogenase gene cluster of *Rhizobium leguminosarum bv. viciae* contains an additional gene (*hypX*), which encodes a protein with sequence similarity to the N10-formyltetrahydrofolate-dependent enzyme family and is required for nickel-dependent hydrogenase processing and activity. Mol Gen Genet 252:237–248

Richard DJ, Sawers G, Sargent F et al (1999) Transcriptional regulation in response to oxygen and nitrate of the operons encoding the [NiFe] hydrogenases 1 and 2 of *Escherichia coli*. Microbiology 145 (Pt 10): 2903–2912

Roberts GP, Kerby RL, Youn H et al (2005) CooA, a paradigm for gas sensing regulatory proteins. J Inorg Biochem 99:280–292

Rosano C, Zuccotti S, Bucciantini M et al (2002) Crystal structure and anion binding in the prokaryotic hydrogenase maturation factor HypF acylphosphatase-like domain. J Mol Biol 321:785–796

Roseboom W, Blokesch M, Bock A et al (2005) The biosynthetic routes for carbon monoxide and cyanide in the Ni-Fe active site of hydrogenases are different. FEBS Lett 579:469–472

Rossmann R, Maier T, Lottspeich F et al (1995) Characterisation of a protease from *Escherichia coli* involved in hydrogenase maturation. Eur J Biochem 227:545–550

Rubach JK, Brazzolotto X, Gaillard J et al (2005) Biochemical characterization of the HydE and HydG iron-only hydrogenase maturation enzymes from *Thermatoga maritima*. FEBS Lett 579:5055–5060

Sandman K, Pereira SL, Reeve JN (1998) Diversity of prokaryotic chromosomal proteins and the origin of the nucleosome. Cell Mol Life Sci 54:1350–1364

Sapra R, Verhagen MF, Adams MW (2000) Purification and characterization of a membrane-bound hydrogenase from the hyperthermophilic archaeon *Pyrococcus furiosus*. J Bacteriol 182: 3423–3428

Sattley WM, Madigan MT, Swingley WD et al (2008) The genome of *Heliobacterium modesticaldum*, a phototrophic representative of the Firmicutes containing the simplest photosynthetic apparatus. J Bacteriol 190:4687–4696

Sauter M, Bohm R, Bock A (1992) Mutational analysis of the operon (*hyc*) determining hydrogenase 3 formation in *Escherichia coli*. Mol Microbiol 6:1523–1532

Sawers G, Heider J, Zehelein E et al (1991) Expression and operon structure of the *sel* genes of *Escherichia coli* and identification of a third selenium-containing formate dehydrogenase isoenzyme. J Bacteriol 173:4983–4993

Sawers RG, Blokesch, M Böck, A (2004) Anaerobic formate and hydrogen metabolism. EcoSal-Escherichia coli and Salmonella: Cellular and Molecular Biology. R. Curtiss III E. i. C. Washington, DC, ASM Press

Schmitz O, Boison G, Bothe H (2001) Quantitative analysis of expression of two circadian clock-controlled gene clusters coding for the bidirectional hydrogenase in the cyanobacterium *Synechococcus* sp. PCC7942. Mol Microbiol 41:1409–1417
Schmitz O, Boison G, Salzmann H et al (2002) HoxE–a subunit specific for the pentameric bidirectional hydrogenase complex (HoxEFUYH) of cyanobacteria. Biochim Biophys Acta 1554:66–74
Schneider K, Schlegel HG (1976) Purification and properties of soluble hydrogenase from *Alcaligenes eutrophus* H 16. Biochim Biophys Acta 452:66–80
Schultz JE, Weaver PF (1982) Fermentation and anaerobic respiration by *Rhodospirillum rubrum* and *Rhodopseudomonas capsulata*. J Bacteriol 149:181–190
Schut GJ, Adams MW (2009) The iron-hydrogenase of *Thermotoga maritima* utilizes ferredoxin and NADH synergistically: a new perspective on anaerobic hydrogen production. J Bacteriol 191:4451–4457
Schwartz CJ, Djaman O, Imlay JA et al (2000) The cysteine desulfurase, IscS, has a major role in in vivo Fe-S cluster formation in *Escherichia coli*. Proc Natl Acad Sci USA 97:9009–9014
Schwartz E, Friedrich B (2006) The H2-metabolizing prokaryotes. The prokaryotes. In: Dworkin M, Falkow S, Rosenberg E, Schleifer KH, Stackebrandt E (eds) A handbook on the biology of bacteria. New York, Springer
Self WT, Grunden AM, Hasona A et al (1999) Transcriptional regulation of molybdoenzyme synthesis in *Escherichia coli* in response to molybdenum: ModE-molybdate, a repressor of the *modABCD* (molybdate transport) operon is a secondary transcriptional activator for the *hyc* and *nar* operons. Microbiology 145(Pt 1): 41–55
Serebryakova L, Zorin N, Lindblad P (1994) Reversible hydrogenase in *Anabaena variabilis* ATCC 29413. Presence and localization in non-N_2-fixing cells. Arch Microbiol 161:140–144
Serebryakova LT, Medina M, Zorin NA et al (1996) Reversible hydrogenase of *Anabaena variabilis* ATCC 29413: catalytic properties and characterization of redox centres. FEBS Lett 383:79–82
Serebryakova LT, Sheremetieva M, Tsygankov AA (1998) Reversible hydrogenase activity of *Gloeocapsa alpicola* in continuous culture. FEMS Microbiology Letters 166:89–94
Serebryakova LT, Zorin NA, Gogotov IN (1992) Hydrogenase activity in filamentous cyanobacteria. Mikrobiologiya 61:107–112
Severin I, Stal LJ (2008) Light dependency of nitrogen fixation in a coastal cyanobacterial mat. ISME J 2:1077–1088
Shalev-Malul G, Lieman-Hurwitz J, Viner-Mozzini Y et al (2008) An AbrB-like protein might be involved in the regulation of cylindrospermopsin production by *Aphanizomenon ovalisporum*. Environ Microbiol 10:988–999
Shelver D, Kerby RL, He Y et al (1997) CooA, a CO-sensing transcription factor from *Rhodospirillum rubrum*, is a CO-binding heme protein. Proc Natl Acad Sci USA 94: 11216–11220
Shomura Y, Komori H, Miyabe N et al (2007) Crystal structures of hydrogenase maturation protein HypE in the Apo and ATP-bound forms. J Mol Biol 372:1045–1054
Sjoholm J, Oliveira P, Lindblad P (2007) Transcription and regulation of the bidirectional hydrogenase in the cyanobacterium *Nostoc* sp. strain PCC 7120. Appl Environ Microbiol 73:5435–5446
Skibinski DA, Golby P, Chang YS et al (2002) Regulation of the hydrogenase-4 operon of *Escherichia coli* by the sigma(54)-dependent transcriptional activators FhlA and HyfR. J Bacteriol 184:6642–6653
Stal LJ, Moezelaar R (1997) Fermentation in cyanobacteria. FEMS microbiology reviews 21: 179–211
Stensjo K, Ow SY, Barrios-Llerena ME et al (2007) An iTRAQ-based quantitative analysis to elaborate the proteomic response of *Nostoc* sp. PCC 7120 under N_2 fixing conditions. J Proteome Res 6:621–635
Stephenson M, Stickland LH (1931) Hydrogenase: a bacterial enzyme activating molecular hydrogen: The properties of the enzyme. Biochem J 25:205–214

Summerfield TC, Sherman LA (2008) Global transcriptional response of the alkali-tolerant cyanobacterium *Synechocystis* sp. strain PCC 6803 to a pH 10 environment. Appl Environ Microbiol 74:5276–5284

Svetlitchnyi V, Peschel C, Acker G et al (2001) Two membrane-associated NiFeS-carbon monoxide dehydrogenases from the anaerobic carbon-monoxide-utilizing eubacterium *Carboxydothermus hydrogenoformans*. J Bacteriol 183:5134–5144

Swem LR, Elsen S, Bird TH et al (2001) The RegB/RegA two-component regulatory system controls synthesis of photosynthesis and respiratory electron transfer components in *Rhodobacter capsulatus*. J Mol Biol 309:121–138

Swem LR, Gong X, Yu CA et al (2006) Identification of a ubiquinone-binding site that affects autophosphorylation of the sensor kinase RegB. J Biol Chem 281:6768–6775

Swem LR, Kraft BJ, Swem DL et al (2003) Signal transduction by the global regulator RegB is mediated by a redox-active cysteine. EMBO J 22:4699–4708

Tamagnini P, Leitao E, Oliveira P et al (2007) Cyanobacterial hydrogenases: diversity, regulation and applications. FEMS Microbiol Rev 31:692–720

Tel-Or E, Luijk LW, Packer L (1977) An inducible hydrogenase in cyanobacteria enhances N_2 fixation. FEBS Lett 78:49–52

Thauer RK, Jungermann K, Decker K (1977) Energy conservation in chemotrophic anaerobic bacteria. Bacteriol Rev 41:100–180

Theodoratou E, Huber R, Bock A (2005) [NiFe]-Hydrogenase maturation endopeptidase: structure and function. Biochem Soc Trans 33:108–111

Theodoratou E, Paschos A, Magalon A et al (2000) Nickel serves as a substrate recognition motif for the endopeptidase involved in hydrogenase maturation. Eur J Biochem 267:1995–1999

Toepel J, Welsh E, Summerfield T et al (2008) Differential transcriptional analysis of the cyanobacterium *Cyanothece* sp. strain ATCC 51142 during light-dark and continuous-light growth. J Bacteriol 190:3904–3913

Toussaint B, Bosc C, Richaud P et al (1991) A mutation in a *Rhodobacter capsulatus* gene encoding an integration host factor-like protein impairs in vivo hydrogenase expression. Proc Natl Acad Sci USA 88:10749–10753

Toussaint B, de Sury d'Aspremont R, Delic-Attree I et al (1997) The *Rhodobacter capsulatus hupSLC* promoter: identification of cis-regulatory elements and of trans-activating factors involved in H_2 activation of *hupSLC* transcription. Mol Microbiol 26:927–937

Troshina O, Serebryakova L,Sheremetieva M et al (2002) Production of H_2 by the unicellular cyanobacterium *Gloeocapsa alpicola* CALU 743 during fermentation. Int J Hydrogen Energy 27:1283–1289

Van Bruggen JJA, Stumm CK, Vogels GD (1983) Symbiosis of methanogenic bacteria and sapropelic protozoa. Arch Microbiol J 36:89–95

Van der Linden E, Burgdorf T, Bernhard M et al (2004) The soluble [NiFe]-hydrogenase from *Ralstonia eutropha* contains four cyanides in its active site, one of which is responsible for the insensitivity towards oxygen. J Biol Inorg Chem 9:616–626

Van der Oost J, Cox RP (1988) Hydrogenase activity in nitrate-grown cells of the unicellular cyanobacterium *Cyanothece* PCC 7822. Arch of Microbiol 151:40–43

Van der Zwaan JW, Coremans JM, Bouwens EC et al (1990) Effect of $^{17}O_2$ and ^{13}CO on EPR spectra of nickel in hydrogenase from *Chromatium vinosum*. Biochim. Biophys. Acta 1041:101–110

Van Praag E, Degli Agosti R, Bachofen R (2000) Rhythmic activity of uptake hydrogenase in the prokaryote *Rhodospirillum rubrum*. J Biol Rhythms 15:218–224

Vaughn JL, Feher V, Naylor S et al (2000) Novel DNA binding domain and genetic regulation model of *Bacillus subtilis* transition state regulator *abrB*. Nat Struct Biol 7:1139–1146

Vignais PM (2008) Hydrogenases and H(+)-reduction in primary energy conservation. Results Probl Cell Differ 45:223–252

Vignais PM., Billoud B. (2007) Occurrence, classification, and biological function of hydrogenases: an overview. Chem Rev 107:4206–4272

Vignais PM, Billoud B, Meyer J (2001) Classification and phylogeny of hydrogenases. FEMS Microbiol Rev 25:455–501

Vignais PM, Colbeau A (2004) Molecular biology of microbial hydrogenases. Curr Issues Mol Biol 6:159–188

Vignais PM, Elsen S, Colbeau A (2005) Transcriptional regulation of the uptake [NiFe]hydrogenase genes in *Rhodobacter capsulatus*. Biochem Soc Trans 33:28–32

Vignais PM, Toussaint B (1994) Molecular biology of membrane-bound H_2 uptake hydrogenases. Arch Microbiol 161:1–10

Volbeda A, Charon MH, Piras C et al (1995) Crystal structure of the nickel-iron hydrogenase from Desulfovibrio gigas. Nature 373:580–587

Volbeda A, Fontecilla-Camps JC, Frey M (1996) Novel metal sites in protein structures. Curr Opin Struct Biol 6:804–812

Volbeda A, Martin L, Cavazza C et al (2005) Structural differences between the ready and unready oxidized states of [NiFe] hydrogenases. J Biol Inorg Chem 10:239–249

Watanabe S, Arai T, Matsumi R et al (2009) Crystal Structure of HypA, a Nickel-Binding Metallochaperone for [NiFe] Hydrogenase Maturation. J Mol Biol 394:448–459

Watanabe S, Matsumi R, Arai T et al (2007) Crystal structures of [NiFe] hydrogenase maturation proteins HypC, HypD, and HypE: insights into cyanation reaction by thiol redox signaling. Mol Cell 27:29–40

Waugh R, Boxer DH (1986) Pleiotropic hydrogenase mutants of *Escherichia coli* K12: growth in the presence of nickel can restore hydrogenase activity. Biochimie 68:157–166

Weyman PD, Pratte B, Thiel T (2008) Transcription of *hupSL* in *Anabaena variabilis* ATCC 29413 is regulated by NtcA and not by hydrogen. Appl Environ Microbiol 74:2103–2110

Wu J, Bauer CE (2008) RegB/RegA, a global redox-responding two-component system. Adv Exp Med Biol 631:131–148

Xu K, Strauch MA (1996) In vitro selection of optimal AbrB-binding sites: comparison to known in vivo sites indicates flexibility in AbrB binding and recognition of three-dimensional DNA structures. Mol Microbiol 19:145–158

Yang F, Hu W, Xu H et al (2007) Solution structure and backbone dynamics of an endopeptidase HycI from *Escherichia coli*: implications for mechanism of the [NiFe] hydrogenase maturation. J Biol Chem 282:3856–3863

Zannoni D (1995) In: Blankenship RE, Miller M, Bauer CE (eds) Anoxygenic Photosynthetic Bacteria.. Kluwer Academic, Dordrecht pp. 949–971

Zhang JW, Butland G, Greenblatt JF et al (2005) A role for SlyD in the *Escherichia coli* hydrogenase biosynthetic pathway. J Biol Chem 280:4360–4366

Zhang Z, Pendse ND, Phillips KN et al (2008) Gene expression patterns of sulfur starvation in *Synechocystis* sp. PCC 6803. BMC Genomics 9:344

Subject Index

P.C. Hallenbeck (ed.), *Recent Advances in Phototrophic Prokaryotes*, Advances in Experimental Medicine and Biology 675, DOI 10.1007/978-1-4419-1528-3,

D

E

F

G

H